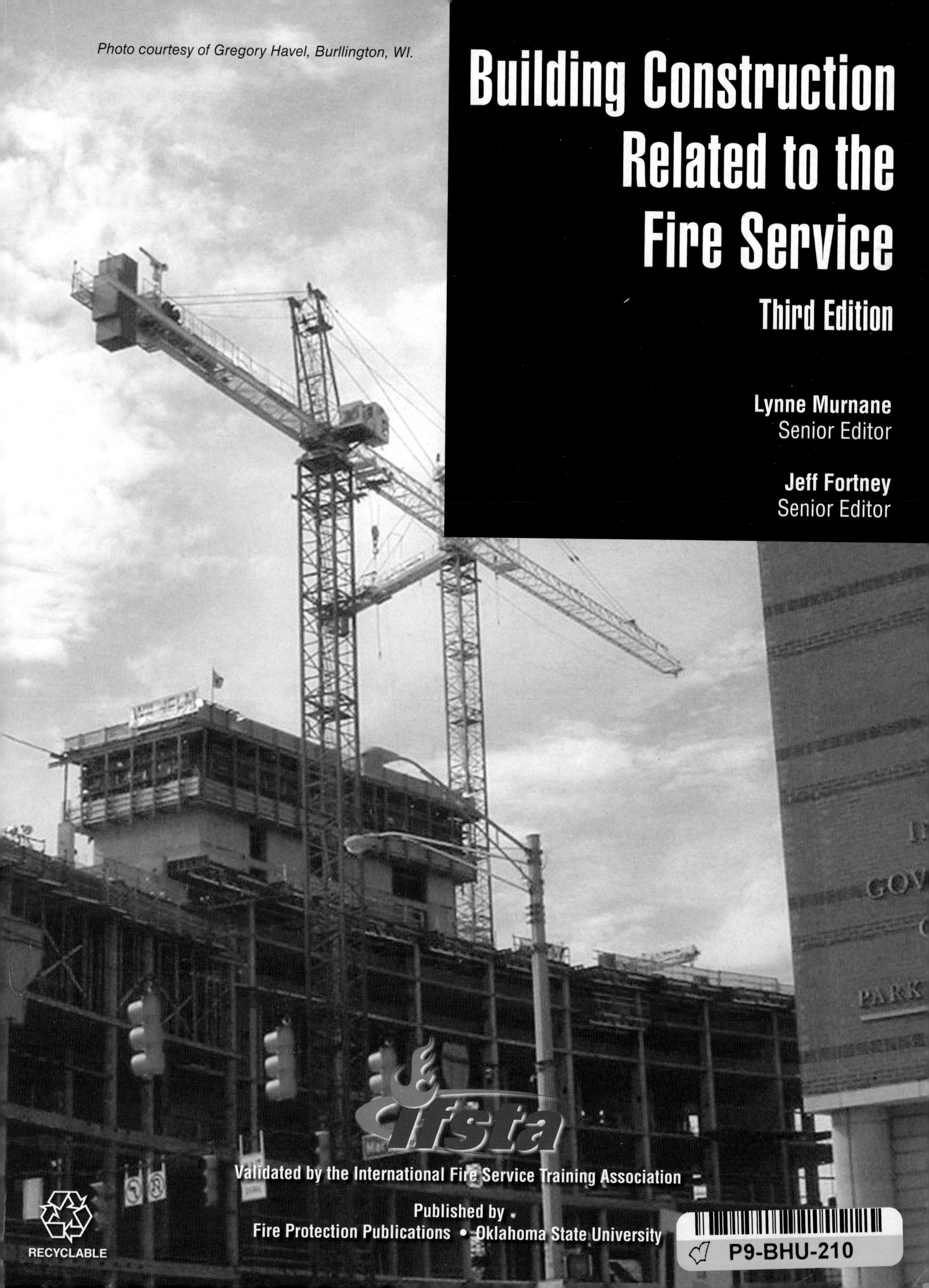

Building Construction Related to the Fire Service

Third Edition

Lynne Murnane
Senior Editor

Jeff Fortney
Senior Editor

Validated by the International Fire Service Training Association

Published by
Fire Protection Publications • Oklahoma State University

The International Fire Service Training Association

The International Fire Service Training Association (IFSTA) was established in 1934 as a *nonprofit educational association of fire fighting personnel who are dedicated to upgrading fire fighting techniques and safety through training.* To carry out the mission of IFSTA, Fire Protection Publications was established as an entity of Oklahoma State University. Fire Protection Publications' primary function is to publish and disseminate training texts as proposed and validated by IFSTA. As a secondary function, Fire Protection Publications researches, acquires, produces, and markets high-quality learning and teaching aids as consistent with IFSTA's mission.

The IFSTA Validation Conference is held the second full week in July. Committees of technical experts meet and work at the conference addressing the current standards of the National Fire Protection Association® and other standard-making groups as applicable. The Validation Conference brings together individuals from several related and allied fields, such as:

- Key fire department executives and training officers
- Educators from colleges and universities
- Representatives from governmental agencies
- Delegates of firefighter associations and industrial organizations

Committee members are not paid nor are they reimbursed for their expenses by IFSTA or Fire Protection Publications. They participate because of commitment to the fire service and its future through training. Being on a committee is prestigious in the fire service community, and committee members are acknowledged leaders in their fields. This unique feature provides a close relationship between the International Fire Service Training Association and fire protection agencies, which helps to correlate the efforts of all concerned.

IFSTA manuals are now the official teaching texts of most of the states and provinces of North America. Additionally, numerous U.S. and Canadian government agencies as well as other English-speaking countries have officially accepted the IFSTA manuals.

ISBN 978-0-87939-371-7 *Library of Congress Control Number: 2009938935*

Third Edition, First Printing, January 2011 *Printed in the United States of America*

10 9 8 7 6 5 4 3 2 1

If you need additional information concerning the International Fire Service Training Association (IFSTA) or Fire Protection Publications, contact:

Customer Service, Fire Protection Publications, Oklahoma State University
930 North Willis, Stillwater, OK 74078-8045
800-654-4055 Fax: 405-744-8204

For assistance with training materials, to recommend material for inclusion in an IFSTA manual, or to ask questions or comment on manual content, contact:

Editorial Department, Fire Protection Publications, Oklahoma State University
930 North Willis, Stillwater, OK 74078-8045
405-744-4111 Fax: 405-744-4112 E-mail: editors@osufpp.org

Chapter Summary

Appendix

Table of Contents

List of Tables

Preface

The information contained in this text is designed to meet the objectives put forth in the model course outlines for ***Building Construction for Fire Protection*** as established by the Fire and Emergency Services Higher Education (FESHE) initiative led by the United States Fire Administration. This initiative brings together leaders in fire service higher education for the purpos e of establishing model core curricula for Associate's, Bachelor's, and Master's level fire service higher education degree programs. It is hoped that this initiative will encourage growth in the fire service higher education field and support commonality of the information taught to all of the students.

IFSTA Building Construction Related to the Fire Service Third Edition Validation Committee

Chair
Don Turno
Fire Protection Engineer/Program Manager
Washington Savannah River
Aiken, SC

Vice Chair
Tonya Hoover
Asst State Fire Marshal
CAL Fire Office of the State Fire Marshal
Sacramento, CA

Secretary
Steve Martin
Deputy Director
Delaware State Fire School
Dover, DE

Committee Members

Michael Boub
Fairfax County Fire and Rescue Dept
Fairfax, VA

Melvin Byrne
Virginia Department of Fire Programs
Ashburn, VA

Dave Coombs
DRC
Columbia River Fire & Rescue
St. Helens, OR

Jon B. Fore
City of Fairburn Fire Department
Sharpsburg, GA

Dave Hanneman
Boise City Fire Department
Boise, ID

Donny Howard
Agent
Oklahoma State Fire Marshal's Office
Oklahoma City, OK

George Jamieson
Regional Training Coordinator
Oregon Dept. of Public Safety
Pendleton, OR

James Lambrechts
San Francisco Fire Department
San Francisco, CA

Committee Members (Concluded)

Dan Lane
International Code Council
Petersburg, NY

Richard Merrell
Fairfax County Fire Rescue Department
Montclair, VA

Randal Novak
Bureau Chief
Iowa Fire Service Training Bureau
Ames, IA

Technical Reviewers

Greg Havel
Burlington Fire Department
Burlington, WI

Dave Bergman
Merced, CA

The following organizations and individuals have also contributed information, photographs, or other assistance that made final completion of this manual possible:

Special thanks to Ron Moore and the McKinney (TX) Fire Department for their special assistance and granting permission for the use of so many photos. We are very grateful.

Gregory Havel
Burlington, WI

Dave Coombs

Martin King
West Allis (WI) Fire Department

Chris E. Mickal
NOFD Photo Unit

Steve Toth

Ron Jeffers

Wil Dane

Elizabeth Titus
Underwriters Laboratories, Inc.

Sturzenbecker Construction Company, Inc.
Gala and Associates

San Diego County Sheriff's Department

APA - The Engineered Wood Association

Loretta Hall, Author
Underground Buildings: More than Meets the Eye

Vermont Timber Works, Inc.

Daniel P. Finnegan
Siemens Building Technologies Inc.

Peter J. Zelonis PE
Structural Engineer

Southern Forest Products Association

Hoover Treated Wood Products, Inc.

The Simon Property Group, Inc.

Penn Square Mall of Oklahoma City, OK
Audrie Thompson
Mis Gaston
Virgil L. Green, Sr.

Oakwood Mall (J. Herzog & Sons, Inc.) Enid, OK

Enid (OK) Fire Department
Ken Helms, Fire Marshall
Kevin Winter, Assistant Fire Marshall
Kenny Hager
Shawn Kuehn
Mark Morris
Kurt Pendergraft

Building Construction Project Team

Project Manager
Lynne Murnane

Technical Writer
Ed Prendergast
Chicago, IL

Technical Writer
Jim Simms
Senior Consulting Engineer
Rolf Jensen & Associates, Inc.
Walnut Creek CA

Senior Editors
Cindy Brakhage
Michael Sturzenbecker

Photographer
Jeff Fortney

Technical Reviewers
Don Turno
Martin King

Production Manager
Ann Moffat

Illustrator and Layout Designer
Ben Brock

Research Technicians
Gabriel Ramirez
Elkie Burnside

The IFSTA Executive Board at the time of validation of the **Building Construction Related to the Fire Service** was as follows:

Introduction

Chapter Contents

Introduction

It is the basic mission of the firefighter, when called upon, to enter a burning structure to rescue occupants and to extinguish the fire. It may be stated, therefore, that the true workplace of a firefighter is inside a building on fire. A burning building is not the typical surrounding in which to practice a profession. The firefighter enters a hostile and unknown environment and if the fire is not controlled, it will only become worse. To extinguish a fire in a building efficiently and safely without injury, the firefighter must be a master of that dangerous environment.

Whenever a fire occurs in a building, a potential exists for each of the following effects:

- Death or injury of the building's occupants and firefighters
- Destruction of the contents of the building
- Ignition of the building itself
- Partial or total collapse of the building
- Extension of the fire to other exposures.

The firefighter must understand building construction to understand the behavior of buildings under fire conditions. However, firefighters cannot perform a detailed engineering analysis of buildings while performing their fire fighting duties on the fireground. Therefore, a fundamental knowledge of buildings is an essential component of the decision-making process in successful fireground operations.

While the idea of firefighters having a basic knowledge of buildings may seem simple or even self-evident, it is made difficult by two fundamental aspects of building construction. First, the field of building construction is constantly changing. Construction methods and materials driven by economics, changing technology, and the needs of society are continually evolving. Second, the life span of a building may be 50 to 100 years or more.

Firefighters face an enormous number of challenges and risk factors during an emergency related to building construction. A few of the common situations encountered are as follows:

- The widespread use of lightweight construction materials such as trusses and lighter-weight building components that leave structures vulnerable to quick collapse during a fire.
- Any number of buildings may have been constructed with materials and methods that may be obsolete.

- Buildings may be constructed with materials that are manufactured to look like something else. For example, a deck may appear to be wood but is actually plastic and therefore will behave like plastic during a fire. What appears to be stone siding on a building may really be polyurethane.
- Any structure may have had several owners with different occupancies and numerous renovations — not all of them legal or structurally sound.
- Standard-looking structural elements may be reinforced, making it difficult for firefighters to gain access.
- Renovations that exceed the capacity of the original building components.
- Unknown changes in structures or storage that result in increased hazards to firefighters.
- Remodeling that results in multiple layers of roofing or floors, leading to increased collapse, fire load, and access hazards.
- Changes in building construction methods that permit lighter building elements and rely on building systems like sprinklers to achieve certain level of protection.

For these reasons and many more, it is vital that today's firefighter have not only a working knowledge of building construction materials and their behavior during a fire, but also pre-incident planning procedures that will identify and lessen the hazards to firefighters during an emergency.

Purpose and Scope

This manual is intended to furnish the reader with basic information about how buildings are designed and constructed. This information will aid in decision making related to fire prevention and fire control. Whether your duties include studying and enforcing fire codes, inspecting buildings, developing pre-incident plans, fighting fires, directing fireground operations, overseeing firefighter safety, or investigating fires, a thorough understanding of building construction principles and practices as these relate to fire behavior and fire load will enable you to make better, safe, and more timely decisions to protect people and property from potential as well as actual fires. Knowledge of various principles discussed in this text is required in a number of NFPA® qualification standards, including NFPA® 1001, 1006, 1021, and 1031.

The information contained in this text is designed to meet the objectives put forth in the model course outline for *Building Construction for Fire Protection* as established by the Fire and Emergency Services Higher Education (FESHE) initiative led by the United States Fire Administration (USFA). This initiative brings together leaders in fire service higher education for the purpose of establishing model core curricula for Associate's, Bachelor's, and Master's level fire service degree programs. It is hoped that this initiative will encourage growth in the fire service higher education field and support commonality of the information that is taught to all of the students.

Key Information

Various types of information in this manual are given in shaded boxes marked by symbols or icons. See the following definitions:

Safety Alert

Provides additional emphasis on matters of safety.

Case Study

A case history analyzes an event. It can describe its development, action taken, investigation results, and lessons learned. Illustrations can be included.

Information Plus

Information Plus Sidebars give additional relevant information that is more detailed, descriptive or explanatory than that given in the text.

A key term is designed to emphasize key concepts, technical terms, or ideas that firefighters need to know. They are listed at the beginning of each chapter and the definition is placed in the margin for easy reference.

Load — Any effect that a structure must be designed to resist. Forces of loads, such as gravity, wind, earthquakes, and soil pressure, are exerted on a building.

Three key signal words are found in the text: **WARNING!, CAUTION,** and **NOTE**. Definitions and examples of each are as follows:

- **WARNING!** indicates information that could result in death or serious injury to fire and emergency services personnel. See the following example:

WARNING!

From a fire fighting and safety standpoint, trusses have the potential for early failure. Lightweight trusses are especially prone to failure.

- **CAUTION** indicates important information or data that fire and emergency service responders need to be aware of in order to perform their duties safely. See the following example:

CAUTION

Firefighters should not step *onto* skylights because they may fall through. In addition, skylights may have been covered with a lightweight material and may not be visible.

- **NOTE** indicates important operational information that helps explain why a particular recommendation is given or describes optional methods for certain procedures. See the following example:

NOTE: Some precast components, such as tilt-up panels, are cast at the site and moved into position.

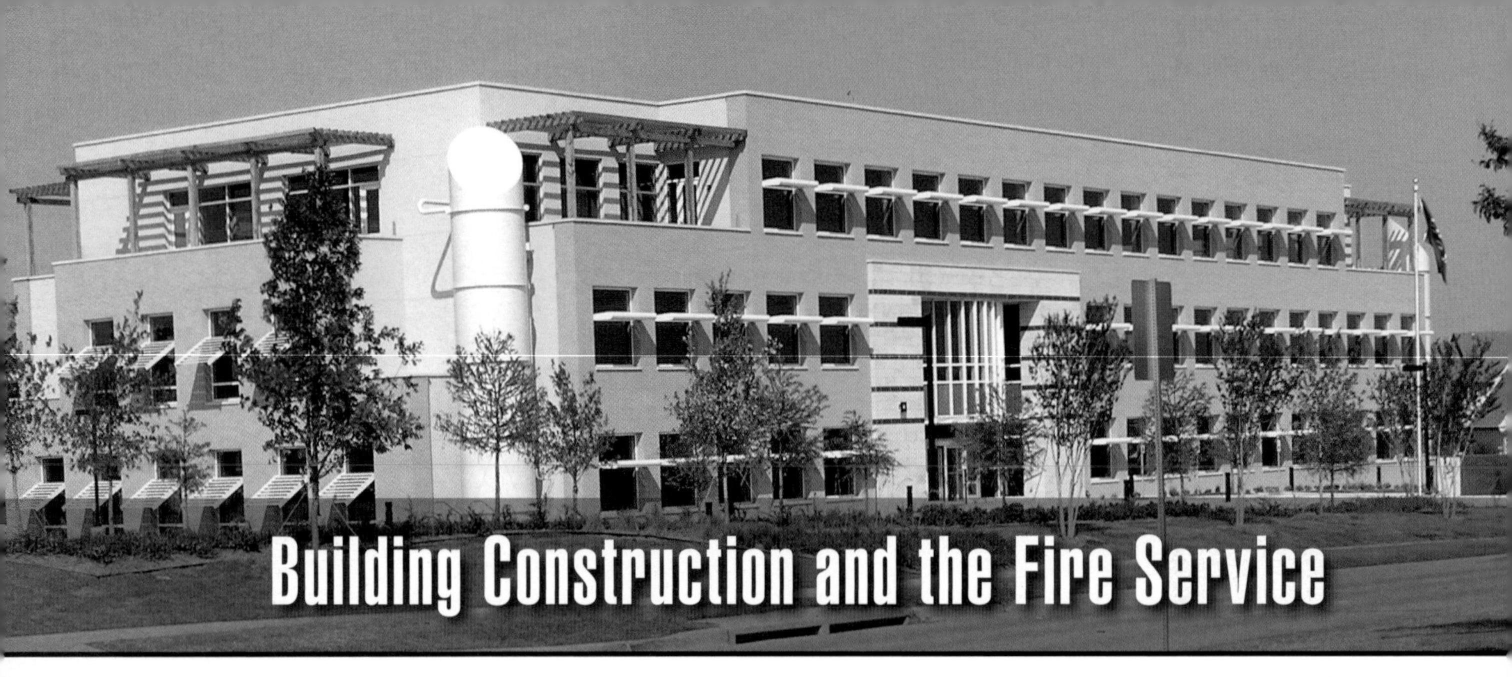

Building Construction and the Fire Service

Chapter Contents

Divider page photo courtesy of McKinney (TX) Fire Department.

Key Terms

FESHE Objectives

Fire and Emergency Services Higher Education (FESHE) Objectives: *Building Construction for Fire Protection*

1. Demonstrate an understanding of building construction as it relates to firefighter safety, building codes, fire prevention, code inspection and firefighting strategy and tactics.

Building Construction and the Fire Service

Learning Objectives

After reading this chapter, students will be able to:

1. Recognize the significance of methods and materials historically used in building construction, as well as the importance of the age of the building itself.
2. Discuss building variables as they relate to the work of firefighters.
3. Explain communication of fire and the ways in which it occurs.
4. Describe factors that affect communication of fire and methods used to protect buildings from exposing fires.
5. Discuss building failure, structural integrity, building systems, and design deficiencies as building design considerations.
6. Explain the principles of design and why buildings are built.
7. Discuss design considerations.
8. Describe the design and construction process.
9. Recognize the role of the building permit process and preincident planning in the construction of a building.

Chapter 1
Building Construction and the Fire Service

Case History

Event Description: In 1991 a fire in western Pennsylvania resulted in the deaths of four volunteer firefighters. The fire began in the basement of a two-story building that housed a furniture refinishing business. Stored in the basement were lacquer, paints, varnish, a spray booth, dip tanks, and furniture. The building was approximately 60 years old and originally had been constructed as an automobile repair garage. It was of masonry construction with concrete floors, and wood joists supported the roof.

Responding firefighters found evidence of fire in the basement but could not initially determine the extent of the fire. As fire fighting operations began, one fire team entered from the front of the building to the first floor; another team entered from the rear of the building and descended to the basement. Approximately 42 minutes after the initial alarm, a 15 x 65 feet (5 m by 20 m) section of the concrete first floor collapsed at the front of the building. The collapsed floor was between the firefighters on the first floor and the front door they had entered. Fire erupted from the basement, cutting off the firefighters' avenue of escape. Subsequent medical examination determined that the men died from burns.

Of special interest in this incident was the construction and condition of the floor that collapsed. Although the floor surface was concrete, it was supported by unprotected steel beams and steel columns. What may have appeared to be a fire-resistive floor slab was, in fact, NOT a fire-resistive assembly. Furthermore, an investigation uncovered that some of the steel at the front of the basement was severely rusted. When the floor assembly was exposed to the heat of the fire, collapse occurred.

Lesson Learned:

— The building's occupancy had changed from its original use. A furniture refinishing operation with its associated fuel and fire hazards was located in a basement where it might not have been expected.

— The belowgrade location of the fire made access difficult. Firefighters were forced to attack a fire of unknown magnitude from above.

— The nature and condition of the first floor construction was not apparent from the outside. Therefore, its behavior when exposed to the fire below could not be anticipated. *Preincident planning is critical,* especially when there is a change of occupancy.

In a scientific sense, the laws of physics and chemistry that govern fire behavior never change. Even though all fires are fundamentally similar or at least share common characteristics, experienced firefighters sometimes say no two fires are alike. Therefore, fighting fires is an eminently practical activity, not a theoretical one. The fire officer confronting fires in buildings must make immediate decisions as each situation unfolds. What makes fire fighting complicated is the variety of circumstances in which fires occur. If the underlying principles of fire behavior and its effect on buildings can be understood, fire fighting operations can be carried out more efficiently and safely.

History of Building Construction

The history of building construction can be traced back over many centuries. The way in which buildings are built is determined by several factors, the most important of which are the technology of the time and economics. The Romans constructed their buildings (quite successfully) based on the materials and methods available 2,000 years ago. Native Americans built teepees using the materials (animal hides) and tools available to them.

In colonial America, buildings were originally built using thatched roofs and post and beam construction. In the 19th century, buildings were built (at least in the cities) primarily using masonry and wood. The invention of the circular saw permitted the shift from log construction to planks and boards.

As the 19th century ended, technology allowed the introduction of cast iron for columns and, later, steel for columns and beams. At the beginning of the 20th century the combined use of steel framing and the elevator gave rise to the high-rise building as it is known today. Steel replaced most heavy-timber framing, which had been used for multistory commercial buildings, because steel is a stronger material than wood.

Economic factors such as material cost, labor cost, and building efficiency also impact building construction. For example, the use of plaster as an interior finish material declined in the second half of the 20th century because it is a more labor-intensive material to apply than gypsum board. Multistory factory and warehouse buildings have become obsolete largely because one-story buildings are generally more efficient for movement of materials. Concerns for energy efficiency have resulted in buildings without openable windows or with no windows at all.

A building constructed in 1925 would not have been constructed with air conditioning and the heating plant would probably have used coal. By contrast, modern buildings are air conditioned and are not likely to use coal as a fuel. Single-family homes built in the 1920's were provided with 30 amp electrical service; today, 200 amp service is typical.

The useful life span of a building can range from 25 to 100 years or more; therefore, almost all communities include buildings that vary widely in age **(Figure 1.1)**. Because construction technology is continually evolving, the varying ages of buildings within a community present a number of challenges to firefighters.

A discussion of building construction for firefighters must include older construction methods as well as new technology. Many buildings built using methods now considered obsolete or undesirable continue to exist in any community.

For example, building codes recognized the need to enclose stairwells and other vertical openings years ago. However, old buildings with open stairwells continue to exist **(Figure 1.2)**. Similarly, although heavy-timber (masonry exterior, mill style) construction has become obsolete, many heavy-timber buildings are still standing and have been converted to other occupancies. For all types of buildings, change in occupancy alone can present significant fire fighting and life safety challenges.

NOTE: Heavy-timber construction is covered in more detail in Chapters 2 and 7.

Figure 1.1 Older buildings often remain in use as newer buildings are built near them. *Courtesy of Ed Prendergast.*

Figure 1.2 The cast-iron staircase in this hotel dates from 1896. *Courtesy of Ed Prendergast.*

Building Variables

In any size community, a fire officer can be faced with difficult situations because building variables affect the course of the fire. These building variables include not only the age of the building but also fire protection systems, occupancy type, fuel load, type of construction, configuration, and the buildings' access and exposures. Some of these variables are technical, such as the thermal properties or strength of building materials. Others, such as occupancy, are more general in nature.

Age

The age of a building is not in itself a hazard, but age is often an indication of potential hazards. The age of a building affects the occurrence and behavior of fire in indirect ways. As a building ages, its general condition deteriorates **(Figure 1.3)**. The physical systems, such as the electrical wiring and heating plant, may require maintenance and updating. Buildings are subject to the prolonged effects of weathering. Roofs, for example, may leak and be resurfaced, possibly leaving damaged wood under the newly repaired roof. Mortar begins to erode and exposed wood may begin to decay. Foundations can settle, resulting in shifting of structural loads.

Over time, the ownership of a building will change and remodeling may take place. Some of the remodeling may be cosmetic, such as concealing defects or creating voids. Older buildings may become less appealing in the real estate market. People with limited resources may have difficulty maintaining the property and needed repairs may have been made in a substandard manner.

Gentrification — Process of restoring rundown or deteriorated properties by more affluent people, often displacing poorer residents.

The effects of age are not uniform. Older neighborhoods sometimes undergo *gentrification* in which older properties are extensively rehabilitated to satisfy the desires of a modern real estate market **(Figure 1.4)**. In these cases the general effects of aging can be negated.

Figure 1.3 A building showing the obvious effects of age.

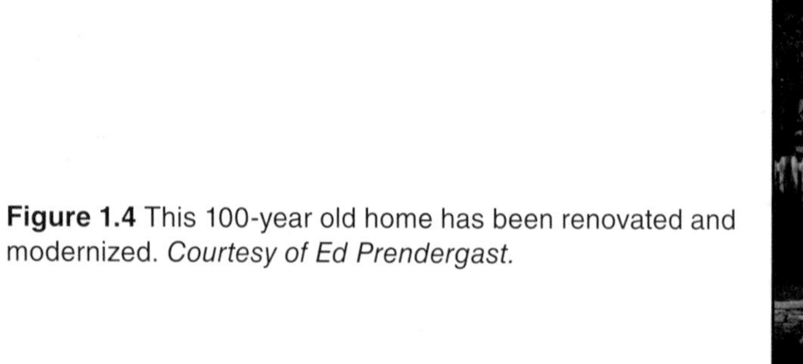

Figure 1.4 This 100-year old home has been renovated and modernized. *Courtesy of Ed Prendergast.*

In some cases older buildings offer some advantages to the firefighter. The design methods used in the past were less precise than those currently available with the use of computers. This necessitated the use of structural safety factors that often resulted in greater structural mass than was absolutely necessary. This greater mass often results in greater structural stability under fire conditions.

Modern design methods permit the more efficient use of materials, which results in less material being used for similar structural members. These more slender and lighter structural members are structurally sound; however, under fire conditions they may fail more quickly than older, heavier members.

Automatic Fire Suppression Systems

An automatic fire protection system, especially an automatic sprinkler system, is the first line of defense in a building. If the sprinkler system is properly designed and maintained, incipient fires will be promptly detected and controlled. Even where there are impediments to complete control of a fire by an automatic suppression system, such systems can provide faster notification of firefighters and a slowing of fire growth.

NOTE: Refer to Chapter 4 for additional information on fire suppression systems.

Occupancy

The occupancy of a building often affects the ways in which building components behave under fire conditions. Because different occupancies contain varying hazards and fuels, a fire will subject structural components to different temperatures, heat releases, and fire durations. For example, an automobile body shop will contain such highly flammable materials as gasoline and diesel oil, plastic (from the interior of the car), rubber tires, flammable liquids, paint, and torches. Some body shops may also work with fiberglass and related solvents. A fire in this environment will develop rapidly and subject the structure to high temperatures.

Fuel Loading — Amount of fuel present expressed quantitatively in terms of weight of fuel per unit area. This may be available fuel (consumable fuel) or total fuel and is usually dry weight.

In contrast, an occupancy such as a bank typically has relatively low amounts of combustibles (low fuel load) with only a few potential sources of ignition. When the amount of fuel is limited, the fire duration and the temperatures developed will not be as great.

Over time, many buildings undergo one or several changes in occupancy. These changes can produce more than new contents or décor: the overall use of a building can change. Developers and architects frequently convert buildings from one occupancy to another to meet changing demands in the real estate market. For example, an old railroad station may be converted to a restaurant or a warehouse may be converted to a residential apartment building **(Figure 1.5)**.

Figure 1.5. These condominium apartments were converted from a cold-storage warehouse. *Courtesy of Ed Prendergast.*

Changes in occupancy frequently result in a significant difference in the amount and type of combustible materials in a building. For example, a new owner or occupant may introduce fuels that exceed the designed capability of an existing sprinkler system. Building and fire inspectors play an important role in detecting hazards associated with changes in occupancy. Very often a change of occupancy is only discovered when a fire or building inspector visits the property in the course of a routine inspection.

Type of Construction

Nothing is more fundamental to a building than the materials from which it is constructed. For example, a wood-frame building reacts to fire conditions very differently than a fire-resistive building. The publishers of model building codes (originally the insurance industry) recognized the fundamentally different fire behavior of different construction types and classified buildings by their construction type. The traditional construction classifications were as follows:

- Fire-Resistive
- Noncombustible
- Masonry or Ordinary
- Heavy Timber
- Wood Frame

Contemporary building codes use numerical designations for the various construction types as follows:

- Type I — Fire-Resistive
- Type II — Protected Noncombustible or Noncombustible
- Type III — Exterior Protected
- Type IV — Heavy Timber
- Type V — Wood Frame

Each classification is further divided into one, two, or three subclassifications. (These are explained in detail in Chapter 2). Wood-frame construction, for example, is divided into two subclassifications depending on whether the wood framing is provided with a protection such as gypsum board.

Figure 1.6 The concrete structural system of this building is clearly identifiable. *Courtesy of Ed Prendergast.*

Because each construction type reacts to fire conditions in a different manner, the firefighter's ability to recognize a construction type is important. Sometimes recognizing the construction type is easy; for example, some wood-frame buildings such as barns or sheds can be classified simply by observing the exterior. Heavy, reinforced concrete structures can usually be identified in a similar manner as well **(Figure 1.6)**.

In many cases, however, the actual construction type is not obvious. A wood-frame structure with a brick veneer can frequently be mistaken for a building with masonry bearing walls **(Figure 1.7)**. Different construction types may also be used in combination. A building may include portions that are wood-joisted masonry along with other portions that are noncombustible. This combination of classifications is especially likely when an older structure has been enlarged or has undergone renovation **(Figure 1.8)**.

Configuration

The *configuration* of a building refers to its general shape or layout. While building code requirements address fire fighting and fire behavior, building designers rarely consider fire fighting needs as a principle element of design. Designers tend to be more concerned with the functionality of a structure and its visual appeal. Design choices such as large undivided areas, ceiling heights, and vertical openings between floors such as stairwells and atriums, can significantly affect fire behavior and the response of building elements to a fire.

Vertical openings of different styles are frequently favored by designers or their clients (the building owners) because of the appearance of openness **(Figure 1.9)**. On the other hand, the potential danger that vertical openings create as a means of fire communication is well-understood by members of the fire service who see fires firsthand.

Figure 1.7 After the brick veneer is applied to this wood-frame building, it will be indistinguishable from a masonry structure. *Courtesy of Ed Prendergast.*

Figure 1.8 A fire-resistive addition being added to an existing heavy-timber building. *Courtesy of Ed Prendergast.*

Figure 1.9 The open atmosphere created by an atrium is a popular feature in hotel design. *Courtesy of Ed Prendergast.*

Building Configuration and Exiting

An example of how building configuration affects fire behavior can be seen in the 1986 fire in the Dupont Plaza hotel in Puerto Rico that resulted in the loss of 97 lives. The hotel was a 20-story, nonsprinklered, fire-resistive structure. A ballroom occupied the first floor. A gambling casino, lobby, and shops occupied the second floor **(Figure 1.10)**.

1. At 3:22 p.m. fire was discovered in the ballroom on the first floor. The fire had been deliberately set in a stack of recently delivered furniture.
2. A glass partition wall separated the ballroom from an adjacent open stairwell that extended to the lobby level above. The glass partition wall failed from the heat of the fire and products of combustion spread up the open stairwell.
3. The gambling casino was provided with two exits but they both led into the lobby where the open stairwell communicated fire from the ballroom level. The open stairwell not only provided a path for the communication of fire through the structure, it also resulted in the blocking of the two exits from the casino.
4. Eighty-five of the victims were located in the casino because there was no safe means of egress for them to escape the fire.[1]

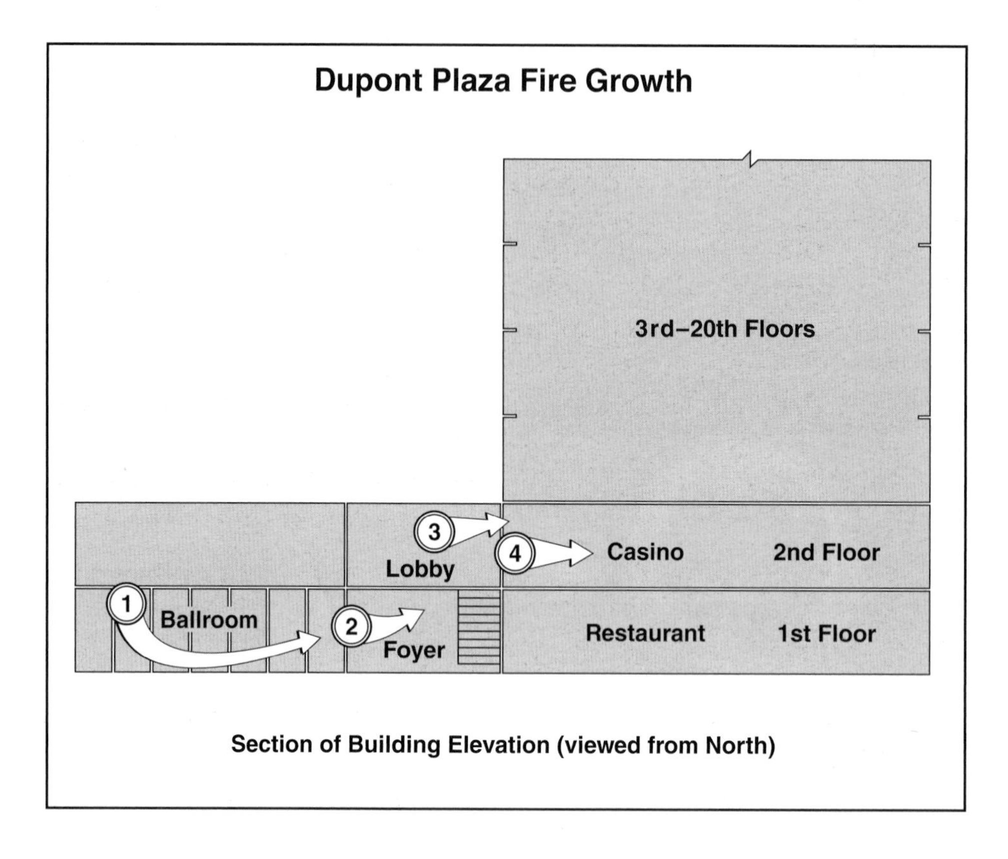

Figure 1.10 A diagram of the growth area of the DuPont Plaza hotel fire. *Source: NIST Engineering Analysis of the Early Stages of Fire Development – The Fire at the Dupont Plaza Hotel and Casino – December 31, 1986.*

Building Access

Access is a primary factor for fire department operations. Buildings are sometimes situated in locations that make fire department access difficult. All aspects of the terrain and built-out environment affect the fire department's ability to access buildings. Examples of terrain features that cause access problems include steep slopes, rivers, and landscaping. Built-out environ-

Figures 1.11 a and b (a) The river adjoining this condominium renders one entire side of the structure inaccessible to emergency vehicles. (b) Security bars are designed to discourage criminal behavior but will also slow access by emergency crews. *Photo a courtesy of Ed Prendergast. Photo b courtesy of Dave Coombs.*

ments that cause access problems include narrow roadways, setbacks, and barriers placed for security. These circumstances do not necessarily affect the fire behavior of the buildings but they do affect fire fighting tactics and must be recognized by the fire officer **(Figures 1.11 a and b)**.

Exposures and Community Fire Defense

Communication of Fire

The National Fire Protection Association® (NFPA) defines a building as an exposure when the heat from an external fire might cause ignition of or damage to the exposed building [(2)]. In tactical fire fighting, the term *exposure* is also used to refer to a structure or an object such as a propane tank or a pile of lumber to which a fire could spread. An exposed building is a building threatened by fire.

Exposure — Structure or separate part of the fireground to which the fire could spread.

Conflagration — Large, uncontrollable fire covering a considerable area and crossing natural fire barriers such as streets; usually involves buildings in more than one block and causes a large fire loss. Forest fires can also be considered conflagrations.

Communication of fire from building to building has been recognized as a serious fire problem for centuries. The earliest provisions of building codes were directed at the danger of *conflagrations*. Conflagrations occurred when fires communicated from building to building. Some of the earliest fire regulations adopted in colonial Boston prohibited thatched roofs to prevent the spread of fire.

As cities evolved and grew, building codes were developed with the aim of preventing the spread of fire between buildings. These provisions included such measures as the following:

- Requiring fire shutters or wire glass to protect the windows of closely spaced buildings.
- Requiring fire walls in an effort to limit the potential size of fires to a level that would not threaten the entire community.
- Imposing limits on the height and area of combustible construction for the same reason.

The problem of communication of fire remains significant, especially in older urban environments with closely spaced combustible construction **(Figure 1.12, p. 18)**. In modern suburban communities, building-to-building fire communication has been somewhat reduced as a result of code

Figure 1.12 These closely spaced combustible garages can pose a fire communication problem. *Courtesy of Ed Prendergast.*

Setback — Distance from the street line to the front of a building.

requirements for building setbacks, limiting building sizes for specific lot sizes, off-street parking, and loading. These requirements result in greater separation between buildings than are found in the inner areas of older communities.

Wildland/Urban Interface — Line, area, or zone where structures and other human development meet or intermingle with undeveloped wildland or vegetative fuels.

In the wildland/urban interface, fire communication to exposures can be from shake (wood) roofs, open vents, exterior sidings, and vegetation. Wildland/urban interface codes are now addressing these problems by regulating types of roofing and siding materials, and requiring the creation of defensible space through fire-resistant vegetation.

Convection

Convection — Transfer of heat by the movement of heated fluids or gases, usually in an upward direction.

Communication of fire from building to building occurs by convection and/or radiation **(Figure 1.13)**. *Convection* is the transfer of heat by the movement of liquids or gases, usually in an upward direction. In a large fire, convective currents can carry flaming debris great distances downwind. Depending on the wind direction, convective plumes are also significant when the exposed building is higher than the exposing building **(Figure 1.14)**.

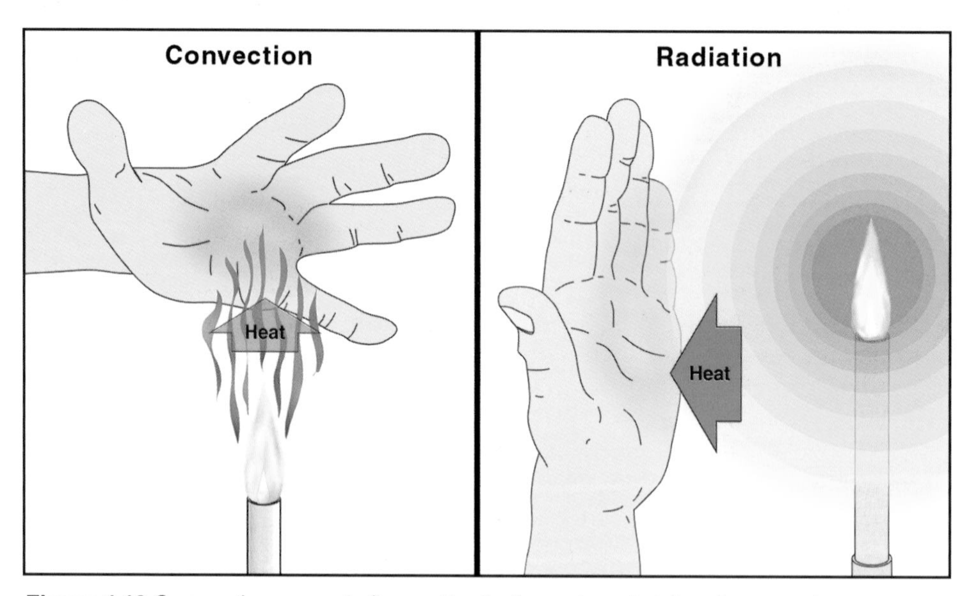

Figure 1.13 Convection spreads fire vertically through a structure by way of open stairs, shafts, and the inside of walls that lack fire stops. Radiation spreads fire in all directions.

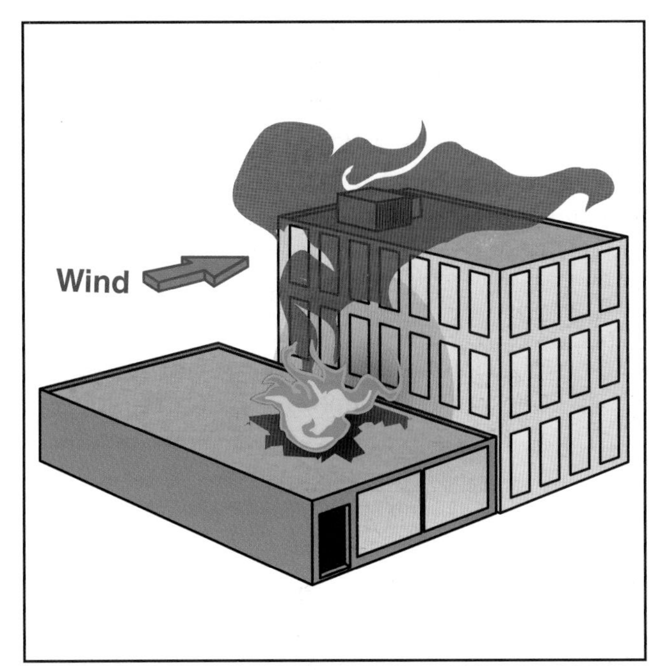

Figure 1.14 A fire in a shorter building can expose a taller neighboring building.

Radiation

Horizontal communication of fire is mainly due to thermal radiation. *Thermal radiation* is the transfer of heat energy through space by electromagnetic waves **(Figure 1.15)**. All bodies emit thermal radiation at a rate dependent on their absolute temperature. When two bodies, such as two buildings, have different temperatures, there will be a net transfer of energy from the body of higher temperature to the body of lower temperature.

Thermal Radiation — The transmission or transfer of heat energy from one body to another body at a lower temperature through intervening space by electromagnetic waves similar to radio waves or X-rays.

In physics, the Stefan-Boltzmann law states that the intensity of thermal radiation is a function of the fourth power of the absolute temperature of the thermal radiation source (T^4). Thus, a small increase in temperature produces a large increase in thermal radiation. For example, if the exterior wood frame wall of a building were to become involved in fire, its temperature would increase from the temperature of the surroundings, which may be 70°F (21°C), to the temperature of the flame, approximately 1,000°F (538°C). This is an increase of 930°F (517°C). A temperature of 1,000°F (538°C) is 1,460R (807K) where R denotes Rankine, the absolute scale for degrees Fahrenheit. Because the thermal radiation is a function of the fourth power of the absolute temperature, the relative increase in thermal radiation emitted in this example can be calculated as follows:

$$\frac{(\text{Higher temperature})^4}{(\text{Original temperature})^4} = \frac{(1{,}460^\circ)^4}{(530^\circ)^4} = \mathbf{57.5}$$

Figure 1.15 Fire spread is often the result of radiant heat. *Courtesy of NIST.*

Therefore, the intensity of the thermal radiation would increase by a factor of more than 57. Simply stated, it is possible for thermal radiation to ignite a building across a wide street.

Factors That Affect Communication of Fire

A heavily involved building can readily communicate fire to an adjacent building. Flame temperature alone is not the only factor in determining if a fire can communicate to an exposed building. The area, namely the height and width, of the exposing flame is also a factor. Flames emitting from a window in a masonry wall can expose an adjacent building. As the area of the opening increases in size, the amount of thermal radiation increases. Similarly, the fully involved wall of a wood-frame building creates an enormous amount of thermal radiation due to the large area of the exposing flame.

NFPA 80A®, *Recommended Practice for Protection from Exterior Fire Exposure*, describes three levels of exposure based on the potential severity of the exposing fire. The standard classifies the levels of exposure as light, moderate, or severe. For example, a lumber yard with its high fire load will obviously pose a greater exposure fire risk than a daycare center. The levels of severity are based on the fire load and the flame spread rating of the wall and ceiling finishes of the exposing building. These parameters establish the severity of the exposing fire.

There are a number of methods used to protect buildings from exposing fires, including the following:

- Clear space between buildings
- Fire suppression systems such as water-based sprinkler systems and fire-retardant distribution systems
- Blank walls of noncombustible construction
- Self-supporting barrier walls between the fire building and the exposure
- Parapets on exterior masonry walls
- Automatic outside deluge systems
- Elimination of openings in exterior walls
- Glass block panels in openings
- Wired glass in steel sash windows
- Automatic fire shutters or dampers on wall openings
- Automatic fire doors on door openings

Because a fully involved building can communicate fire across wide streets, the possibility of communication of fire should always be included in preincident planning.

Building Design Principles and Considerations

Building Failure

In engineering, failure is said to occur when a structure or part "is no longer capable of performing its required function in a satisfactory manner." Depending on what a structure is designed for, failure could mean excessive vibration, deflection, noise, or wear.

To the firefighter, building failure usually means structural collapse. In the broader sense, however, it can mean that the building or part of it is no longer performing its required function in a satisfactory (designed) manner. Thus, the communication of fire through a fire-rated barrier can be viewed as a failure. The unsatisfactory performance of a fire protection system is also a failure. Potential sources of building failure under fire conditions can include some or all of the following:

- Structural integrity
- Building systems
- Design deficiencies

Structural Integrity

The collapse of a building under fire conditions is a result of the loss of a building's structural integrity **(Figure 1.16)**. The structural integrity of a building under fire conditions is related to the fire resistance and combustibility of the materials of which it is constructed. Combustible materials may possess some initial fire resistance, as with heavy plank flooring, and be able to act as a barrier to fire, but ultimately they will be consumed. Depending on their physical dimensions, noncombustible materials such as steel or glass may also retain structural integrity at first but will fail from the effects of heat.

Figure 1.16 Buildings constructed of combustible materials will collapse under heavy fire. *Courtesy of District Chief Chris E. Mickal, NOFD Photo Unit.*

Fire-resistive materials possess the ability to maintain structural integrity (See Chapter 2). Structural integrity permits effective interior attacks and, therefore, is of fundamental importance to the firefighter.

Building Systems

In the modern world, buildings function as total systems to provide a healthy, productive, and comfortable environment for the occupants. Specific building systems include heating, ventilation, and air conditioning (HVAC); electrical power; communications; plumbing; and transportation such as elevators and conveyors. Improper or inadequate design of these systems can contribute to building failures under fire conditions. For example, it has long been recognized that the ductwork and circulating fans of a ventilation system can contribute to the spread of products of combustion throughout a building **(Figure 1.17, p. 22)**. Good design practice requires that provisions be built into a system to prevent the spread of combustion products. These provisions would include such measures as smoke detectors to initiate the shutdown of units or to operate dampers in ducts.

Electrical systems are an essential aspect of all buildings. Their design and installation should include fire stopping at the openings where conduits penetrate floor slabs and firewalls. Electrical systems should also include a provision for emergency power that is protected from fire and highly unlikely to fail as a backup system. If both electrical systems fail, fire fighting operations become much more difficult. For example, a fire penetrating a single utility

Figure 1.17 An HVAC system can draw the products of combustion into the ducts and transport them throughout the building.

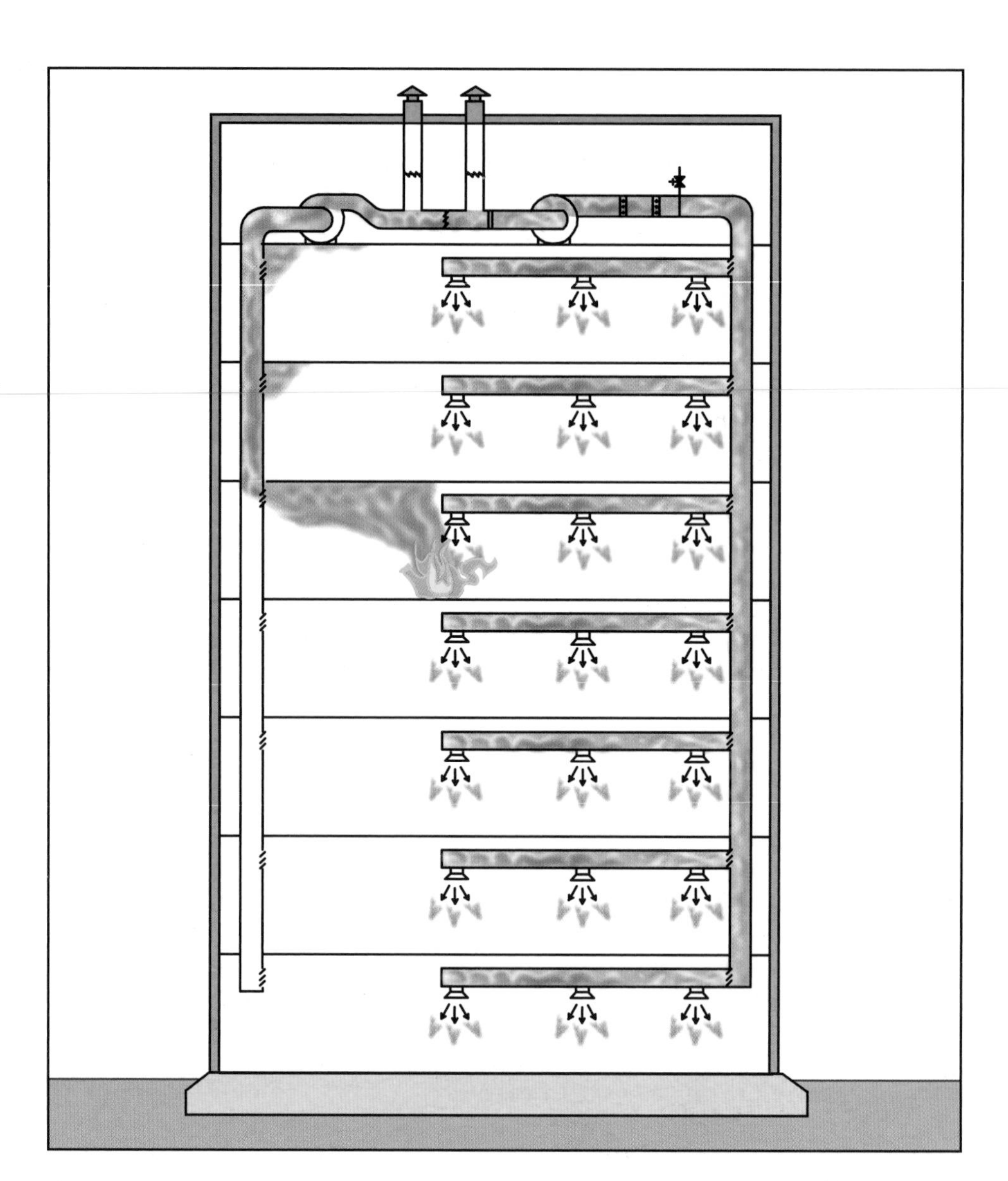

closet at the 1991 Meridian Plaza fire in Philadelphia resulted in the failure of the primary and the emergency power to the entire high-rise building, resulting in loss of power to fire pumps and elevators.

NOTE: Building systems are discussed in greater detail in Chapter 4. The Meridian Plaza fire is described in Chapter 12.

Design Deficiencies

The term *design deficiencies* in this text refers to a failure to provide a level of fire safety appropriate to the ultimate use of the building. Incorrect assumptions and oversimplifications about the activities planned for a building frequently creep into the design process. One very basic aspect of building safety is the provision of an adequate number of exits **(Figure 1.18)**. This number is determined by the number of persons likely to occupy a building, subject to building code provisions for occupancy classifications. This aspect can be difficult to plan. An exhibition hall might be used for an industrial trade show with a restricted attendance; the same hall might also be used for a rock concert with a highly dense and volatile crowd. Obviously the difference in these two situations is enormous in terms of emergency evacuation.

Design deficiencies can also occur in fire protection systems. An example of such a deficiency would be failure to provide adequate water supply for a sprinkler system. This can occur when the magnitude of a likely fire (the fire load) is underestimated.

In the design process, heavy reliance is placed on codes and standards but those who write building codes cannot anticipate every situation that may arise. Codes are also subject to the political process in which those involved have competing interests. Codes can only provide a "reasonable" level of protection for the most commonly encountered situations. In specific cases, a code may not provide an adequate level of safety or compliance with the provisions of a code may not be possible. In these situations, a fundamental technical analysis of the problem posed is the best course of action. However, competing priorities, including economics, frequently discourage measures that exceed the safety provisions outlined in codes.

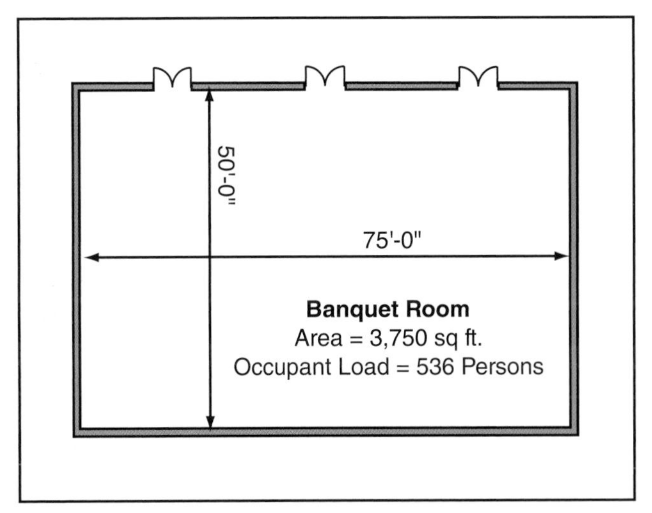

Figure 1.18 Providing adequate exits for expected occupant load is a crucial part of fire safety design.

Principles of Design

Buildings are not designed to collapse, catch fire, burn, or restrict egress; however, that does not mean that fire and life safety are the most prominent design considerations when owners and designers plan their projects. The following sections of this chapter explore building design principles, processes, and products. The text will also help give the reader an understanding of the way structures, systems, and space are integrated to make a building useful, livable, and profitable.

In the field of architecture, an enormous amount has been written and said about buildings and the way they are designed. One of America's most prominent architects, Frank Lloyd Wright, called buildings "machines for living." A few years earlier, Chicago architect Louis A. Sullivan, known for his stately commercial buildings, intoned his design philosophy by stating, "form follows function."

The truth recognized by these famous architects was simple: Buildings are a complex of many diverse elements and systems. Fire service personnel must understand the essential elements of building design and construction in order to predict how the structure, occupants, and contents will be affected by fire and how best to mitigate those effects.

Why Buildings Are Built

An enormous amount of energy, resources, and talent are expended in the design and construction of buildings. Ironically, a fair amount of energy is also expended in tearing them down. A basic philosophical question asks why humans build buildings. Why do buildings exist in the first place? Exploring this question gives insight into how buildings are designed, what the designer is trying to accomplish, and how buildings are constructed.

Security From Physical and Social Forces

Buildings are built because people need them. In primitive societies, simple structures made from animal skins or tree limbs provided shelter from the elements. As societies evolved, buildings were designed to provide people and their property with security from both physical and social forces. They also provide security and privacy for the activities carried out within a society, such as in office buildings and factories **(Figure 1.19)**.

Figure 1.19 This library is an example of a building designed for a specific purpose.

Investments

Because energy and resources are expended in the construction of buildings, they become repositories of wealth. Buildings are frequently constructed as investment vehicles. Some building construction is undertaken without an occupant being known. These buildings are referred to as "spec buildings." Office plazas, shopping malls, light industrial buildings, and condominiums may be developed primarily as investments rather than for specific needs. Obviously, real estate investment is not foolproof and is subject to the economic forces of the market.

Spec Building — (short for speculation). Building built without a tenant or occupant.

Real estate can outlive its useful economic life. Fire investigators are aware that when buildings become obsolete and unmarketable, they frequently become a target for arson.

Cultural Desires

Buildings take on cultural as well as functional characteristics. The design of a building frequently serves as an expression of taste or to convey a certain image **(Figures 1.20 a-c)**. Projections of an owner's wealth or a corporation's sense of permanence are frequently incorporated into the design. Because buildings are ultimately used by people, designers strive for designs that are appealing and comfortable as well as those that enhance human endeavor.

Design Considerations

Before studying the details of building construction, it is useful to examine the overall design process. The log cabin illustrates how various considerations impact building design. Log cabins were originally built because logs were an available and useful building material. Today, log cabins are erected mainly for appearance rather than for the structural or economic advantage of using logs as a building material.

Figure 1.20 a-c The appearance of a building often conveys its intended use. *Photo a Courtesy of Ed Prendergast.*

In earlier times, the appearance of a building was determined by the materials used in its construction. Today, building design is used to achieve a pleasant aesthetic appearance, so the desired appearance will sometimes dictate the material used. When a decision is made to design and construct a building, the designer has many factors to consider in reaching a satisfactory result. These include the following:

- Fiscal resources (cost)
- Building use
- Aesthetic tastes
- Building codes
- Safety
- Accessibility
- Climate
- Infrastructure
- Soil conditions
- The physical laws of engineering
- The owner's needs and desires

Cost

The overall design of a building, as well as the individual details, is determined by available funds, and there are few situations where construction funds are unlimited. A school design is an example of the way funds affect the overall design of a building. Communities usually desire a school that is attractive architecturally and is equipped with good classrooms, laboratories, and athletic facilities. Available tax dollars, however, may force the architect to design a less costly facility. Thus, the size of the library may be reduced or music rooms may be eliminated.

With respect to the details of design, an owner may be willing to install marble flooring in an office building lobby but may in turn reduce the size of an emergency generator. The reason given would likely be that the lobby decor is visible and marketable while the generator is hidden from view and, presumably, will be used infrequently. Cutting costs in this way may lead to difficulties during an emergency if inexpensive or inadequate systems fail.

Building Use

A building's design must facilitate its end use. Grain silos, aircraft hangars, fire stations, and movie theaters are common examples of buildings whose fundamental design is determined by their use **(Figure 1.21)**. For example, It would be hard to imagine a one-story grain silo or a movie theater with windows.

Figure 1.21 The use of this fire station has determined its design. *Courtesy of Ed Prendergast.*

The end use of a building also dictates subtle requirements. In the design of mercantile buildings, the floor plan is usually laid out to maximize the customers' exposure to the product while minimizing the opportunities for shoplifting. In hospitals, the nurses' station is centrally located so the staff can observe the corridors and have a minimum travel distance to the patients' rooms.

The desire to renovate older buildings in some urban areas can result in unique efforts at change of occupancy. A few typical examples include:

- A church converted into condominiums
- A storm window manufacturing plant into a community theater
- Parking facility converted to a flea market
- Two-story factory converted to a shopping mall
- Multistory industrial buildings converted to condominiums with parking

The fact that a building has been designed for a specific occupancy can create problems when it is later converted to a different use. For example, a building's fundamental design might be unsuitable for a proposed occupancy under the provisions of the building code. A building or fire official could be placed in the uncomfortable position of stopping a desired building rehabilitation, but the fundamental issue would be that the building was never designed for the proposed function in the first place!

NOTE: Refer to the case history at the end of Chapter 2 for an example of an emergency that was affected by a change in occupancy.

Aesthetics

Although normally of little interest to the fire service, *aesthetics*, or the art in building design, is a major force in architecture. Architects are creative professionals, and their work affects the quality of the community environment. Architectural styles are subject to change, just as are the styles of other products. For example, prominent architect Mies van der Rohe introduced a distinctive style of building (International Style) in the early 1930's **(Figure 1.22)**. The style was characterized by very clean rectangular lines and was used in many buildings in that era. More recent designs, however, are very different from the geometric simplicity used by van der Rohe **(Figure 1.23)**.

Aesthetics — Branch of philosophy dealing with the nature of beauty, art, and taste.

Aesthetics is such an important concern to an architect that it frequently clashes with fire safety concerns. For example, an architect may object to the placement of a fire command panel in a hotel lobby because it detracts from the lobby decor. An architect may attempt to conceal a fire department connection (FDC) or a pump test header because of its "clunky" appearance. Architects frequently want to place sprinklers above decorative ceilings, which may impede sprinkler operation. Interestingly, in other cases the architect will leave piping and other mechanical equipment exposed to achieve a rugged atmosphere **(Figure 1.24)**.

Figure 1.22 Simple and clean architectural lines represent a particular style. *Courtesy of Ed Prendergast.*

Figure 1.24 Exposed piping and ductwork are relatively common design features.

Figure 1.23 Some buildings are designed to make a strong impression through their distinctive design. *Courtesy of Ed Prendergast.*

Building Codes

Before a building is constructed, a unit of local government — such as a city, county, or state, typically requires that a building permit be obtained. Before the permit is issued, the proposed design must meet the provisions of the local building code. A *building code* is a body of law that determines the minimum standards that buildings must meet in the interest of community safety and health. Because the code is a body of law, it imposes restrictions on designers that may conflict with their creative desires. For example, a code typically restricts the maximum size of wood-frame construction or requires the use of fire-retardant roofing materials and disallows others.

Building Code — Body of local law, adopted by states, counties, cities, or other governmental bodies to regulate the construction, renovation, and maintenance of buildings.

Although building designers and developers may view codes as burdensome governmental regulation, their importance to the fire service cannot be overstated. It is the building code that determines, by law, the kinds of buildings in which the firefighter must ultimately fight fires. The procedures for issuing building permits varies from jurisdiction to jurisdiction. In some jurisdictions fire officials are actively involved in the permitting process; in others they are not.

NOTE: For a more detailed description of the permitting process, see the IFSTA **Plans Examiner for Fire and Emergency Services** and **Fire Inspection and Code Enforcement** manuals.

Although jurisdictions can write and adopt their own codes, they typically adopt all or a portion of a "model code" package as their building and fire code. The practice of adopting a model code saves the many government entities the task of writing a fairly complex document and provides a fundamental degree of uniformity among jurisdictions. Jurisdictions are free to amend a model code by appropriate legislative act to suit local conditions and frequently do so.

International Building Code

Currently the most widely used model building code in the US is the *International Building Code* (IBC) published by the International Code Council Incorporated. The *International Building Code* is the successor to three earlier model codes:

- The *Uniform Building Code* (UBC) published by the International Council of Building Officials (ICBO).
- The *Standard Building Code* published by the Southern Building Code Congress International.
- The *BOCA National Building Code* published by the Building Officials and Code Administrators International.

The International Building Code has a companion fire code, the *International Fire Code*. The building and fire codes complement each other and are intended to be used together. Because they complement each other, both the fire and building codes must be used together during facility design. This interdependence underscores the desirability for building and fire officials to work together in applying and enforcing building and fire codes.

Some jurisdictions may adopt and use codes developed by the National Fire Protection Association (such as NFPA® 1, *Uniform Fire Code*, and NFPA® 5000, *Building Construction and Safety Code*. A commonly adopted NFPA®

code is the *Life Safety Code* (NFPA® 101). This code is often adopted by a governmental agency with a specific area of responsibility, such as a state health department.

In Canada, a widely used building code is the *National Building Code of Canada* published by the National Research Council of Canada. This document is published in the form of a model code and is adopted or adapted by most of the provinces and territories in Canada.

Safety

All designers have a fundamental responsibility to provide a safe end product. A significant portion of building codes are devoted to building fire safety. Specific fire safety provisions include requirements for the following elements:

- Structural fire resistance
- Flammability of interior finishes
- Adequacy of means of egress
- Enclosure of vertical openings
- Fire protection systems
- Exposure protection
- Occupancy separation

In addition to fire safety, a building must have adequate strength to prevent structural collapse. This means more than simply not collapsing under its own weight. The structural system must be adequate to withstand environmental forces such as wind, snow, and earthquakes **(Figure 1.25)**. Occupant safety must also be considered in the following areas:

- Design of stairs and walking surfaces
- Balcony railings
- Overhead obstacles
- Electrical wiring
- Elevator operation

In addition, the widespread use of air conditioning with recirculation of interior air has given rise to concerns over the quality of the air breathed by building occupants and indoor air pollution.

NOTE: For more information on the effects of wind and structural stability on structures, see Chapter 3.

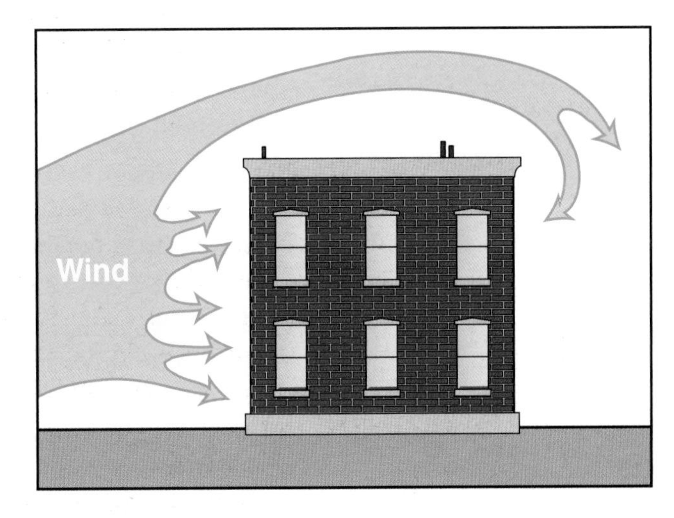

Figure 1.25 Wind can create a negative pressure on the roof and downwind side of a building. These forces must be taken into account during the design phase.

Accessibility

In 1990, the Americans With Disabilities Act (ADA) was signed into law in the US. This act requires that public facilities be accessible to persons with disabilities. The following impairments qualify as disabilities under the Act:

- Vision impairments
- Hearing impairments
- Learning
- Speech
- Neuromuscular impairments
- Mental illness

To provide accessibility for persons with impairments, the Act requires the removal of architectural barriers. This imposes on the architect an additional set of requirements in the design process. Some of the specific building elements that must be designed to accommodate individuals with restricted abilities include the following:

- Building entrances **(Figure 1.26)**
- Parking and passenger loading zones
- Elevators
- Drinking fountains
- Toilet facilities
- Alarms (visible and/or audible)
- Telephones
- Automated teller machines
- Means of egress

For example, manual fire alarm stations must be not more than 4½ feet (1.37 m) and not less than 3½ feet (1.1 m) above the floor level so they can be reached from a wheelchair **(Figure 1.27)**.

Area of Refuge — (1) Area where persons who are unable to use stairs can temporarily wait for instructions or assistance during an emergency building evacuation. (2) Space in the normal means of egress protected from fire by an approved sprinkler system, by means of separation from other spaces within the same building by smokeproof walls, or by virtue of location in an adjacent building.

People who are unable to use the stairs for emergency evacuation must be provided with alternate protection such as an area of refuge. An *area of refuge* is a protected area where a person can remain temporarily until someone else provides assistance or instructions. Areas of refuge can take several forms but may include such arrangements as a stairway landing in a smokeproof enclosure, a balcony located adjacent to an exterior stair, or a protected vestibule adjacent to an exit enclosure. Areas of refuge may not be required in buildings equipped with an automatic sprinkler system.

Where required, the areas of refuge must be equipped with two-way communication so individuals can call for assistance. Firefighters must be prepared to respond to persons who may be calling for assistance.

Utilities

The availability of public utilities must be considered early in the design process. For example, virtually all buildings require some form of sewer service. If a building is to be constructed where public sewers do not exist, a septic tank may be needed. If the project is large - such as a residential subdivision — it may be necessary or desirable to construct a sewer treatment facility.

Figure 1.26 A wheelchair ramp is an example of a design feature required by the Americans with Disabilities Act (ADA). *Courtesy of Ed Prendergast.*

Figure 1.27 Manual pull stations must be placed so that they are accessible by someone in a wheelchair.

Availability of water is a basic consideration in the design of fire protection systems. The primary concern is the amount of water that will be needed for the flow rate and duration of the fire protection system. A test may be necessary to determine if existing water mains can supply the required flow and pressure.

If the existing water mains cannot supply the required flow, it would be necessary to either increase the size of the mains or provide for on-site storage and fire pumps. In the case of the development of a subdivision, it may be necessary to extend water mains into the subdivision. This is a cost that may be borne by the developer or shared with the local community. If public water mains are not readily accessible, a well and storage tank may be necessary. In any case, the provision of water to a project contributes to the cost of the project.

Climate

Climatic concerns affect other aspects of building design. In northern regions, for example, it may be desirable to incorporate snow-melting equipment into the sidewalk pavement. In areas where there is a heavy annual rainfall, buildings may be designed with overhangs or enclosed walkways between them to protect pedestrians **(Figure 1.28)**.

The heating and cooling requirements of buildings are determined by the temperature variations of the region. Buildings in northern climates require greater heating

Figure 1.28 An elevated walkway between buildings protects pedestrians from traffic and inclement weather. *Courtesy of Ed Prendergast.*

capabilities, which in turn dictates the amount of space that must be dedicated to a heating plant and the choice of fuels. In contrast, buildings in the southwestern part of the United States may require only incidental heating but larger cooling capacity. In all climates, energy conservation dictates the use of insulating materials.

Energy Conservation (Green Design)

Green Design — Term used to describe the incorporation of such environmental principles as energy efficiency and environmentally friendly building materials into design and construction.

The increased awareness of the sometimes negative impact of human activities on the environment has resulted in efforts to make buildings more energy efficient **(Figure 1.29)**. This overall concept is known as *green design*. Efforts are made, for example, to reduce the greenhouse gas emissions from buildings arising from burning heating fuel. Other efforts include increasing energy efficiency by using less electrical power. In areas, efforts are made to reduce water consumption at a building site by curtailing landscaping that requires watering.

These efforts can have an indirect impact on fire fighting. Concerns with energy efficiency, for example, result in buildings with dual-pane windows and tighter fitting doors and seals. These affect the speed with which firefighters can ventilate a building and the rate of combustion within a building. When doors and windows are tightly fitted, the flow of air to a fire is reduced. This results in a fire burning in an oxygen-poor environment. Under these circumstances, when ventilation does occur, a rapid development of the fire can result. It is also possible in some cases that a fire can be so starved for oxygen that it burns itself out.

Soil Conditions

The conditions and properties of soil vary and must be evaluated before the foundation of a building is designed. Soil strength, strain resistance, and stability are properties of importance in foundation design. Soil properties are affected by factors such as frost action, water content, seismic shock, organic decomposition, and disturbance during construction. These soil properties affect building design in such fundamental matters as building height and in determining whether or not a basement is practical.

Physical Laws of Engineering

Regardless of the purpose of a building, it must be designed so it can actually be built. The proposed design of a building and the loads and forces exerted upon it must be matched to the strength of building materials and the mathematics of structural mechanics. The professions of architecture and structural engineering must be joined to accomplish the end product. Buildings that encompass exotic or innovative designs pose great challenges to the engineer and architect **(Figure 1.30)**. Some very well-known landmarks serve as testimony to the human ability to solve very difficult problems. A few examples include the Golden Gate Bridge, Sears Tower, and the St. Louis Gateway Arch.

Owner's Needs and Desires

In the final analysis, the building and its design belong to the owner. The architect works for the owner so it is appropriate for the owner, who will pay for and occupy the building, to have the final word in many design matters. Although

Figure 1.29 Green buildings are designed to conserve energy and may feature water collectors, solar panels, and rooftop gardens. *Courtesy of McKinney (TX) Fire Department.*

Figure 1.30 Architecture and engineering are combined in the design of this gymnasium. *Courtesy of Ed Prendergast.*

the person paying the bills is entitled to some indulgence, the owner's input is not necessarily a matter of whimsy. For example, the design of a restaurant may be influenced by the specific atmosphere desired or the owner's practical knowledge of restaurant operations.

Often, all the design considerations confronting the architect cannot be satisfied. For example, a conflict can exist between the maximum surface flammability permitted by a code and the desired grandeur of a particular interior finish material. The design process involves compromises and a prioritization of objectives. Cost is always a major concern and frequently necessitates a reduction in plans. The final building design always involves a balance of what is desired, what is needed, and what is practical.

The Design and Construction Process

The design and construction of a building is a process that begins as an idea and may end with a massive structure covering acres of space and housing thousands of people. Getting from the idea to the finished product may take weeks or it may take years. The process involves creative talent, technical knowledge, legal expertise, management skills, and financial resources.

Concept

The process begins when either the developer or the owner perceives a need. This need could be a need for a new hospital, school, warehouse, or air terminal. The owner may contact an architect or solicit proposals from several architectural firms. The architects and their associates will have meetings with the owner to determine the owner's needs and desires. If the building is not large, as in the case of an ordinary house, the process may begin with a contractor. The contractor may retain an architect, who will produce preliminary plans for presentation to the owner. Once the idea of the building appears on paper, it begins to be something concrete. The owner has the opportunity to evaluate the design and changes can be made.

The owner may contract with a single firm to undertake both the design and construction of a building. Such an arrangement is known as a "design-build" project. The company undertaking the total project in this manner is usually a general contractor with its own architects and engineers on staff. It is also possible for separate contracting and design firms to affiliate through a joint venture for a particular project.

The Design Product

The design and construction process is a serious and expensive undertaking. Many people are involved and many decisions must be made. The architect usually functions as the prime designer who has the responsibility and control to choose major aspects of the building and to eliminate alternatives [6].

Some buildings may be designed using tried and proven techniques involving predesigned, prefabricated components. This reduces the time and cost of the design work. Other buildings involve unique architecture and newly developed methods and materials. A design may also involve old materials used in innovative ways **(Figure 1.31)**. Within the overall design process are areas of specialization. The erection of buildings involves several branches of engineering and all of the building trades. In addition to architecture, the major technical specialties and their functions include the following:

- **Civil engineering**
  - Water supply
  - Sanitary sewers
  - Surveying
  - Site preparation and excavation
  - Roadways
  - Storm water drainage
- **Structural engineering**
  - Determination of loads
  - Foundation design
  - Structural behavior
  - Structural members
  - Structural erection
- **Mechanical engineering**
  - Heating, ventilation, and air conditioning
  - Pumping systems
  - Elevators
  - Plumbing systems
- **Electrical engineering**
  - Lighting
  - Power
  - Communications

Figure 1.31 Laminated wood arches are used to support the roof of a subway station and to provide a design element. *Courtesy of Ed Prendergast.*

- **Fire protection engineering**
  - Automatic sprinklers **(Figure 1.32)**
  - Standpipes
  - Fire alarm
  - Smoke control
  - Code compliance

The finishing touch in the design process is landscaping. It is not uncommon for a local code to require a specified type or amount of landscaping in the interest of community beautification. The details of this work will fall to yet another professional: the landscape architect.

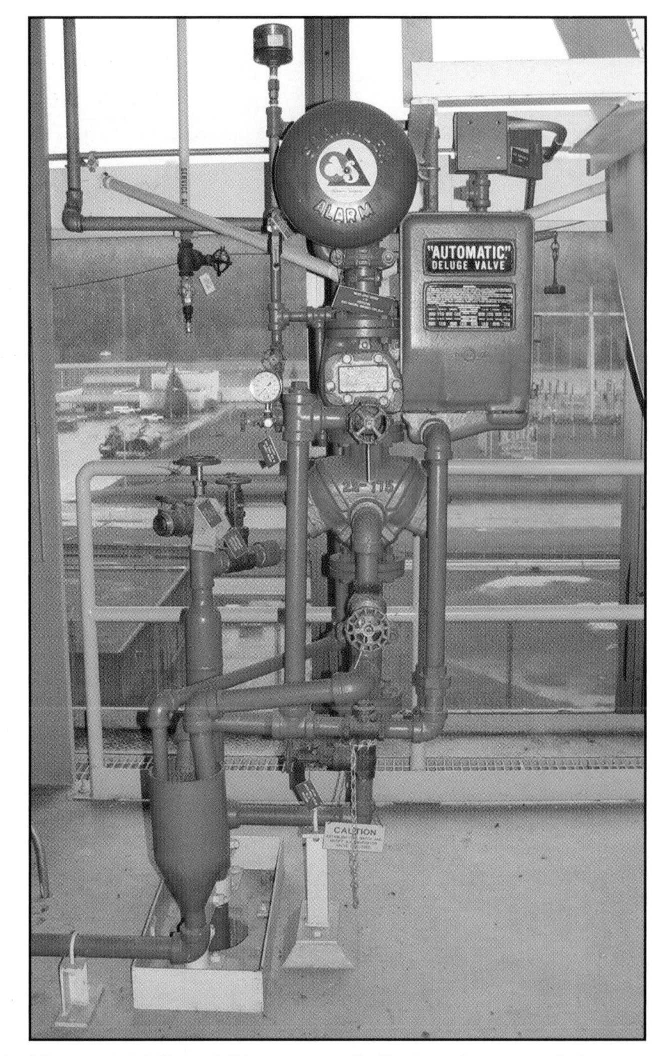

Figure 1.32 A sprinkler system is likely to be a crucial part of any commercial building. *Courtesy of Dave Coombs.*

Financing

After the initial design has been selected, the owner must secure financing for construction. The financial institution providing the construction loan will have certain requirements, which may include a market analysis to evaluate the economic feasibility of the project. In the case of such projects as an office building or a shopping center, the lending institution will want to know that tenants have expressed a willingness to rent space when the building is completed. The lending institutions' technical requirements will necessitate a review of the architect's design drawings and may include such engineering documents as land surveys, preliminary budgets, and soil test reports.

Documentation and Permits

When financing has been secured, the engineering design of the building can proceed. Meetings are held to make final modifications and design decisions. At this time, the details of the building become more specific. These details involve everything within the scope of the final building and extend to such common items as toilet facilities, door handles, and lighting fixtures. Bids are received from subcontractors who will perform specialized portions of the construction, such as the electrical wiring, and contracts are signed.

Board of Appeals — Group of people, usually five to seven, with experience in fire prevention, building construction, and/or code enforcement legally constituted to arbitrate differences of opinion between fire inspectors and building officials, property owners, occupants or builders.

A building permit is obtained from the local building department. Normally, plans of the building are submitted to the building department for review as part of the permit process. If the building official notes conditions that do not comply with the code, further changes may be necessary. Unusual designs or circumstances may require the use of compensatory measures or equivalencies rather than strict compliance with a prescriptive code.

When a proposed design is rejected by a building official, an architect may make use of an appeals process. Building codes provide for an appeals process to resolve differences in interpretation of the specific provisions of a code or to review an alternative means of complying with the code. The appeals process usually involves a *Board of Appeals* constituted as provided for in the building or fire codes.

It is very important for building and fire officials to cooperate in the review and permit process. When the fire department is involved in the plans process, fire safety issues can be addressed before construction begins. Uncovering and correcting problems before the start of construction improves the efficiency and cost effectiveness of the process **(Figure 1.33)**.

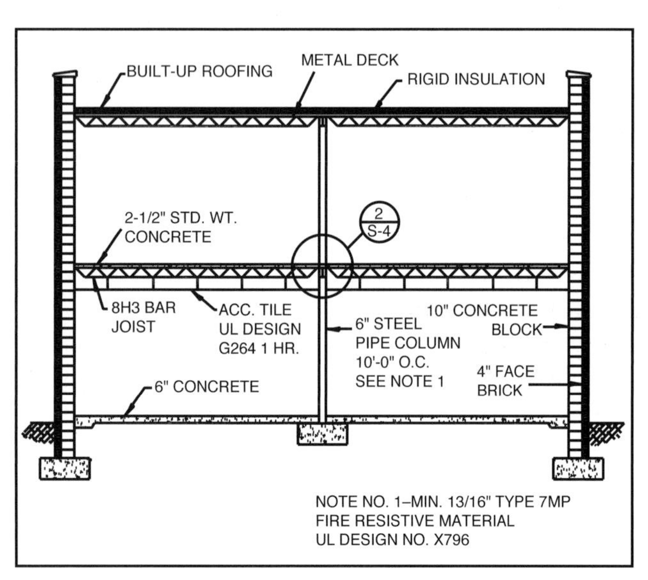

Figure 1.33 Fire and code enforcement officials must have the opportunity to evaluate building plans before construction begins. The details of structural fireproofing shown here are just one aspect that needs to be reviewed.

In some jurisdictions, the fire authorities have legal authority to review building plans. In jurisdictions where the fire department does not have formal authority, it is very useful to establish a cooperative relationship between the fire department and the building officials. In addition to providing for fire safety compliance, a working relationship with the building officials provides the fire department with information on the type of construction being undertaken in its jurisdiction.

In the case of large projects such as a high-rise building that will take many months to construct, the building department may issue a permit for the initial phase of work so construction can begin while the final details of design are being completed. A permit may be issued for the foundation so that the foundation excavation and construction can begin, thus saving both time and money.

Renovation and Remodeling

Another important aspect of the permit process concerns permits for building renovation and remodeling. Over the course of their useful lives, buildings often undergo various renovations. If this work is not carried out properly, unsafe conditions could result **(Figure 1.34)**.

Structural modifications of buildings are of particular concern to firefighters. Building departments require that structural modifications be designed by structural engineers and performed by licensed contractors. Minor renovations such as adding an electrical receptacle also need to be performed by licensed contractors. A jurisdiction may not require the submittal of plans for minor renovations; however, an inspector will come out to ensure that the work was completed in a proper manner.

Without oversight by a building department, buildings can be weakened if work is performed without regard to proper methods. Examples of these structural problems include the following:

- Removal or penetration of bearing walls
- Modification of beams or trusses
- Structural overloading of roofs
- Creation of mezzanine floors in attic spaces
- Rooftop additions
- Remodeling that creates additional voids, such as suspended ceilings
- Illegal remodeling or overloading that poses extreme hazards

Figure 1.34 This collapse resulted when workers – working without a permit – attempted to enlarge a nightclub by removing part of a load-bearing wall.

Subdivision of existing spaces through the creation of partitions can be hazardous, creating maze-like floor plans for emergency responders.

Depending on the extent of the renovation, the owners may have to bring the building up to current code and may be reluctant to do so because of additional costs. Unwillingness to spend this additional money is an incentive for doing work without permits. Situations, therefore, will be encountered where renovations violate the provisions of a building code. An example of this is the construction of corridor partitions that do not have the required fire resistance. A contractor may use non-fire-rated materials for an exit corridor enclosure where a fire-rated enclosure is required. Other examples include the use of interior finish materials that exceed the flammability requirements of the code or fire-rated windows that have been replaced with non-fire-rated windows. In each of these examples compliance with the specific requirements of a building code would be more costly.

Construction

The actual construction process requires coordination and scheduling. Some tasks must be performed before others; for example, the structural steel must be in place before the roof can be constructed. The need for coordination also extends to suppliers, who need to know when materials will be needed at the job site. If a hotel is being constructed, for example, a plumbing supplier may have a contract to supply 1,000 bathtubs. If the bathtubs are not ready when they are needed, construction will be delayed. If the bathtubs arrive before they can be installed, storage will become a problem.

One method of managing a construction project is a technique known as "fast tracking." In a fast-track project, the design and construction phases overlap. As the early design phases are completed, construction is begun while later phases are still undergoing design. Because this method can greatly reduce total construction time, it is very attractive to owners.

The timing of other components may affect the fire department. For example, the street must be paved and water supply in *before* combustibles are on site to start the project. In winter this may be difficult because of weather conditions, yet if the road and access does not get done, the builders will still want to go ahead and build the structures.

Inspection/Testing

Inspection of the construction takes place during and upon completion of a project. Inspections are performed by the architect or the architect's representative and by the building department. Inspection involves the verification that proper materials and construction techniques are used. Building construction is a dynamic process and it is not uncommon for problems to arise or mistakes to occur over the course of construction. As construction nears completion, the architect makes a final inspection.

It is critical that knowledgeable inspectors work closely with architects and contractors during construction. There have been cases of structural failure due to seemingly minor changes to architects' or engineers' plans for ease of construction.

Hyatt Regency Collapse

A tragic example of structural failure occurred in Kansas City, Missouri in July of 1981 at the Hyatt Regency Hotel. Extensive renovation to the hotel had been completed in July of 1980. The hotel complex included an atrium 50 feet (15 m) high that was used for various functions.

The atrium was spanned by three walkways at the second, third, and fourth floor levels. The walkway for the fourth floor was located above the walkway for the second floor. On the evening of July 17th, the second and fourth level walkways collapsed while a dance competition was being held in the atrium. The collapse resulted in the deaths of 144 persons and injuries to an additional 200 people.

In the original design for the hotel, the walkways for the second and fourth floors were to be supported by a series of continuous steel rods from a truss system at the roof of the atrium. The rods were to pass through box beams used to support the walkway at the fourth level and continue to the box beams used to support the second floor level walkway. The steel rods were to support the weight of both the second and fourth floor walkways. **Figure 1.35** illustrates the manner in which the original design was to support the walkways.

The box beams were supported by a steel nut and washer at the bottom of the box beam. The contractor made a change in the final design. Instead of having one continuous rod supporting both walkways, two rods were used. One rod terminated at the underside of the box beam supporting the fourth-level walkway. A second rod extended down from the top of the fourth level box beam to the underside of the box beam to support the second level walkway. This change was actually reviewed by the structural engineers but they evidently did not realize the significance of the change.

The design change resulted in a significant alteration of the stresses within the box beams at the fourth level walkway to the extent that the nut and washer assembly pulled out of the box beam it was supporting.

The owner's representative will also inspect the building. The general contractor is notified of any defects or discrepancies noted or details that remain to be completed.

The local building department also inspects the building to ensure compliance with the building code. Inspections by the building department usually take place during construction as well as upon completion.

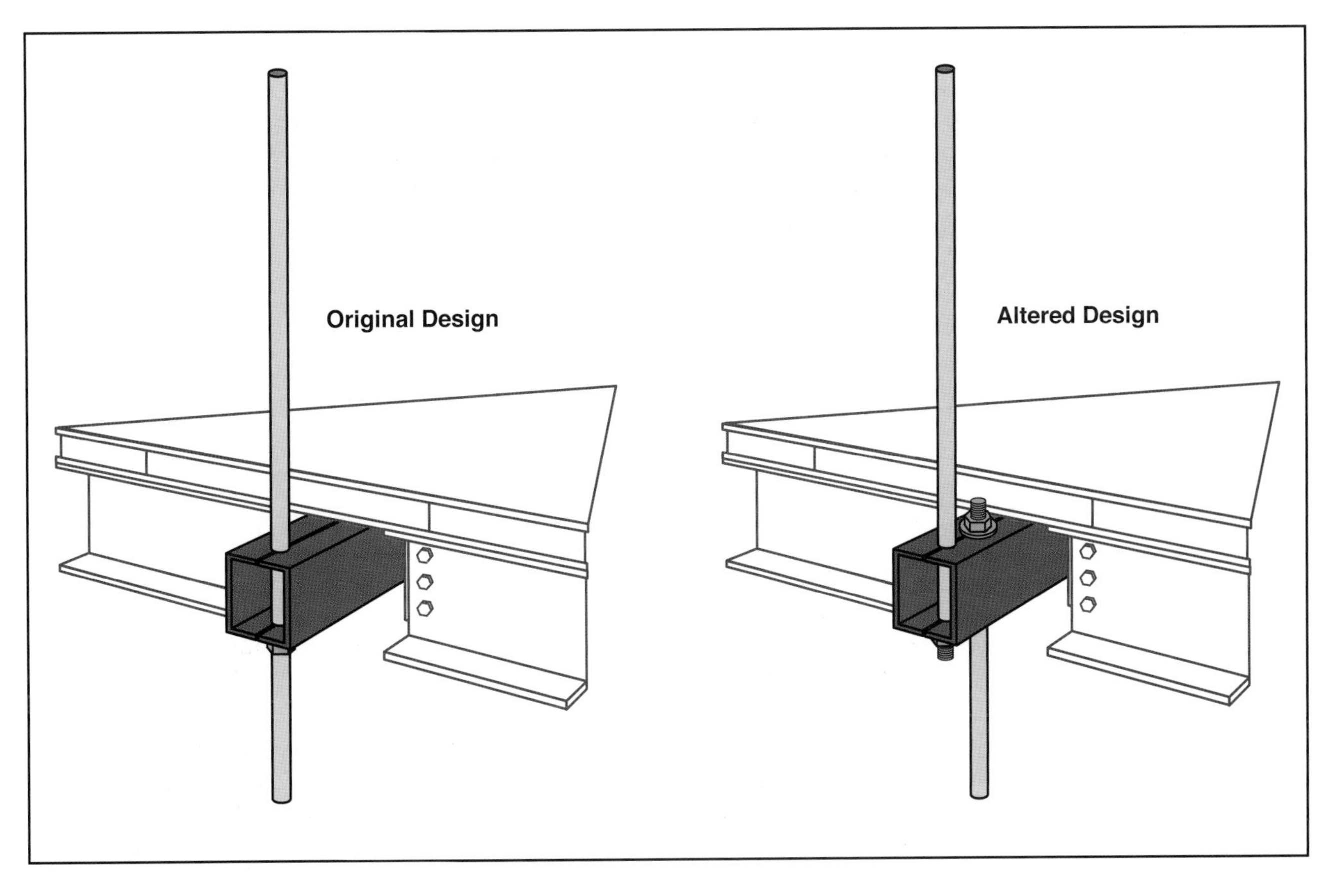

Figure 1.35 The change from the original design for the Hyatt Regency placed additional stress on the walkway supports, which eventually failed.

Testing is performed on certain materials, systems, and components such as concrete, fire pumps, and emergency generators. The fire department is usually involved in the testing of the fire protection systems in new construction. For example, the fire marshal may witness the testing of the automatic fire suppression systems. Fire personnel who must operate in buildings under fire conditions have a basic interest in the reliability of the automatic sprinklers, standpipe systems, fire alarm systems, and smoke control systems.

The primary role of the fire inspector is to ensure proper installation and operation of the fire protection systems. In addition, the involvement of a fire department representative in new construction provides the fire department with first-hand information useful in preincident planning. It should be noted that the role of the fire inspector is to *witness* system tests, not to actually perform the tests. This is to avoid issues of liability should a system component fail during the test. The actual system tests are carried out by representatives of the installing contractor.

Documentation of fire protection system test results should be maintained by the fire prevention bureau. This serves several purposes. One is to establish that the systems were installed in accordance with the fire code and that the system operated properly. The other is to facilitate re-inspection and subsequent testing over the life of the systems. As years go by, personnel changes occur both in the fire department and in building management. Documentation provides for continuity of the system as personnel change.

Building Construction and Preincident Planning

There are several ways tactical firefighters can develop knowledge of building construction in their jurisdiction. One way is to obtain information through the building permit process. As buildings are permitted for construction, information can be passed from the building department to the fire department. This method is particularly useful in the case of large structures such as hospitals, shopping malls, exhibition halls, and high-rise buildings. This information can be shared between fire companies and the fire prevention bureau. In the case of smaller buildings such as single-family dwellings and small commercial buildings, however, it is impractical and probably overwhelming to try and track information through the paper flow of permits.

Preincident Planning — Act of preparing to handle an incident at a particular location or a particular type of incident before an incident occurs. Also called Prefire Planning, Preplanning, Prefire Inspection, or Preincident Inspection.

A second means of developing knowledge of building construction is through preincident (pre-fire) planning. *Preincident planning* is a valuable tool in accomplishing control of emergencies in individual buildings. The more information that is known in advance, the more effectively and expeditiously a fire or other emergency can be controlled. Preincident plans include information regarding occupancy, industrial process, hazardous materials, fire protection systems, building access, and utilities.

Geographic information systems (GIS) are another means of obtaining, analyzing, and using data based on location. GIS has multiple uses across many disciplines and is also becoming more widely used in building construction applications. For large-scale construction projects, GIS allows a large project to be divided into smaller areas or sectors. This makes the project more manageable and allows for logistical, planning, and tracking functions to take place with greater efficiency. By pairing global positioning systems (GPS) with survey information, GIS applications also provide a means for determining the exact locations necessary for placement of critical building elements such as foundations and structural steel.

Figure 1.36 A fire at a large construction site presents many additional hazards for firefighters, particularly access and structural stability.

Fundamental to the preincident planning process is information about a building's construction. This includes the construction type (i.e., wood frame, fire-resistive), location of fire walls, vertical openings, roof construction, exits, and any smoke control systems that may be provided.

Because buildings periodically undergo renovation or remodeling, preincident planning cannot be a one-time occurrence. For this reason, periodic updates to preincident plans are necessary. The frequency of updates will vary with the nature of the occupancy. For example, the preincident plans for large hospitals and shopping malls could be updated more often than a free-standing restaurant. Having current preincident plans ensures accuracy and familiarity with individual sites.

It is also critical to preplan during construction because buildings probably are never closer to "falling down" than they are during construction or renovation **(Figure 1.36)**. Fire emergencies during this stage result in emergency response during the worst time for accessibility and building stability.

Summary

The laws of physics and chemistry that govern fire behavior never change but the buildings in which fires occur vary greatly. The buildings in any community are products of many circumstances and design variables over which firefighters have little or no control. The task of the tactical firefighter is to understand the design, construction, and functioning of a building so emergency operations of all types can be carried out effectively. The structural firefighter must be a student of building construction. The building on fire is the workplace of the firefighter.

Review Questions

1. What types of building configuration issues can significantly affect fire behavior?
2. How is fire communicated by radiation?
3. What are three potential sources of building failure under fire conditions?
4. How do the accessibility requirements of the Americans With Disabilities Act (ADA) affect building construction?
5. What is *Green Design*?

References

1. Thomas J. Klem, "92 Die in Arson Fire at Dupont Plaza Hotel," *NFPA Fire Journal,* Vol. 81, No.3 May/June 1987, pp.74-82.
2. NFPA® 80A, *Recommended Practice for Protection of Buildings from Exterior Fire Exposure,* National Fire Protection Association®, Quincy, MA. 2001 ed.
3. Ibid.
4. Richard M. Phelan, *Fundamentals of Mechanical Design,* McGaw Hill, 3rd ed. p. 96.
5. NFPA® *Fire Protection Handbook, 19th ed.* Sect 12, Chap 10.
6. James Ambrose, *Building Construction,* 2nd ed. John Wiley and Sons..
7. Matthys Levy and Mario Salvadori, *Why Buildings Fall Down,* W.W. Norton and Company.

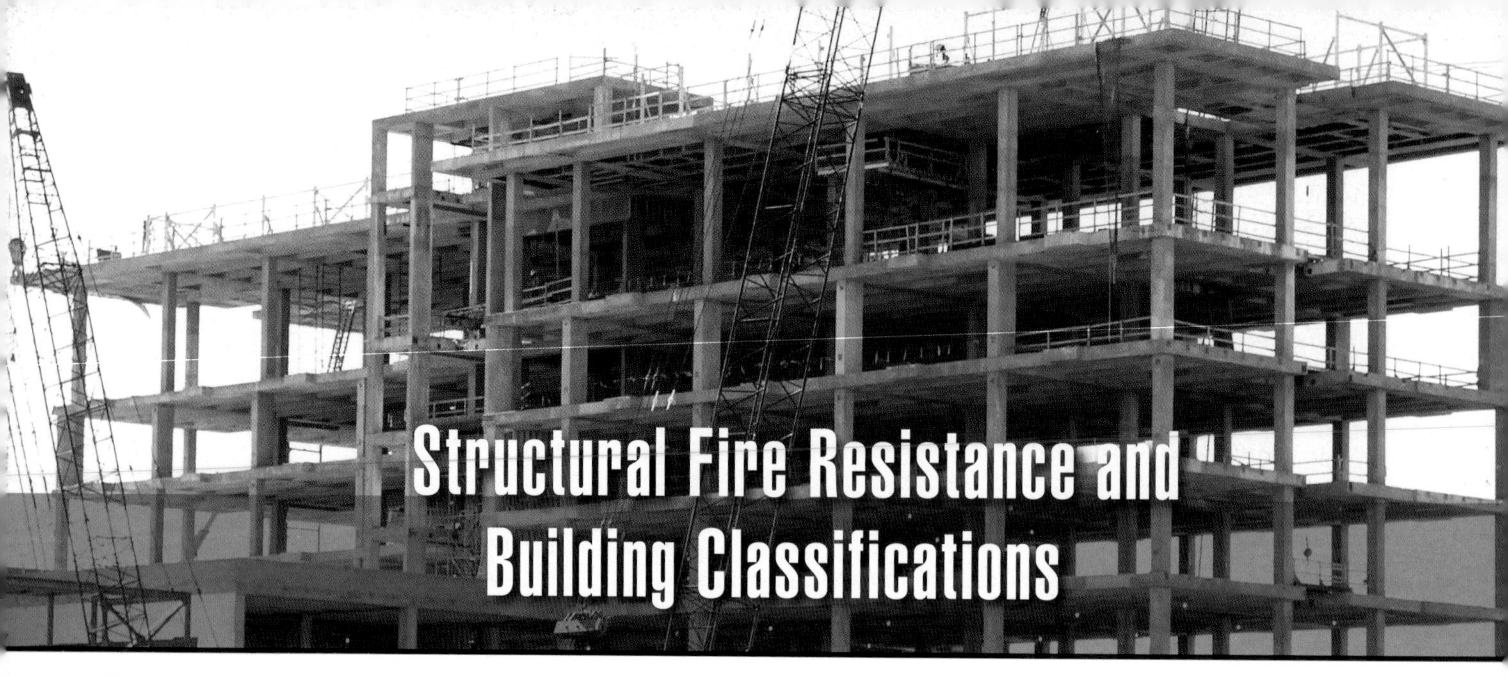

Chapter Contents

Divider page photo courtesy of McKinney (TX) Fire Department.

Key Terms

FESHE Objectives

Fire and Emergency Services Higher Education (FESHE) Objectives: *Building Construction for Fire Protection*

1. Demonstrate an understanding of building construction as it relates to firefighter safety, building codes, fire prevention, code inspection and firefighting strategy and tactics.
2. Classify major types of building construction.
3. Analyze the hazards and tactical considerations associated with the various types of building construction.
7. Classify occupancy designations of the building code.

Structural Fire Resistance and Building Classifications

Learning Objectives

After reading this chapter, students will be able to:

1. Define fire resistance.
2. Discuss methods of determining fire resistance and the limitations of each method.
3. Identify fire testing organizations and discuss the significance of fire test results.
4. Recognize the role of analysis in determining fire resistance.
5. Discuss the basic building classifications as they relate to fire resistance.
6. Discuss the concept of fire load and its impact on building construction types.
7. Explain occupancy classifications as they relate to fire risks.

Chapter 2
Structural Fire Resistance and Building Classifications

Case History

Event Description: We were operating at a two-story residential structure fire with a heavy fire load. The building was a few years old and constructed with TCI floor joists and trusses. The owner finished the garage, which from the outside still looked like a garage and screwed Plexiglas® panels over the windows to seal from air leaks, which hindered our efforts. Within 10 minutes of the first-due engine arriving, approximately 10 firefighters started an interior attack. Some progress was made; however, those of us who were outside could see and hear that the attic area was heavily involved. We evacuated the structure and within minutes the roof collapsed and we had a partial second-floor collapse. Had the firefighters been allowed to continue making their progress we would have had serious injuries if not a LODD.

Lesson Learned: We learned some important lessons about modern lightweight construction and its effects on fire in these buildings. We have had a building boom here for years and have a large number of these homes. We have trained on these issues before and will be revisiting the issues again. It was an eye-opening experience to say the least.

Source: National Fire Fighter Near-Miss Reporting System.

Chapter One briefly described the building design process. To the firefighter, however, the most significant characteristic of a building is not its architectural style but how it behaves under fire conditions. In the field of fire protection, buildings are classified according to the manner in which they behave under fire conditions. For example, wood-frame and reinforced concrete buildings behave very differently under fire conditions so are classified differently. (See Chapters 7 and 10 for more information).

All building codes classify buildings by construction type. These classifications are based on two attributes of building construction: fire resistance and combustibility. Fire resistance determines the likelihood of structural collapse under fire conditions. The combustible nature of a building's structural system will impact the rate of fire growth.

These building classifications are fundamental from both fire fighting and fire safety standpoints. For example, fire-resistive buildings permit firefighters to make a more aggressive interior attack than they can make in non-fire resistive buildings. Fire-resistive buildings also provide an increased degree of occupant safety.

In addition to classifying buildings by construction type, building codes classify buildings by their occupancy. These occupancy classifications reflect the differing life safety issues posed by different occupancies. Health care occupancies, for example, present different fire safety issues than warehouses. Taken together, the building construction and occupancy classifications are used in the building codes to establish limitations on the permissible heights and areas of buildings.

This chapter discusses fire resistance and test methods for determining degrees of fire resistance, fire testing organizations, basic building classifications and construction types, fuel loads, and occupancy classifications.

Fire Resistance

From a fire protection standpoint, one of the most basic properties of building materials is their degree of fire resistance. Fire resistance is a function of the properties of all materials used, including combustibility, thermal conductivity, chemical composition, density, and dimensions **(Figure 2.1)**. *Fire resistance* is the ability of a structural assembly to maintain its load-bearing capacity and structural integrity under fire conditions. Fire-resistive construction is not prone to structural failure under fire conditions. In the case of walls, partitions, and ceilings, fire resistivity also means the ability to act as a barrier to fire.

Fire Resistance Rating — Rating assigned to a material or an assembly after standardized testing by an independent testing organization that identifies the amount of time a material or assembly of materials will resist a typical fire as measured on a standard time-temperature curve.

The fire resistance of structural components can be evaluated quantitatively and is known as the *fire resistance rating*. Fire resistance ratings are expressed in hours and fractions of hours. The fire resistance ratings for structural components are incorporated into the construction classification of buildings in building codes. Building codes will have requirements for the fire resistance of structural elements such as the following:

- Beams
- Columns
- Walls and partitions
- Floor and ceiling assemblies
- Roof and ceiling assemblies

Figure 2.1 An example of noncombustible construction. *Courtesy of McKinney (TX) Fire Department.*

For example, a building code will typically require that columns supporting the floors in a fire-resistive building have a fire resistance rating of 3 hours. The walls enclosing an exit stairwell, which may or may not be load bearing, typically must have a fire-resistive rating of 1 or 2 hours to protect the stairwell. In addition, fire doors and windows will have a fire-resistance rating.

Determination of Fire Resistance

There are three means by which the fire resistance of structural assemblies can be determined:

- Conducting standard fire resistance testing in a laboratory
- Performing analytical calculations to determine the resistance to a standard fire test exposure
- Employing analytical structural fire engineering design methods based on real fire exposure characteristics.[1]

Of these three, the most commonly used method of determining fire resistance is by laboratory test. The laboratory test is also incorporated into building codes.

The earliest known fire tests on building materials were conducted in Germany in 1884-86.[2] In the United States the first known fire tests were conducted in Denver, Colorado in 1890, with subsequent tests in New York City in 1896.

A standardized test method for floor construction was adopted by the American Society for Testing Materials (ASTM) in 1907. A method for the fire testing of walls and partitions was adopted in 1909. The standards that were developed were presented to the NFPA® in 1914 for consideration. Revised versions of the standards were adopted by the NFPA®, ASTM, and the American National Standards Institute (ANSI). The standard was adopted by the NFPA® in 1917 and has evolved as an NFPA® standard ever since. Today the standard test is described in NFPA® 251, *Standard Method of Tests of Fire Endurance of Building Construction and Materials.* It is also designated as ASTM E-119.

NOTE: See Chapter 5 for more information on testing interior finishes.

Fire Resistance Test Method

The standard fire-resistance test is widely used in fire protection to establish the required performance standards in building codes. The primary means used to determine a fire resistance rating is to subject the component to be evaluated to the heat of a standard fire in a test furnace **(Figure 2.2, p. 48)**. In the standard fire test, the furnace temperature is regulated to conform to a standard time-temperature curve. In other words, the temperature in the test furnace is raised along a time scale. Key points on the curve are shown in **Figure 2.3, p. 48**. A temperature of 1,000°F (538°C) is reached at five minutes and 1,550°F (843°C) after 30 minutes. At one hour the temperature is 1,700°F (927°C).

The fire resistance of structural systems is affected by the manner in which they are used in the field. Some structural components, such as partition walls between individual residential units, do not support loads other than their own weight. Other components, such as columns, support the weight of the building. Therefore, in the test the structural elements are loaded in a manner that will approximate the working stresses expected in the design. The test results will be classified as either load bearing or non-load bearing.

In addition, the fire resistance ratings for floor and ceiling assemblies are developed for both restrained and unrestrained assemblies. This is because end restraints affect the extent to which an assembly may expand or rotate at its ends when exposed to high temperatures, affecting its ability to support a load.

Figure 2.2 Testing material inside furnace. *Courtesy of Underwriters Laboratories Inc.*

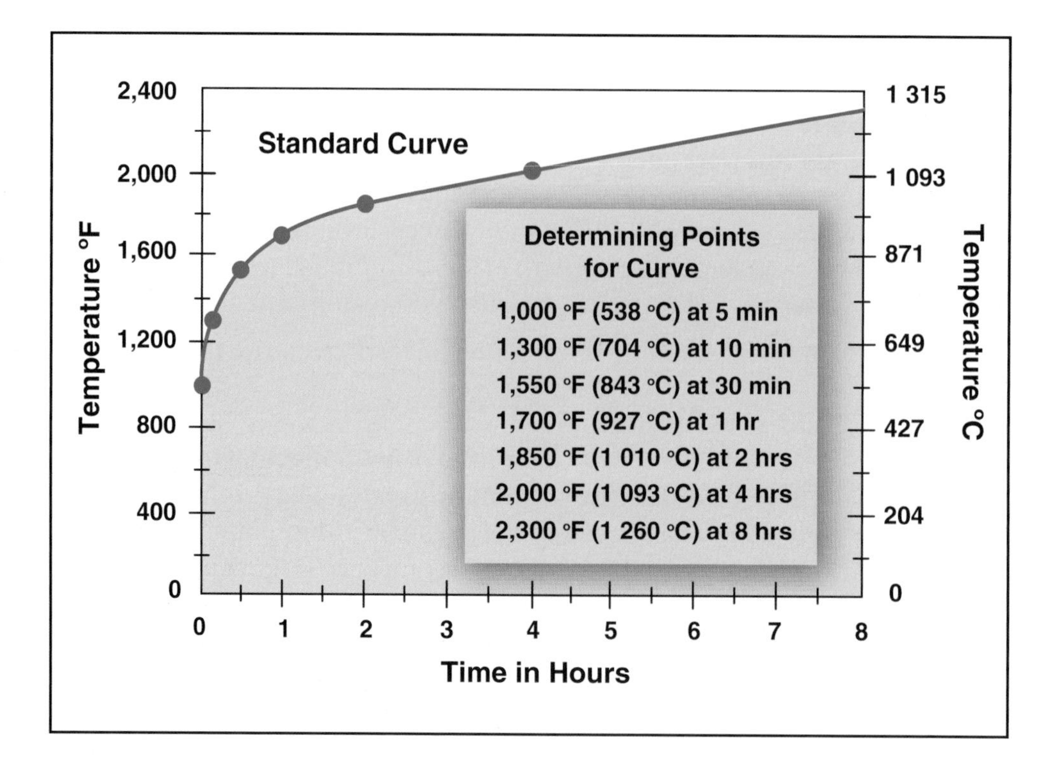

Figure 2.3 Furnace temperatures are regulated to conform to an established time-temperature curve.

When a structural specimen is tested, the test is continued until the specimen fails or the specified fire endurance for which the specimen being tested is reached. Thus, if a partition wall were being tested to obtain a 2-hour fire rating, the test would be stopped at two hours if failure had not occurred. Normally, assemblies are not tested beyond four hours because this is the maximum time required by the building codes.

The specific failure criteria depend on the specimen being tested -- such as a column or a wall assembly -- and whether the test specimen is restrained or unrestrained. The primary points of failure for the test are as follows:

- Failure to support an applied load
- Temperature increase on the unexposed side of wall, floor, and roof assemblies of 250°F (121°C) above ambient temperatures

- Passage of heat or flame through the assembly sufficient to ignite cotton waste
- Excess temperature on steel members*

* **NOTE:** The failure point temperature will depend on how the steel is used. Different maximum temperatures are also stipulated for single points as well as an average temperature.

Underwriters Laboratories, Inc. (UL) — Independent fire research and testing laboratory with headquarters in Northbrook, Illinois that certifies equipment and materials. Equipment and materials are approved only for the specific use for which it is tested.

In addition, certain wall and partition assemblies are subjected to the application of a hose stream to duplicate the impact and thermal shock of water that might occur during fire fighting operations.

Although an assembly may fail at any point during the test, fire resistance ratings for test specimens, including fire doors and windows, are expressed in standard intervals such as 15 minutes, 30 minutes, 45 minutes, 1 hour, 1½ hours, 2 hours, 3 hours, and 4 hours. Thus, if a given assembly failed one hour and ten minutes into a test, its fire rating would be one hour.

Not all field conditions can be duplicated in the laboratory, including construction variables. Obviously, size restrictions do not permit testing entire buildings. For example, the test furnace used by Underwriters Laboratories, Inc. (UL) for the testing of beams, floor, and roof assemblies has approximate plan dimensions of 14 x 17 feet (4.26 m x 5.18 m) **(Figure 2.4)**. The standard test procedures replicate the behavior of materials in relatively small compartments. The behavior of identical materials or assemblies in larger configurations could vary from the standard test results due to the effects of thermal expansion in larger members[(3)].

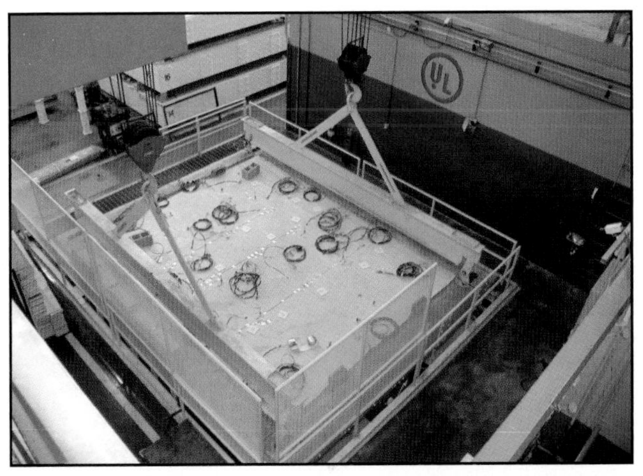

Figure 2.4 Fire resistance test for larger assemblies. *Courtesy of Underwriters Laboratories, Inc.*

It must be clearly understood that the fire-resistance ratings are established using a standard laboratory test fire. In actual situations, rated assemblies may perform satisfactorily for longer or shorter periods of time. This variation is due partly to differences in workmanship and materials encountered in the real world that may vary substantially from those used in test specimens.

It must be remembered that the standard time-temperature curve may not duplicate the situations encountered in real fires. The standard test fire assumes an endless fuel supply and adequate ventilation to produce increasing temperatures. This obviously would not be the case with a fire occurring in a light-hazard occupancy with limited ventilation. In contrast, a fire involving flammable liquids could produce higher temperatures more quickly than in the standard fire. For this purpose a more severe test fire is described in ASTM Standard 1529, *Standard Test Methods for Determining Effects of Large Hydrocarbon Pool Fires on Structural Members and Assemblies.* This test would be useful for evaluating the structural members used in such occupancies as petroleum refineries.

The limitations of the standard fire tests do not mean that they are without value in the field of building fire safety. On the contrary, the use of the fire ratings developed over the years has contributed significantly to the safety of individual buildings and collectively to the fire safety of communities. The E-119 test is the only method currently universally accepted by building codes.

It is important to remember that the standard test evaluates the ability of structural assemblies to carry a structural load and to act as a fire barrier. The test does *not* provide the following information:

- Information about performance of assemblies constructed with components or lengths other than those tested
- Evaluation of the extent to which the assembly may generate smoke, toxic gases, or other products of combustion
- Measurement of the degree of control or limitation of the passage of smoke or products of combustion
- Fire behavior of joints between building elements such as floor-to-wall or wall-to-wall connections

Fire Stop — Solid materials, such as wood blocks, used to prevent or limit the vertical and horizontal spread of fire and the products of combustion in hollow walls or floors, above false ceilings, in penetrations for plumbing or electrical installations, in penetrations of a fire-rated assembly, or in cocklofts and crawl spaces.

NOTE: Joint systems for floor-to-wall and wall-to-wall connections are tested in accordance with UL Standard 2079, "Standard for Fire Tests of Joint Systems."

- Measurement of flame spread over the surface of the tested material
- The effect on fire endurance of openings in an assembly such as electrical outlets and plumbing openings unless specifically provided for in the construction tested.[4]

Although all of the above limitations are important, the last is of particular interest to building and fire prevention inspectors. It is not uncommon over time for fire-resistive assemblies to be penetrated, which often occurs when buildings undergo renovation **(Figure 2.5)**. Penetrations of fire-resistive assemblies may be made for ductwork, plumbing, electrical, and communication purposes and not be adequately firestopped. When the continuity of an assembly is destroyed, it cannot function as a fire barrier.

Figure 2.5 Fire walls need to be kept intact if they are to serve as a barrier. *Courtesy of McKinney (TX) Fire Department.*

Fire Testing Organizations

The furnaces used to determine fire-resistance ratings are very large and materials are tested at high temperatures. Consequently, the test is beyond the capability of local fire and building departments. Laboratories equipped for such work must conduct the testing. The following are some of the organizations that perform fire-resistance testing:

- Underwriters Laboratories, Inc.
- Underwriters Laboratories of Canada
- Building Research Division of the National Research Council of Canada
- Southwest Research Institute
- Intertek Testing
- University of California at Berkeley, Forest Products Laboratory
- Armstrong Cork Company
- National Gypsum Company

Some of these organizations use their furnaces primarily for research and product development.

Fire Test Results

To make the results of fire-resistance testing available and useful to engineers, architects, and building officials, the test results are published by the testing laboratories. Probably the best known of the laboratories is Underwriters Laboratories, Inc. Underwriters Laboratories annually publishes a *Fire Resistance Directory,* which lists assemblies that have been tested and their fire-resistance ratings.

Fire Resistance Directory — Directory that lists building assemblies that have been tested and given fire-resistance ratings. Published by Underwriters Laboratories.

Figure 2.6 is a sample of a listing for a 1-hour-rated floor and ceiling assembly that uses a concrete floor, steel joists, and a gypsum wallboard ceiling. **Figure 2.7** illustrates a listing for a 3-hour-rated column using a center steel column protected by three layers of gypsum wallboard. Notice that the listings specify all the details of the assemblies. Deviation from the materials or dimensions specified would alter the test results. The attachment of the gypsum wall board to the steel column in Figure 2.6 is especially important to ensure its long-term integrity. Therefore, inspectors should not allow deviations from the listings in the field.

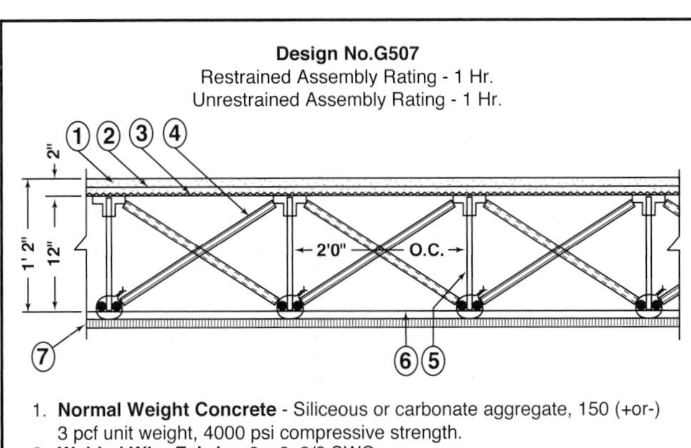

1. **Normal Weight Concrete** - Siliceous or carbonate aggregate, 150 (+or-) 3 pcf unit weight, 4000 psi compressive strength.
2. **Welded Wire Fabric** - 6 x 6, 8/8 SWG.
3. **Metal Lath** - 3/8 in. rib, 3.4 lb/sq yd expanded steel; tied to each joist at every other rib, and midway between joists at side lap with 18 SWG galv steel wire.
4. **Bridging** - 3/4 in., 16 USS gauge box channels or min 1/2 in. diam steel bars.
5. **Steel Joists** - Type 12J4 min size; spaced 24 in. OC and welded to end supports.
6. **Furring Channel** - 3/4 in. 0.30 lb furring channel or 7/8 in. 24 MSG nailing channels, 16 in. OC, fastened to each joist with double tie of galv 18 SWG wire, double furring at each butt joint of wallboard.
7. **Wallboard, Gypsum*** - 5/8 in. thick, secured to furring channels with No.6 flathead sheet-metal screws spaced 8 in. OC or to nailing channels with fetter ring barbed nails 1-1/4 in. long with 11 SWG shanks and 3/8 in. heads, spaced 6 in. OC. Joint treatment not required for this rating, except for tapered, rounded-edge wallboard where edge joints are covered with paper tape and joint compound.

    Celotex Corp. - Type B, C or FRP.
    Continental Gypsum Company - Type CG5-5.
    G-P Gypsum Corp. - Type 5 or C.
    James Hardie gypsum Inc. - Types Fire X, Max"C".
    Republic Gypsum Co. - Type RG-1 or Rg-3.

*Bearing the UL Classification Marking.

Figure 2.6 Specifications for one particular 1-hour rated floor and ceiling assembly from the UL *Fire Resistance Directory.*

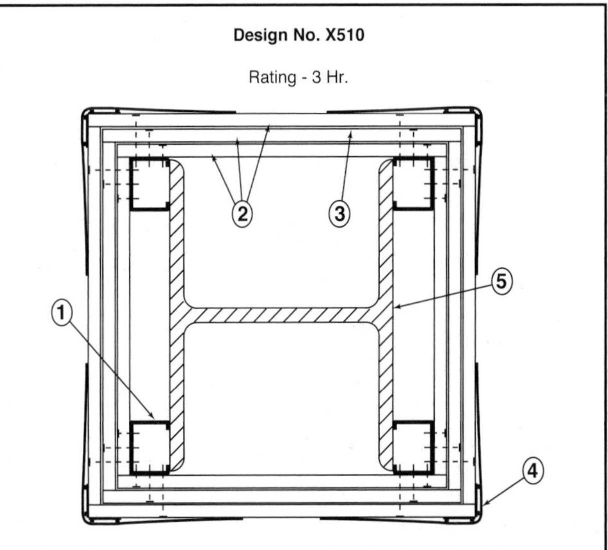

1. **Steel Studs** - 1-5/8 in. deep by 1-7/16 in. wide with 1/4-in. folded flange in legs, fabricated from 25 MSG galv steel. Studs cut 1-2 in. less in length than assembly height and attached to inner layer wallboard with 1 in. long self-drilling, self-tapping screws, spaced vertically 24 in. OC.
2. **Wallboard, Gypsum*** - Three layers of 5/8 in. thick wallboard. Second layer attached to steel studs with 1-5/8 in. long, self-drilling, self-tapping screws, spaced vertically 12 in. OC. Outer layer attached to studs with 2-1/4 in. long, self-drilling, self-tapping screws, spaced vertically 12 in. OC.

    National Gypsum Co., Charlotte, NC - Types FSW, FSW-G.
    Standard Gypsum Corp. - Types SGC, SGC-G.
    Weyerhaeuser Co., Dierks Div. - Type DDN1.

3. **Tie Wire** - Two strands of 18 MSG soft wire spaced vertically 24 in. OC used to secure second layer of wallboard only.
4. **Corner Beads** - 24 MSG galv steel, 1-1/4 in. legs. Attached to wallboard with 6d by 1-3/4 in. nails spaced 12 in. OC at each leg. Covered with joint compound treatment. Nom. 3/32 in. thick gypsum veneer plaster may be applied to the entire surface of Classified veneer baseboard.
5. **Steel Column** - Min. size of column, W10X49, with outside dimensions of 10X10 in. with a flange thickness of 9/16 in., a web thickness of 5/16 in., and a cross-sectional area of 14.4 sq in.

*Bearing the UL Classification Marking.

Figure 2.7 This design provides 3-hour rated protection for a column as specified in the UL *Fire Resistance Directory.*

Analytical Calculation of Fire Resistance

Because testing materials in a furnace is costly and because some structural members may not match those that have been previously tested, mathematical equations have been developed to predict the behavior of materials under test conditions without the need for actual testing. These equations have evolved into mathematical models that utilize the mechanical and thermal properties of materials at high temperatures.

In 1997, the American Society of Civil Engineers (ASCE) and the Society of Fire Protection Engineers (SFPE) jointly developed a standard for the calculation of fire resistance of structural elements. That standard, known as ASCE/SFPE 29, *Standard Calculation Methods for Structural Fire Protection,* provides the methods for calculating fire-resistance ratings that are equivalent to the results obtained from the standard fire test.[5] These calculation methods are limited to use with structural steel, plain and reinforced concrete, timber and wood, concrete masonry, and clay masonry.

The calculation methods are based in part on the years of data that have been obtained in the laboratory testing of materials. They may not provide accurate results when applied to materials that have not been used in the actual tests. For example, the most commonly used structural steel is designated A7 or A36. This is the steel that has been used in compiling the test data. If a high-strength steel such as A242 were to be used the calculated results would not necessarily be accurate.

Calculation

As an example of the calculating method, the following equation can be used to calculate the fire endurance of steel beams and columns protected by light insulation:

$$R = \{[(C_1 \times M) + C_2] \times I\} \div D$$

Where R = fire endurance in minutes

M = mass of the member in lb/ft

D = heated perimeter in inches

I = thickness of protection in inches

C_1 and C_2 are constants that are empirically derived for the insulating units. In metric units the equation is:

$$R = \{[(0.672 \times C_1 \times M) + (0.039 \times C_2)] \times I\} \div D$$

Where R = fire endurance in minutes

M = mass of the member in kg/m

D = heated perimeter in mm

I = thickness of protection in mm

C_1 and C_2 are the same as for U.S. units.

The constant C_1 is calculated from the equation:

$$C_1 = 1200 \div r$$

where r is the insulating material density in lb per cubic foot. For insulating materials such as mineral fibers, vermiculite, and perlite that have densities between 20 and 50 lb per cubic foot, C_2 has a value of 30. For common insulating materials such as cement pastes or gypsum with similar densities, $C_2 = 72$.

The above equation is a relatively simple example of the calculation method. More detailed methods must be used for other materials and structural components.

Analytical Design Using Real Fire Exposures

The NFPA® 251 standard time-temperature test is the most commonly used method of satisfying building code requirements for structural fire resistance. As was noted earlier, however, the standard time-temperature curve may not reflect a given real fire situation. The NFPA® 251 test could be too severe or not severe enough for a given situation. Efforts have been undertaken in recent years to calculate fire resistance based on a time-temperature curve that reflects a more realistic fire occurrence for a given set of circumstances. It should be noted that in some cases this would be a less severe fire exposure than provided in the NFPA® 251 time-temperature curve. Consequently, fire resistance ratings determined analytically using a different time-temperature curve must be interpreted very cautiously.

Noncombustible Materials

Determining whether a material is combustible or noncombustible may at first seem very obvious. Because of the variety of materials used in building construction, however, the meaning of the term *combustible material* must be clearly established. This is especially important where materials are used in combination or have been treated in some manner to alter their properties. Building codes contain explicit criteria for determining what constitutes a combustible material **(Figures 2.8 a and b)**. Fundamentally, a *noncombus-*

Noncombustible — Incapable of supporting combustion under normal circumstances.

Figures 2.8 a and b A concrete structure is classified as noncombustible; the lightweight wood construction that makes up the fast-food structure is classified as combustible. *Photo B courtesy of Dave Coombs.*

tible material is one that "in the form in which used and under the conditions anticipated, will not ignite, burn, support combustion, or release flammable vapors, when subjected to fire or heat."[6] The most commonly used test for determining combustibility is ASTM E 136, *Standard Test Method for Behavior of Materials in a Vertical Tube Furnace at 750°C.*

Basic Building Classifications

In the fields of fire protection and building code enforcement, buildings are grouped into five major classifications. These classifications are commonly designated as follows:

- Type I, Fire-Resistive
- Type II, Noncombustible or protected noncombustible
- Type III, Exterior protected (masonry)
- Type IV, Heavy timber
- Type V, Wood frame

Masonry — Bricks, blocks, stones, and unreinforced and reinforced concrete products.

Building codes do not make use of the descriptive terms *fire-resistive, noncombustible, masonry,* or *wood-frame* because they do not fully define the construction types. The codes make use of the numerical designations only. The descriptive terms are included here because they were used in the past and may still be used to describe buildings.

The building classifications used in the building codes are based on the materials used in construction and the hourly fire-resistance ratings required for the structural components. With the exception of Type IV, Heavy Timber, the major classifications are further divided into two or three subclassifications.

NFPA® 220, *Standard on Types of Building Construction,* details the requirements for each of the classifications and subclassifications. In NFPA® 220, each classification is designated by a three-digit number code. For example, Type I construction can be either 4-4-3 or 3-3-2. The digits are explained as follows:

- The first digit refers to the fire-resistance rating (in hours) of exterior bearing walls.
- The second digit refers to the fire-resistance rating of structural frames or columns and girders that support loads of more than one floor.
- The third digit indicates the fire-resistance rating of the floor construction.

In Type IV construction the designation 2HH is used. The structural members so indicated are of heavy timber with minimum dimensions greater than those used in Type III or Type V construction. The highest requirements for fire resistance are for Type I construction, with lesser requirements for other types of construction.

The International Building Code (IBC) makes use of construction classifications similar to NFPA® 220, although the requirements for individual structural members differ. **Table 2.1, p. 56,** shows the basic fire-resistance rating requirements for the five construction types and subclassifications in the IBC. Note that the basic requirements in Table 2.1 are permitted to be reduced.

Building codes use the types of construction and building occupancy, in connection with sprinkler systems and separations, to establish limits on the heights and areas of buildings. For example, a building code might restrict wood-frame (Type V-A) schools to one story. An architect who wanted to build a two-story school would have to provide an automatic sprinkler system or resort to a type of construction with greater fire resistance. A 15-story apartment building would be required to be of Type I-A construction in the IBC (NFPA® 4-4-2). However, an 11-story or lower apartment building could be of Type I-B construction (NFPA®3-3-2).

NOTE: For more information, see the IFSTA **Fire Inspection and Code Enforcement** manual.

Building code regulations vary from state to state and province to province. It is important that you understand the code process in your local jurisdiction.

Type I Construction

In Type I or fire-resistive construction, the structural members are of noncombustible construction that has a specified fire resistance (see Table 2.1). Steel, for example, is a noncombustible material but it is not fire-resistive and must be protected to attain fire resistance. Type I construction is divided into two subclassifications. In NFPA® 220 they are 4-4-2. In Table 2.1 the subclassifications are I-A and I-B.

Generally, bearing walls, columns, and beams are required to have a fire resistance of two to four hours, depending on the code and the construction classification. Floor construction is required to have a fire resistance of two or three hours. The roof deck and construction supporting the roof must have a fire resistance of one to two hours. In addition, interior partitions enclosing stairwells and corridors are required to be fire-resistive as specified by the local code, usually one or two hours. Partitions that separate occupancies or tenants may also be required to be fire resistive depending on the code.

As previously noted, fire resistance provides structural integrity during a fire. However, some building codes contain a provision to omit the fire-resistive rating for a roof construction for some occupancies when the roof is located more than 20 feet (6.1 m) above the floor **(Figure 2.9, p. 57)**. This exception can create a situation in which a building is classified under a building code as a Type I (fire-resistive) building but is actually constructed as a Type II (noncombustible) building. This difference can bring about troubling consequences in the event of a fire. For example, a typical event staged in an exhibition hall, such as a dog show, introduces limited amounts of combustible materials. On the other hand, an event such as a boating show might introduce a considerable amount of combustible material. If a fire were to break out in an exhibition hall when it contained a higher fuel load, the roof construction could be subject to failure.

Table 2.1
Fire-Resistance Rating Requirements for Building Elements (Hours)

Building Element	Type I		Type II		Type III		Type IV	Type V	
	A	B	A[e]	B	A[e]	B	HT	A[e]	B
Structural Frame[a]	3[b]	2[b]	1	0	1	0	HT	1	0
Bearing Walls									
Exterior[g]	3	2	1	0	2	2	2	1	0
Interior	3[b]	2[b]	1	0	1	0	1/HT	1	0
Nonbearing Walls and Partitions Interior[f]	0	0	0	0	0	0	See Section 602.4.6*	0	0
Floor Construction Including Supporting Beams and Joists	2	2	1	0	1	0	HT	1	0
Floor Construction Including Supporting Beams and Joists	1½[c]	1[c, d]	1[c, d]	0[c, d]	1[c, d]	0[c, d]	HT	1[c, d]	0

For SI: 1 foot = 304.8 mm
HT = Heavy Timber

a. The structural frame shall be considered to be the columns and the girders, beams, trusses, and spandrels having direct connections to the columns and bracing members designed to carry gravity loads. The members of the floor panel or roof panels which have no connection to the columns shall be considered secondary members and not a part of the structural frame.

b. Roof supports: Fire-resistance ratings of structural frame bearing walls are permitted to be reduced by 1 hour where supporting a roof only.

c. Except in Group F-1, H, M, and S-1 occupancies, fire protection of structural members shall not be required, including protection of a roof framing and decking where every part of the roof construction is 20 feet or more above any floor immediately below. Fire-retardant-treated wood members shall be allowed to be used for such unprotected members.

d. In all occupancies, heavy timber shall be allowed where a 1-hour or less fire-resistance rating is required.

e. An approved automatic sprinkler system in accordance with Section 903.3.1.1* shall be allowed to be substituted for 1-hour fire-resistance-rated construction, provided such system is not otherwise required by other provisions of the code or used for an allowable area increase in accordance with Section 504.2*. The 1-hour substitution for the fire-resistance exterior of walls shall not be permitted.

f. Not less than the fire-resistance rating required by other sections* of this code.

g. Not less than the fire-resistance rating based on fire separation distance.

* Section numbers refer to sections in the *2006 International Building Code®*.

Courtesy of the International Code Council®, International Building Code®, 2006, Table 601.

The fire-resistive compartmentation provided by partitions and floors tends to retard the spread of fire through a building. These features provide time for occupant evacuation and interior fire fighting. In a Type I structure, firefighters are able to launch an interior attack with greater confidence than they are in a building that is not of fire-resistive construction. *This does not mean that fire fighting will be easy in a fire-resistive building.* The fuel loads may be quite high and produce a severe fire, but the building is less likely to collapse on the firefighters.

Figure 2.9 The walls of this church are Type I construction (fire-resistive), but part of the roof covering is of combustible materials. *Courtesy of McKinney (TX) Fire Department.*

As a practical matter, building codes usually permit a limited use of combustible materials in Type I construction. A code may also allow the use of fire retardant-treated wood in roofs or interior partitions (See Chapter 7). Combustible materials typically are permitted for such uses as the following:

- Roof coverings
- Interior floor finishes
- Interior wall finishes and trims
- Doors and door frames
- Window sashes and frames
- Platforms
- Nailing and furring strips
- Light-transmitting plastics
- Foam plastics subject to restrictions

Figure 2.10 Reinforcing concrete increases its structural strength.

Although Type I (Fire-resistive) construction is desirable or even essential, it does not provide total fire protection because the contents will contribute most of the fuel for a fire. These contents can enable very serious fires to develop in a Type I building. Loss of life can occur in Type I buildings and the fire-resistive components of a building do not provide for fire extinguishment.

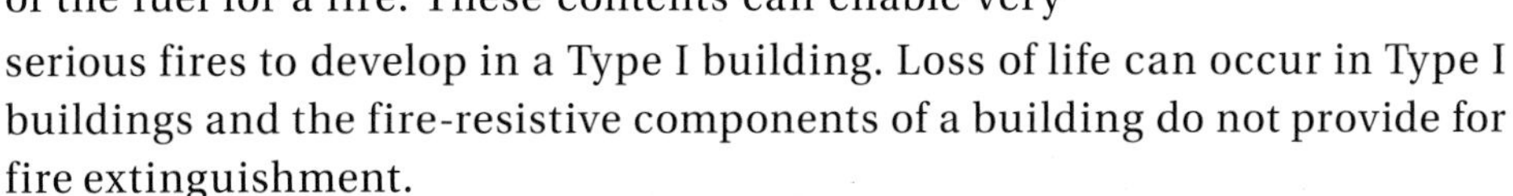

The two most common methods of constructing Type I buildings are by using reinforced concrete or a protected steel frame **(Figure 2.10)**. Several techniques can be used in designing a reinforced concrete building. Concrete is an inherently noncombustible material with good thermal insulating properties. These two attributes result in a material that is fundamentally fire-resistive, although the degree of fire resistance will vary with the specific type of concrete assembly. Reinforced concrete can fail if it is subjected to a very intense fire of long duration. Reinforced concrete structures can also be damaged by an explosion.

NOTE: See Chapter 10, Concrete Construction, for more information on concrete structures.

In the case of a steel-frame building, it must be remembered that unprotected steel has no fire resistance. When steel is used in fire-resistive designs, it must be protected by an insulating material. The combination of the structural strength of steel and the insulation produce a fire-resistive structural assembly. Several insulating materials are commonly used to protect the steel. The thickness of the insulating material can be varied to achieve different fire ratings.

Type II Construction

Type II construction (noncombustible) can be either protected or unprotected. In unprotected construction, the major components are noncombustible but have no fire resistance. The use of unprotected steel is the most common characteristic of unprotected, noncombustible construction **(Figure 2.11)**. An example of this is the use of unprotected steel columns or steel beams for roof support.

In Type II construction, structural steel is frequently provided with a degree of fire resistance that is less than that required for Type I construction. This has sometimes been referred to as protected noncombustible construction. In Table 2.1, Type II construction has two subclassifications, designated II-A and II-B. In II-A construction the structural components are required to have one-hour fire resistance. In II-B construction structural components are unprotected. As in the case of Type I construction, the structural members contribute little or no fuel. A fire in a vacant noncombustible building will also be of limited magnitude.

Materials other than steel can be used in Type II construction. A material such as concrete block can be used for the walls of a Type II building with steel beams or trusses used to support the roof. Glass and aluminum can be used but their structural role is limited. Building codes allow the use of combustible material in Type II construction for applications similar to those in Type I construction.

An unprotected, noncombustible building cannot be expected to provide structural stability under fire conditions. Firefighters should anticipate failure of unprotected steel from the heat of burning contents. The point at which unprotected members will fail, however, depends on the following factors:

- Ceiling height of the building
- Size of the unprotected steel members
- Intensity and duration of the exposing fire

Protected, noncombustible construction provides a degree of structural fire protection similar to Type I, which will depend on the degree of fire resistance provided.

Type III Construction

Type III construction has been commonly referred to as "ordinary construction." Type III construction is frequently constructed with exterior walls of masonry, but from a technical standpoint any noncombustible material with the required fire resistance can be used for the exterior walls **(Figure 2.12)**. Interior structural members including walls, columns, beams, floors, and roofs are permitted to be partially or wholly combustible.

Figure 2.11 Unprotected steel is most commonly used in unprotected, noncombustible construction. *Courtesy of McKinney (TX) Fire Department.*

Figure 2.12 An example of an older Type III structure. *Courtesy of Dave Coombs.*

The interior structural members of Type III construction may be protected or unprotected. Referring to Table 2.1, it can be seen that Type III construction has two subclassifications:

- Type III A construction is required to have a one-hour fire-resistive rating for interior members
- Type III B construction has no fire resistance requirements for the interior members.

Figure 2.13 Gypsum is an extremely common interior covering.

When the structural components of Type III construction are required to have a fire rating (such as for IBC Type III A and NFPA® Type III 2-1-1 construction) they can be protected by several means. Probably the most common are the use of plaster in older buildings and gypsum board in newer buildings **(Figure 2.13)**. In NFPA® Type III 2-0-0 and IBC Type III B, unprotected steel is sometimes used to support combustible members. For example, unprotected steel trusses could be used to support a combustible roof deck. The presence of the combustible members would result in the construction being classed as Type III.

The dimensions of the wood used in Type III construction are permitted to be smaller than those required in Type IV construction. In Type III construction it would not be uncommon to use nominal 2- x 10-inch (50 mm x 250 mm) joists for floor construction. But in Type IV construction, the minimum dimensions would be nominal 6 x 10 inch (150 mm x 250 mm) for floor construction.

NOTE: Nominal dimensions are not exact. See Chapter 7, Wood Construction, for more information.

A fundamental fire concern with Type III construction is the combustible concealed spaces that are created between floor and ceiling joists and between studs in partition walls when they are covered with interior finish materials. These spaces provide combustible paths for the communication of fire through a building. Fire can enter these spaces when openings exist in the interior finish materials or when a fire is of sufficient magnitude to destroy the material. It is essential, therefore, that the concealed spaces in Type III construction be properly firestopped.

In older Type III construction it is not unusual to encounter dropped ceilings that have been installed during renovations. These ceilings are typically installed several inches or several feet beneath existing ceilings with the original ceiling left in place. Dropped ceilings can facilitate hidden fires and make it difficult for firefighters to find the seat of a fire.

Firefighters cannot assume a level of structural stability where the structural components are combustible. When combustible materials become involved in a fire they will be consumed and collapse. Further, when the structural support provided by the interior beams, columns, and floors is lost, the exterior masonry walls can lose their stability and may collapse **(Figure 2.14)**.

Type IV Construction

Type IV construction is commonly known as heavy-timber or "mill" construction. Like Type III construction, the exterior walls are normally of masonry construction and the interior structural members are combustible. There are two important distinctions between Type III and Type IV construction:

- In Type IV construction the beams, columns, floors, and roofs are made of solid or laminated wood with dimensions greater than in Type III construction.
- Concealed spaces are not permitted between structural members in Type IV construction **(Figure 2.15)**.

Type IV construction was used extensively in factories, mills, and warehouses in the 19th and early 20th centuries. It is not commonly used in new construction for multistory buildings, although many buildings of this type

Figure 2.14 Loss of interior supports can lead to collapse of walls.

Figure 2.15 In Type IV construction concealed spaces are not permitted between structural members. *Courtesy of McKinney (TX) Fire Department.*

remain in use. Many old Type IV warehouse and industrial buildings have been converted to residential use. Today, heavy-timber wood frame construction is encountered primarily where it is desired for appearance.

NOTE: For more information about heavy-timber construction, including hazards that can be left in older gentrified buildings, see Chapter 7, Wood Construction.

The primary fire hazard associated with Type IV construction is the massive amount of fuel presented by the large structural members in addition to the building contents **(Figure 2.16)**. In addition, fire hazards can be present if the interior of the building is not cleaned of old oils and residues left from previous use.

The greater mass of the heavy-timber components provides greater structural endurance under fire conditions than the structural members used in Type III buildings. The larger timbers are slower to ignite and burn. If the members have not been exposed to a prolonged fire, it is possible to sandblast away the charring and continue their use after a fire.

However, the wood members are combustible and ultimately will be consumed in a fire. The exterior masonry walls can then become unstable and collapse because of the loss of the interior bracing provided by the floors and columns. See Chapter 7 for more discussion on the hazards of Type IV construction.

Type V Construction

In Type V construction, also commonly known as wood-frame construction, all major structural components are permitted to be of combustible construction. The basic method of construction in a Type V building consists of using a wood frame to provide the primary structural support **(Figure 2.17)**. Many Type V structures are required to have a 1-hour fire resistance for the structural members. This is typically accomplished by protecting the combustible frame members with plaster or fire-rated gypsum board.

Figure 2.16 Because of the larger size of structural members, an enormous amount of fuel is available to feed a fire. *Courtesy of Dave Coombs.*

Figure 2.17 Type V construction.

A fundamental problem posed by Type V construction is the creation of combustible concealed voids and channels more extensive than are found in Type III construction. These concealed spaces provide avenues for extension of fire within a building. Because it is inherently combustible, a Type V building can become totally involved and completely destroyed in a fire **(Figure 2.18)**.

Because a heavily involved wood-frame building also poses a threat to adjacent structures, building codes impose restrictions on the maximum allowable heights and areas of Type V buildings. The building codes may also require a separation distance between a Type V building and an adjacent property line.

Light-Frame Construction — Method for construction of wood-frame buildings. Replaced the use of heavy timber wood framing.

Several different methods can be used to construct a Type V building. In modern practice, wood-frame buildings are most often constructed using a method known as light-frame construction. This technique was developed approximately 150 years ago. At that time, development of the water-powered sawmill made it possible to cut thinner boards from logs than had been possible using hand saws. Light-frame construction eliminated heavy posts and beams and made use of smaller studs, joists, and rafters. This change permitted a building to be erected faster and more cheaply. Light-frame construction is so commonly used that one author has described it as "the common currency of small residential and commercial buildings in North America today." [7]

Mixed Construction

Some buildings are allowed to have a mixed use and a mixed construction type. This situation may be found where a different type of structure is built on top on an existing one. These mixed situations maybe present special challenges for emergency responders. For example, in some jurisdictions it is permissible to have Type V construction over Type I construction up to a height of 70 feet (21.3 m) **(Figure 2.19)**. An example might be a single-story commercial occupancy of Type I construction with a concrete deck on top and Type V construction above it. Another example would be a three-story parking garage with Type V construction above.

Figure 2.18 Type V fully involved. *Courtesy of District Chief Chris E. Mickal, NOFD Photo Unit.*

Figure 2.19 Mixed construction. Note this example of Type V residential construction over Type I commercial. *Courtesy of McKinney (TX) Fire Department.*

Building codes contain height limitations for the different construction classifications. In the situation of mixed construction, however, emergency responders could be confronted with a fire in Type V construction that exceeds the normal height restrictions for that type of construction.

Fire or Fuel Load

When discussing fire protection, the terms fire load or fuel load are used frequently. Both terms refer to the maximum heat that would be released if all the available fuel in a building was consumed.[8] The fire load is the product of the weight of the combustibles multiplied by their heat of combustion. It is typically expressed in terms of pounds per square foot (or kg per sq meter). The fire load will vary depending on the heat of combustion of the fuel. Thus, the fire load in a shoe store is different from the fire load in a warehouse full of plastics.

Fire Load — The amount of fuel within a compartment expressed in pounds per square foot obtained by dividing the amount of fuel present by the floor area. Fire load is used as a measure of the potential heat release of a fire within a compartment. Also known as fuel load.

Ordinary combustibles such as wood, paper, and similar materials have heats of combustion between 7,000 and 8,000 BTU per pound (16,282 to 18,608 J/kg). Therefore an occupancy with a fire load of 10 pounds per square foot (48.6 kg/sq m) would produce 70,000 to 80,000 BTU per square foot (791,305 to 904,348 J/ sq m) if it were totally consumed.

The fire load can be used as an estimate of the total potential heat release or thermal energy to which a building may be subjected if all combustibles become fully involved in fire. Buildings with combustible structural components (Types III, IV, and V) have an inherently greater fire load than Types I or II because the structural framing materials contribute a significant amount of fuel to a fire.

A fire load does not translate into an equivalent structural load. For example, a warehouse of Type IV (heavy-timber) construction containing iron radiators would have a light fire load but a very large structural load. Conversely, a toy store would have a moderate structural load but a high fire load because most toys are combustible with combustible packaging.

Occupancy Classifications

Building codes classify buildings according to their occupancy as well as their construction type. Classifying buildings according to occupancy facilitates the administration of a code. Grouping all the various uses to which a building can be put (the occupancies) into a relatively small number of classifications allows for the use of less cumbersome code language.

The occupancy classifications assign building occupancies into groups with broadly similar fire risks. For example, occupancies in which crowds of people gather, such as night clubs, present a potential threat to life because of crowd density. In any occupancy, crowd density could slow the rate at which the occupants would be able to flee in an emergency. By contrast, in a warehouse the occupants are scattered throughout the building and are more mobile. Building codes, therefore, treat occupancies in which crowds of people can gather differently from occupancies that hold relatively few people **(Figures 2.20 a and b, p. 64)**.

Figures 2.20 a and b The threat to life in a crowded theater would be much greater than in a warehouse with relatively few persons, so occupancy codes treat the occupancies differently.

International Building Code (IBC) Classifications

The building codes group building occupancies into occupancy classifications. The International Building Code (IBC) contains ten major occupancy classifications:

- Assembly — Group A
- Business — Group B
- Educational — Group E
- Factories — Group F
- High Hazard — Group H
- Institutional — Group I
- Mercantile — Group M
- Residential — Group R
- Storage — Group S
- Utility and Miscellaneous — Group U

Ten classifications is a relatively small number in which to group all the potential uses for a building. Not surprisingly, considerable variation of hazards can exist within groups. For example, Group R, residential, includes hotels, apartment buildings, monasteries, and single-family dwellings. Obviously a hotel with 1,500 rooms and highly transient occupants presents a far greater fire safety risk than a monastery, even though both are classified under the same grouping. Therefore, the codes further divide the major occupancy classifications into a number of subclassifications to describe occupancy groups with more similar characteristics.

As an example, in the IBC, residential occupancies (occupancy group R) are subdivided into four subgroups: R-1, R-2, R-3, and R-4. The subgroups are characterized as follows:

- R-1 — Occupancies with primarily transient occupants including hotels and motels.
- R-2 — Occupancies with primarily permanent occupants and more than two dwelling units, such as apartment buildings, dormitories, and nontransient hotels **(Figure 2.21)**.

- R-3 — Occupancies with primarily permanent occupants and not more than two dwelling units. In addition, adult facilities that provide accommodations for five or fewer persons of any age for less than 24 hours and child care facilities for five or fewer persons of any age for less than 24 hours.
- R-4 — Occupancies used as assisted-living facilities with five to sixteen occupants.

Figure 2.21 A typical multifamily dwelling.

Similarly, assembly occupancies are subdivided into five subgroups: A-1, A-2, A-3, A-4, and A-5 as follows:

- A-1 — Assembly occupancies, usually with fixed seating such as motion picture theaters and concert halls.
- A-2 — Assembly occupancies used for consuming food and drink such as night clubs and restaurants.
- A-3 — Assembly occupancies used for worship, recreation, or amusement such as bowling alleys, churches, dance halls, and exhibition halls.
- A-4 — Assembly occupancies intended for viewing indoor sporting events such as tennis courts and arenas.
- A-5 — Assembly uses for outdoor activities such as bleachers, grandstands, and stadiums.

The IBC contains a total of 26 subgroups within the 10 major occupancy classifications. The IBC also makes separate provision for one-and two-family dwellings not more than three stories high. Although these buildings would be classified as Rs in the IBC they are governed by a separate code, the International Residential Code.

NFPA® Classifications

NFPA® 5000, Building Construction and Safety Code, and NFPA® 101 Life Safety Code®, make use of 12 major occupancy classifications:

- Assembly
- Educational
- Day care
- Health care
- Ambulatory health care
- Detention and correctional
- Residential
- Residential board and care
- Mercantile
- Business
- Industrial
- Storage

The occupancy classifications of the both the NFPA® and the IBC are typical; however, other building codes may use other occupancy classifications. One code, for example, classifies both correctional facilities and health care facilities as institutional occupancies.

Mixed Occupancies

Buildings frequently contain occupants that represent more than one occupancy classification. For example, a building might contain mercantile and residential occupancies. The different occupancies will pose different hazards to each other. A mercantile occupancy could endanger the occupants of a nightclub if a fire occurred in the mercantile occupancy located in the same building. If an infant care center were located in the same building with a restaurant, a fire in the restaurant could endanger the children in the infant care center.

To alleviate this problem, building codes will require fire-resistive separations between various occupancies. Thus an infant care center and a restaurant could be required to be separated by a 2-hour fire-resistive separation when located in the same building.

The specific requirements for occupancy separation will depend on the local building code. Required separations can vary from one to three hours and not all occupancies would require a separation. Not all occupancies would require separations. Furthermore, a building code may permit a reduction in the required occupancy separation if a building is sprinklered.

Change of Occupancy

Buildings frequently undergo a change of occupancy. As was noted in Chapter 1, a change of occupancy can create serious problems. The following case study illustrates the point.

A Nightclub Disaster

In the early morning hours of February 16, 2003 a crowd later estimated at 1,150 persons was occupying the E2 nightclub on the south side of Chicago. At approximately 1:00 a.m. a disturbance occurred as a result of a fight between two patrons. A security guard discharged mace in an attempt to quell the disturbance. The crowd began to surge toward the front entrance in an effort to flee the effects of the mace. Twenty-one persons were crushed to death in the stairway that led from the second floor where the club was located to the street-level main entrance.

The E2 nightclub was located on the second floor of a two-story building that had been constructed as an automobile dealership. The building had first and second floors of reinforced concrete and exterior walls of masonry. The roof was wood deck supported by wood bowstring trusses. A mezzanine used for VIP suites had been constructed in the space between the upper and lower chords of the trusses. At the time of the incident the first floor was occupied by a restaurant that was a separate operation from the nightclub.

The net area of the second floor was 4,260 square feet (395.7 sq m). The mezzanine had a net area of 1,491 square feet (138.5 sq m). Under the provisions of the Chicago building code the calculated occupant load was 1,040 persons for the second floor and mezzanine. The building code required that exits be provided for this number of people.

There were three exits from the second floor. These consisted of two rear stairways and a front stairway that served as the main entrance. The exits had not been modified from the time the building was originally occupied as an automobile dealership (more than 70 years). The combined capacity of the three existing stairs under the provisions of was only 240 persons; therefore, the existing exits were inadequate by a total of 800!

In order to provide enough exits to handle the occupant load, it would have been necessary to construct additional stairways from the second floor to grade level: a costly undertaking. The operators of the club had not done this. At the time of the incident city building inspectors had cited the nightclub for code violations and the case was in court.

Summary

How a building will react under fire conditions is of critical importance to firefighters. The fire behavior in a building is determined in large measure by the materials of which it is constructed and by its structural fire resistance. The structural fire resistance of building components is determined most often through laboratory testing. Building codes classify construction into major types depending on the construction material used and the structural fire resistance. The codes also classify buildings according to their occupancy. Occupancies within the individual occupancy groups present roughly similar fire risk factors.

Review Questions

1. What is fire resistance?
2. Explain the value of standard fire tests.
3. What is a noncombustible material?
4. What considerations do firefighters need to take into account when they encounter dropped ceilings?
5. What are the major occupancy classifications contained in the *International Building Code (IBC)*?

References

1. NFPA® *Fire Protection Handbook* 19th ed., Sect 12, Chap 4.
2. Ibid.
3. *Handbook of Building Materials for Fire Protection*, Charles A. Harper, McGraw-Hill, 2004.
4. NFPA® 251, *Standard Methods of Tests of Fire Resistance of Building Construction and Materials*, 2006 Edition
5. NFPA® *Fire Protection Handbook*, 19th ed., Sect12, Chap 4.
6. *International Building Code*, 2006 edition
7. *Fundamentals of Building Construction, Materials and Methods,* Edward Allen and Joseph Manno, John Wiley and Sons, Fourth Edition, 2004.
8. NFPA® *Fire Protection Handbook*, 19th ed., Sect. 12, Chap. 5.
9. NFPA 5000®, *Building Construction and Safety Code,* 2006
10. NFPA® 101, *Life Safety Code®* , 2006

Chapter Contents

Divider page photo courtesy of McKinney (TX) Fire Department.

Key Terms

Axial Load ... 88	Inertia ... 77
Beam ... 90	Kinetic Energy ... 73
Bearing Wall ... 98	Live Load ... 82
Cantilever ... 87	Load ... 72
Column ... 93	Membrane Structure ... 101
Concentrated Load ... 82	Seismic Forces ... 75
Damping Mechanism ... 79	Shell Structure ... 102
Dead Load ... 81	Static Load(s) ... 84
Dynamic Load ... 84	Surface System ... 101
Eccentric Load ... 89	Tension ... 87
Equilibrium ... 85	Torsional Load ... 89
Failure Point ... 88	Truss ... 95
Gravity ... 72	

FESHE Objectives

Fire and Emergency Services Higher Education (FESHE) Objectives: *Building Construction for Fire Protection*

4. Explain the different loads and stresses that are placed on a building and their interrelationships.
5. Identify the principle structural components of buildings and demonstrate an understanding of the functions of each.

The Way Buildings are Built: Structural Design Features

Learning Objectives

After reading this chapter, students will be able to:

1. Explain the various loads exerted on a building resulting from environmental sources.
2. Distinguish between the classifications of loads based on origin and movement.
3. Recognize and discuss the internal forces resulting from the loads and forces applied to a structural member.
4. Describe the basic structural components.
5. Describe the basic structural systems.

Chapter 3
The Way Buildings are Built: Structural Design Features

Case History

Firefighters were called to a structure fire in a house that was under construction. They arrived to find a two-story wood frame house approximately 3,500 square feet, with light smoke puffing from the eaves. The house was dry-walled, but rails were not in place on the stairs and second floor. When firefighters opened the door, they found fire extending to the second floor. The initial attack crew fought fire on the first floor and the second-in crew ascended to the second floor to stop fire extension. On the second floor, a firefighter fell from a bridge over the foyer that did not have any rails. The firefighter fell approximately 12 feet to the first floor, breaking an arm. The impact also broke a 2 x 12 floor joist. Only inches from where the firefighter landed, the fire had burned through the floor. The firefighter could easily have fallen an additional 8 to 10 feet to a concrete floor. Firefighters reported the smoke conditions were not very bad and the heat was tolerable.

Lesson learned: This house was under construction, there was no life hazard, and there were no exposure hazards. The fire was small enough to warrant an interior attack, but situational awareness tells us that a house under construction may not have railings to prevent falls from the second floor.

Source: National Fire Fighter Near-Miss Reporting System.

With regard to buildings, the interest of the firefighter is necessarily focused on the way a building reacts when it is involved in a fire. Evaluating these reactions is necessary to facilitate fire control and rescue. The interest of the structural designer, however, extends to a broader and more basic variety of circumstances. Although the architect and engineer may be required by building codes to provide for the fire safety of a building — as was noted in Chapter 1 — fire safety is only one of many considerations for the designer.

To design an adequate structure, the engineer must first determine the type and magnitude of the forces to which the structure will be subjected. Making this determination is the most critical aspect of engineering design. The ability to understand and evaluate these forces is what distinguishes a casual knowledge of buildings from a professional knowledge. For example, it is commonly understood that wind exerts a force on a building. As the size of a building increases, the actual force exerted by the wind becomes substantial and must be considered so that adequate bracing can be designed. If a building is to be located in a region where it will be subjected to hurricanes, the force of hurricane winds must be determined so adequate roof anchoring can be designed. Simply knowing that the wind exerts a force is not enough. What matters is understanding the *magnitude* of the force.

This chapter discusses loads and the ways they impact building stability, the strengths and weaknesses of such structural components as beams, trusses, arches, and columns, and basic structural systems such as framing and bearing walls.

Loads: the Sources of Force

Load — Any effect that a structure must be designed to resist. Forces of loads, such as gravity, wind, earthquakes, and soil pressure, are exerted on a building.

It is necessary to understand the nature of the loads exerted on a building before they can be evaluated. A *load* is defined as any effect (or force) that a structure must *resist* ([1]). Loads arise from several sources such as gravity, wind, earthquakes, and soil pressure and can be classified in several ways.

Gravity

Gravity — Force acting to draw an object toward the earth's center; force is equal to the object's weight.

Gravity creates a force on a building through the weight of the building components and all of its contents. In addition, gravity can exert a force (added weight) if snow, ice, and water accumulate on the roof or water accumulates inside the building due to flooding **(Figure 3.1)**. A knowledge of gravity is important to understand the cumulative effects of these loads on the supporting structure. In this illustration, the roof and the snow are supported by the second-story exterior bearing walls. The second-story walls as well as the roof and snow are in turn supported by the first-story bearing walls. These first-story walls also support the second floor and the contents of the second floor. All this weight, including the additional weight of the first-story walls, is supported by the foundation. Finally, the total weight of the building exerts a force on the soil beneath it.

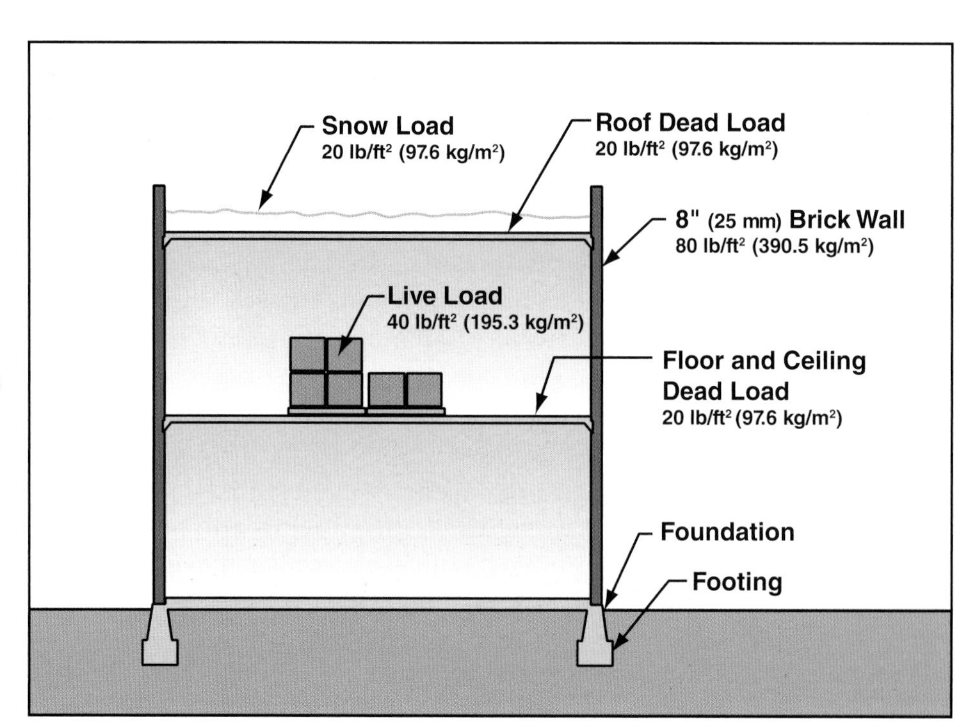

Figure 3.1 Effects of gravity forces on a building.

Wind

The air that makes up the atmosphere is a gas that, like all substances, has mass. When air is in motion, it possesses *kinetic energy* according to the following fundamental relationship:

$$E = \frac{1}{2}mv^2$$

where E = energy

m = mass of a body

v = velocity

We feel this kinetic energy as wind. When the wind encounters a fixed object, it exerts a force. Wind exerts the following basic forces on a building: **(Figure 3.2)**

- *Direct pressure* - the impact effect the wind has on a surface. This force may be reduced by streamlining the surface encountered.
- *Aerodynamic drag* - when wind encounters an object, its fluid nature causes it to flow around the object. This exerts a drag effect on the object.
- *Negative pressure* - a suction effect produced on the downwind side of the building resulting in an outward pressure.

These fundamental forces push on a building in the direction of the wind. In addition, there are secondary effects that may be produced, including the following:

- *Rocking effects* - a back-and-forth effect due to variations in the velocity of the wind.
- *Vibration* - wind passing over a surface such as a roof may cause vibration of the surface depending on the velocity of the wind and the harmonic characteristics of the surface.
- *Clean-off effect* - the tendency of wind to dislodge objects from a building.

Kinetic Energy — The energy possessed by a moving object.

Building design can increase or reduce the effect of wind on a building. The "clean-off effect" noted above is of particular concern where a building has projections such as canopies and parapets **(Figure 3.3, p. 74)**. Conversely, providing a design with a smooth contour can reduce the force on a building.

In designing buildings to withstand the force of wind, the primary effect considered is the force due to direct pressure. This force increases with the velocity of the wind and can be evaluated by the basic equation:

$$p = Cv(2)$$

where v = velocity

p = static pressure

C = .00256 a numerical constant that accounts for the air mass and simplifying assumptions of building behavior.

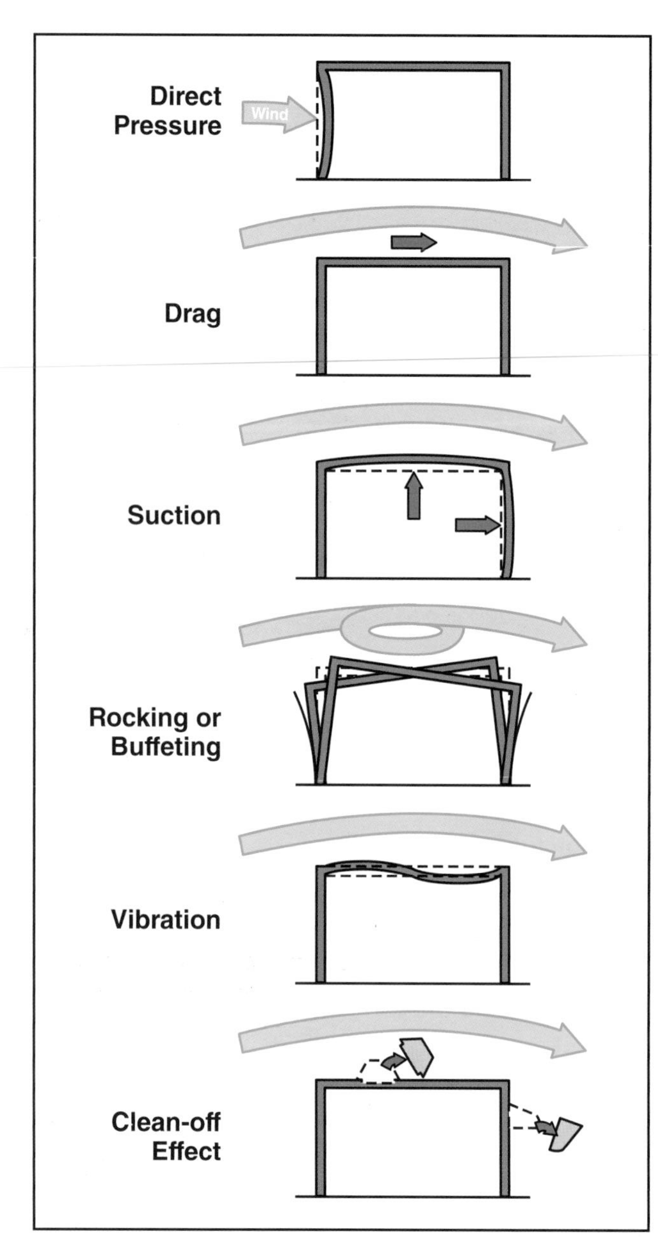

Figure 3.2 Effects of wind.

Figure 3.3 Damage to a masonry parapet caused by wind. *Courtesy of Ed Prendergast.*

For example, a wind velocity of 10 mph (16 km/h) creates a pressure equal to the following:

$p = .00256 \times 10^{(2)} = .256$ psf (3.8 $N/m^{(2)}$)

However, if hurricane conditions occur and a speed of 100 mph is reached, the wind pressure would become:

$p = .00256 \times 100(2) = 25.6$ psf (1.2 kPa)

This is 100 times greater than the pressure at 10 mph (16 km/h).

If the wall of a two-story building were 20 feet high and 60 feet long (6.1 m by 18.3 m), the total basic force exerted by the above pressure would be:

$20 \times 60 \times 25.6 = 30{,}720$ pounds (137,625 N) **(Figure 3.4)**

This calculation involves the basic wind pressure and has assumed a wind blowing perpendicular to the building wall. The actual design pressure used by engineers must be adjusted to account for building height, surrounding terrain, and specific features relating to the shape of the building. In addition, designers must consider the combined effects of wind forces and localized conditions, such as closely spaced buildings in an urban area.

The force resulting from wind is particularly dangerous when it occurs against an unbraced wall. Situations of this type can occur at construction sites. In the course of construction, a wall may have been erected that was not provided with temporary bracing or did not have roof framing or lateral bracing in place. If a strong wind occurs, these walls can be blown over and injure or kill workers **(Figure 3.5)**. Fire personnel should also be alert for unbraced walls at demolition sites and at fire-damaged buildings where interior structural supports have collapsed or been destroyed.

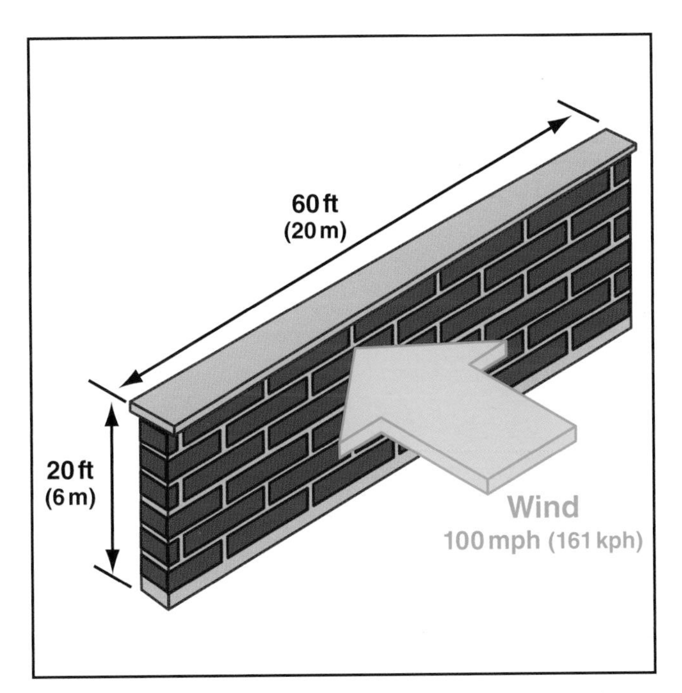

Figure 3.4 Hurricane-force winds of 100 mph (161 kph) can exert 30,720 pounds (136 649 newtons) of force against a two-story high wall.

Figure 3.5 This wall at a construction site was toppled by wind. *Courtesy of Ed Prendergast.*

Earthquakes

Earthquakes occur as a result of slippage between the tectonic plates that make up the earth's surface. The movement of the tectonic plates produces vibrations at the earth's surface, usually along a fault line. The vibrational motion of the surface subjects buildings to forces known as *seismic forces* that can be very destructive, as history has shown **(Figure 3.6)**.

Seismic Forces — Forces developed by earthquakes. Seismic forces are the some of the most complex forces exerted on a building.

Earthquakes can occur anywhere on earth; however, major earthquakes occur most frequently in parts of the world known as *fault zones* or *zones of high probability* [(2)]. As in the case of wind loads, not all geographic areas are subject to the same degree of risk from earthquakes. In some areas, the design professional needs to give little or no special consideration to seismic loads because the structural provisions for wind or gravitational loads are adequate for likely seismic loads. In other areas seismic considerations are essential and fundamental aspects of building design.

Figure 3.6 An earthquake caused extensive damage to this store in Washington state. *FEMA News Photo.*

Zones of high earthquake probability include the Pacific Coast of the United States and Canada, Hawaii, central Utah, and southern Illinois. All the model building codes provide seismic maps to help designers and code officials determine the zones in which their communities are located **(Figures 3.7 a and b USGS seismological map, p. 76)**.

Buildings located in fault zones can be subjected to seismic loads that are far more complex than those previously described for wind. For example, the vibrational motion produced by earthquakes can be three-dimensional. The shifting of the earth's surface can also produce forces that are either torsional or resonant in nature. *Torsional forces* are produced in a structural member when it is twisted. *Resonant forces* are movements of relatively large amplitude resulting from a small force applied at the natural frequency of a structure. The magnitude of the forces developed within a building during an earthquake depends on several factors, including the following:

- Magnitude of the vibratory motion
- Type of foundation
- Nature of the soil under the building

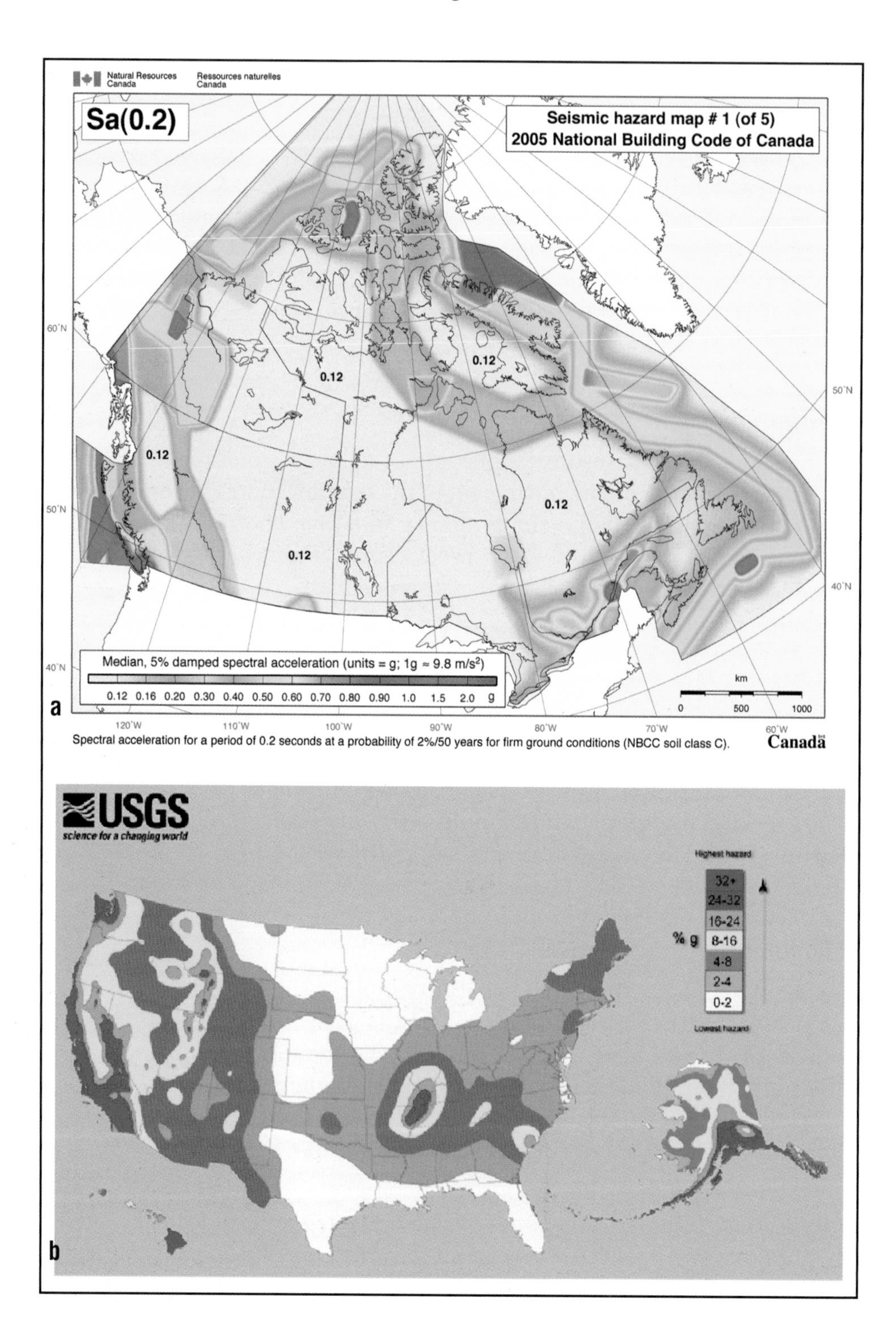

Figures 3.7 a and b Seismic maps of the United States and Canada showing areas that are more and less prone to earthquakes.

- Stiffness of the structure
- Presence of damping mechanisms within the building

One phenomenon associated with earthquakes is *soil liquefaction*. Soil liquefaction occurs where the soil is of a loose, sandy nature and is saturated with water. These conditions frequently occur in low-lying land near bodies of water. Normally the individual particles of sand exert forces on each other that give the soil its adhesive properties. When the sand is subjected to rapidly applied force, as in an earthquake, the sand particles attempt to rearrange themselves and move closer; however, the entrapped water prevents the particles from moving closer together. There is also a simultaneous increase in the water pressure that reduces the forces between the sand particles. This action results in a reduction of the strength of the soil leading to shifting of foundations and serious structural damage.

Although the movement of the ground beneath a building can be three-dimensional, the horizontal motion is the most significant force **(Figure 3.8)**. As the ground moves under a building, *inertia* tends to keep the upper portion of the building momentarily in its initial position. A shear force develops internally between the upper and lower portions of the building. If an earthquake generates additional ground motion, more complex forces can occur and be transmitted up through the height of the building. This motion can produce a swaying motion in a building, which the structural system must be designed to overcome. Low buildings are less susceptible to this type of motion than tall buildings.

Inertia — The tendency of a body to remain in motion or at rest until it is acted upon by force.

The basic architecture of a building affects the way in which a building reacts to an earthquake. Buildings with geometric irregularities are inherently more susceptible to damage from earthquakes than buildings having a symmetrical design **(Figure 3.9)**. Because the tall and short sections have different heights, their responses

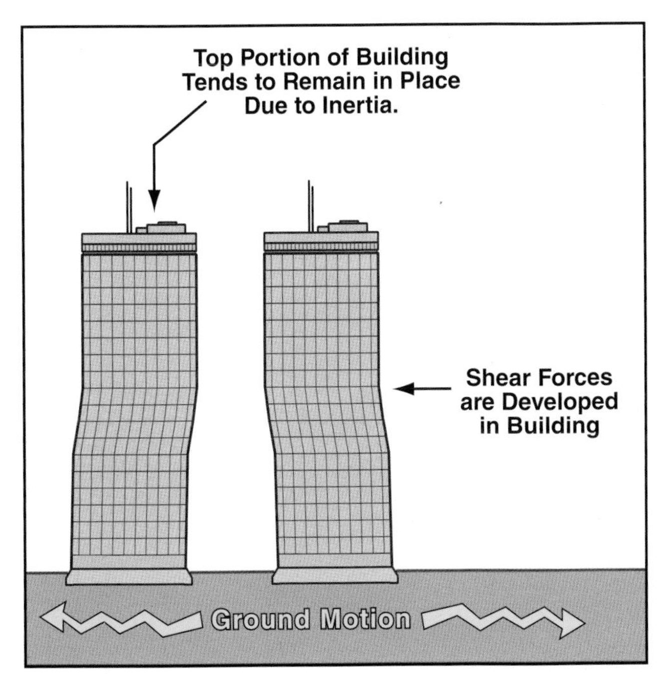

Figure 3.8 The basic action of an earthquake on a multistory building.

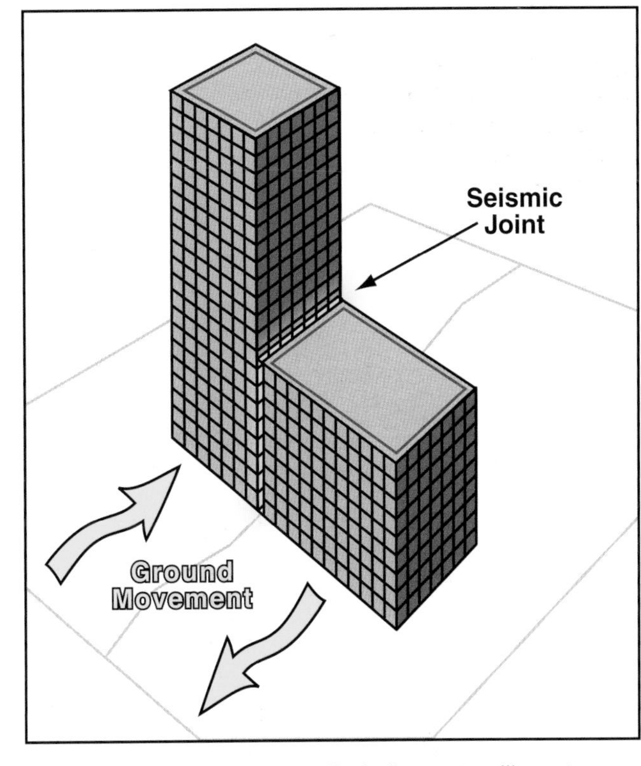

Figure 3.9 Buildings having dissimilar parts will react differently in an earthquake.

to the vibrations of the surface will be different. This difference in their responses can produce damaging forces at the junction of their two parts. To overcome this difference, a seismic joint can be designed into the structure between the two sections. The required width of the joint increases based on the height of the building.

NOTE: Maintaining the fire rating of seismic joints can be difficult and has been ruled as a main contributor of fire spread in the 1980 fire at the MGM Grand in Las Vegas that killed 87 people and injured 650, including 14 firefighters.

The severity of the force developed depends on the acceleration of the ground beneath the building, not necessarily the total movement. Building codes contain provisions for seismic design based on the duration and magnitude of the ground acceleration for the geographic area in which the building is located.

The codes also recognize that some buildings are more critical than others and require stronger seismic bracing. This category includes buildings with such occupancies as large places of public assembly, large office buildings, schools, and institutions. Other buildings essential for community recovery after an earthquake that also must be designed for greater seismic loads include the following:

- Fire and police stations
- Hospitals
- Communication centers
- Emergency preparedness centers
- Generating stations

Several design methods can be employed to protect buildings against the forces of earthquakes. One method is to increase a building's stiffness through the use of shear walls and cross bracing **(Figure 3.10)**. Increasing the stiffness of a building reduces its susceptibility to ground motions that have a relatively long (meaning slow) vibrational period.

Increasing a building's stiffness may not be suitable for all locations. When a building is located in an area where short vibrational periods are expected, the use of a flexible design is appropriate. Furthermore, the shear walls and cross bracing must be symmetrically located in a building so undesirable torsional forces do not develop within it.

Another method for increasing a building's stability is to use continuous structures with a high degree of redundancy in their structural frames. In a structure with redundant structural members, more members would have to fail for collapse to occur **(Figure 3.11)**. In addition, structures with continuity at their joints have a greater ability to absorb energy at the joints compared to structures in which the joints are free to rotate. (See the discussion on beams in the Structural Components section later in this chapter.) Thus, a poured-in-place concrete structure is easier to design for earthquake resistance than a structure using precast concrete.

NOTE: Concrete construction methods are discussed further in Chapter 10.

Figure 3.10 This building shows cross bracing that has been retrofitted to a building in earthquake-prone California. *Courtesy of Tanya Hoover.*

Figure 3.11 Buildings designed with redundant structural frames are less likely to collapse if one member fails.

Buildings can also be provided with damping mechanisms to control building motion. Damping mechanisms operate on a principle similar to the manner in which a door closer controls the speed of a door. Damping mechanisms are typically installed at the connections between columns and beams, and they absorb energy as the structure begins to move.

Damping Mechanism — Structural element designed to control vibration.

One method that has received considerable attention in recent years is *base isolation*. The basic concept of base isolation is to isolate the building from the horizontal movement of the earth's surface. The following two isolation methods have been used:

- ***Elastomeric bearings*** - Create a layer between the building and the foundation, which has a low horizontal stiffness **(Figure 3.12)**. The bearings are made of either natural rubber or neoprene. The bearings change the fundamental vibrational frequency of the building. Several buildings in the U.S. have been built or retrofitted with these bearings, including the City Hall of Oakland, California.
- ***Sliding Systems*** - Make use of special plates sliding on each other (a less common method of building isolation). These systems isolate the building from the horizontal shear force created by an earthquake.

Figure 3.12 Elastomeric bearings used in base isolation. *Courtesy of San Diego County Sheriff's Department.*

Soil Pressure

Like the forces caused by wind and earthquakes, the force generated by soil pressure must be evaluated in the design process. As with wind and earthquakes, the forces associated with soils are difficult to determine accurately and may only be estimated.

Soil exerts a horizontal pressure against a foundation. The magnitude of the pressure depends on the type of soil (clay, sand, rock, etc.), its degree of cohesion, and its moisture content. The pressure exerted by the soil against the foundation is known as the *active soil pressure*. The force of the foundation against the soil is known as the *passive soil pressure*. In determining the force created by the active soil pressure, the soil is assumed to behave like a fluid **(Figure 3.13)**. Thus, the pressure would vary from zero at the top of a foundation wall to some maximum pressure at the base, which would depend on the depth and density of the soil.

The basic equation used to determine soil pressure is:

$$p = Cwh$$

where p = pressure

h = depth of soil

w = density of soil

C = numerical constant that depends on the physical properties of the soil.

In **Figure 3.13** the wall is 8 feet (2.44 m) in height and the soil is soft clay with a density of 100 pounds per cubic foot (1.6 kg/L). The numeric constant for the soil is 0.4. Therefore, the pressure at the base of the wall is 0.4 x 100 x 8 = 320 pounds per square foot (1.56 N/m2). This is the pressure at the *base* of the wall. The average pressure is:

320/2 = 160 pounds per square foot (0.73 N/m2)

Because the wall is 8 feet (2.44 m) in height, the force exerted by the soil is 8 x 160 = 1,280 pounds per square foot (1.9 N/m2) of wall length. If the wall were 50 feet (15 m) in length, the total force against the wall would be 64,000 pounds (284,672 N).

The previous example is a very simple one that illustrates the basic concept of soil pressure. Situations encountered in actual practice are frequently more complicated. In **Figure 3.14**, for example, a foundation wall must withstand a soil mass that is not horizontal. This situation changes the force acting on the wall and the equations that are used to analyze it.

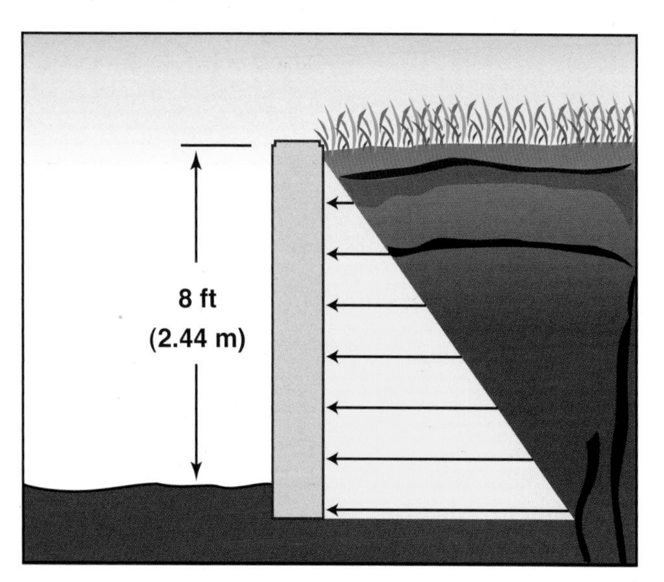

Figure 3.13 Soil pressure against a foundation wall.

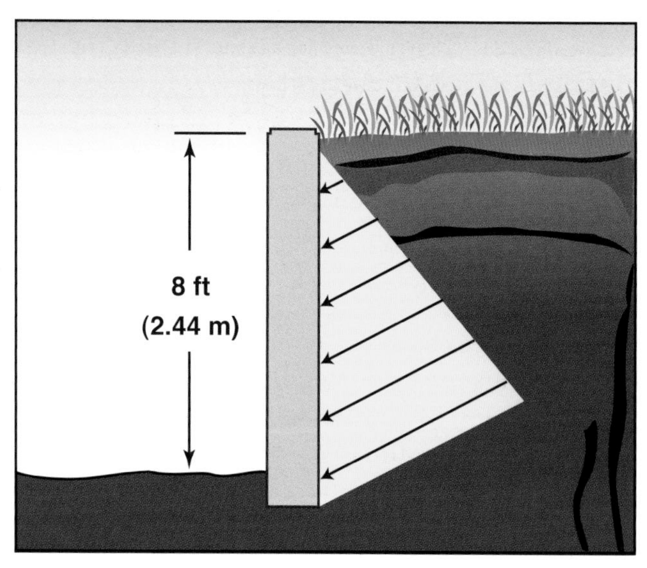

Figure 3.14 If the soil configuration supported by a foundation wall is not horizontal, the soil pressure and resultant force will change.

Firefighters typically have little interest in foundations, but by examining the previous example, they can gain insight into the possibility of foundation shifts over time and consequent shifts of forces in the building being supported. Firefighters frequently become involved in rescue operations because of trench collapses or construction excavation accidents, so these shifts have real-life application.

Other Forces

In addition to the forces already discussed, other forces arising from temperature change, vibration, and shrinkage also may be exerted on a building.

Temperature. As the temperature outside a building changes, the structural members at the periphery of the building expand and contract. The structural members inside the building are less subject to expansion and contraction because the interior temperature of a building is more constant. The differing rates of expansion and contraction between the structural members cause the members to exert forces on each other.

Vibration. Vibration can arise from sources within a building such as rotating machinery. An example would be blower motors for the ventilation system. Vibration can also arise from sources outside a building such as the passing of a freight train.

Shrinkage. Shrinkage can occur in wood structural members as the lumber dries over time. This can result in tensile forces at connections between the members (See Chapter 7).

Live Loads and Dead Loads

The forces on a building resulting from gravity, which were described earlier, are classified into two types: dead loads and live loads. A *dead load* is the weight of any permanent part of a building. This includes the weight of the building components such as roofs, floor slabs or decks, interior walls, stair systems, exterior walls, and columns. It also includes permanent equipment such as heating plants, elevator hoists, pumps, and water supply tanks in earthquake zones. A dead load has the characteristic of being fixed in location and accurately known **(Figure 3.15)**. For example, the beam in shown in Figure 3.15 weighs 142 pounds per foot (211 kg/m0 of length. If it is 20 feet (6.1 m) long, its total weight is 2,840 pounds (1 288 kg). If the beam is used to support a portion of the floor, it is likely to remain in place for the life of the building and its weight will not change.

Dead Load — Weight of the structure, structural members, building components, and any other feature permanently attached to the building that is constant and immobile. Load on a structure due to its own weight and other fixed weights.

Although dead loads usually remain the same, they can change, such as when an air conditioning unit is installed on the roof of a building. The dead load represented by the weight of the roof can also increase over time when additional layers of roofing material are added in the course of resurfacing. Another example of changing loads is the addition of large plantings and even gardens on rooftops.

Figure 3.15 Dead weight load.

Live Load — Force placed upon a structure by the addition of people, objects, or weather.

A *live load* is any load that is not fixed or permanent. Although live loads can include wind and seismic loads, the term is usually applied to building contents, occupants, and the weight of snow or rain on the roof. Usually the actual weight and distribution of the contents are not known exactly. In addition, live loads vary by occupancy. Therefore, building codes specify minimum live loads to be used in the design process for different occupancies. **Table 3.1** shows some of the uniformly distributed live loads required by the International Building Code. Building codes specify that when the actual live load for a given occupancy is known and exceeds the values contained in the code, the actual load must be used in the design calculations.

Distribution of Loads

Concentrated Load — Load that is applied at one point or over a small area.

It must be emphasized that the loads shown in Table 3.1 are *uniformly distributed loads* applied over a large area. A *concentrated load* is one that is applied at one point or over a small area **(Figure 3.16)**. Concentrated loads produce

Table 3.1
Minimum Uniformly Distributed Live Loads
International Building Code®

Occupancy	Pounds per Square Foot (psf)	Kilograms per Square Meter (kgsm)
Assembly Areas and Theaters		
Fixed Seats	60	293
Lobbies	100	488
Movable Seats	100	488
Stages	125	610
Catwalks	40	195
Balconies (exterior)	100	488
On One-and Two-Family Residences Not Exceeding 100 ft^2 (m^2)	60	293
Bowling Alleys	75	366
Dining Rooms and Restaurants	100	488
Fire Escapes	100	488
Gymnasiums	100	488
Manufacturing		
Light	125	610
Heavy	250	1221
Residential		
Uninhabitable Attics without Storage	10	49
Uninhabitable Attics with Storage	20	98
Habitable Attics and Sleeping Areas	30	146
All Other Areas except Balconies	40	195
Stores		
Retail, First Floor	100	488
Retail, Upper Floors	75	366
Wholesale, all Floors	125	610

high localized forces and nonuniform loads in the supporting structural members. Building codes require that a specified minimum concentrated load be used in the structural analysis when it creates greater load effects than the uniform load.

Figure 3.17 illustrates the differences in the structural loads created by a uniformly distributed load and a concentrated load of the same magnitude. In the upper figure, each column supports a load of 1,000 pounds (454 kg). In the lower figure, a concentrated load positioned 4 feet (1.2 m) from the left side produces a load of 1,600 pounds (726 kg) on the left column and 400 pounds (181 kg) on the right column. Thus, the column on left side would have to be designed to support four times the load of the column on the right side.

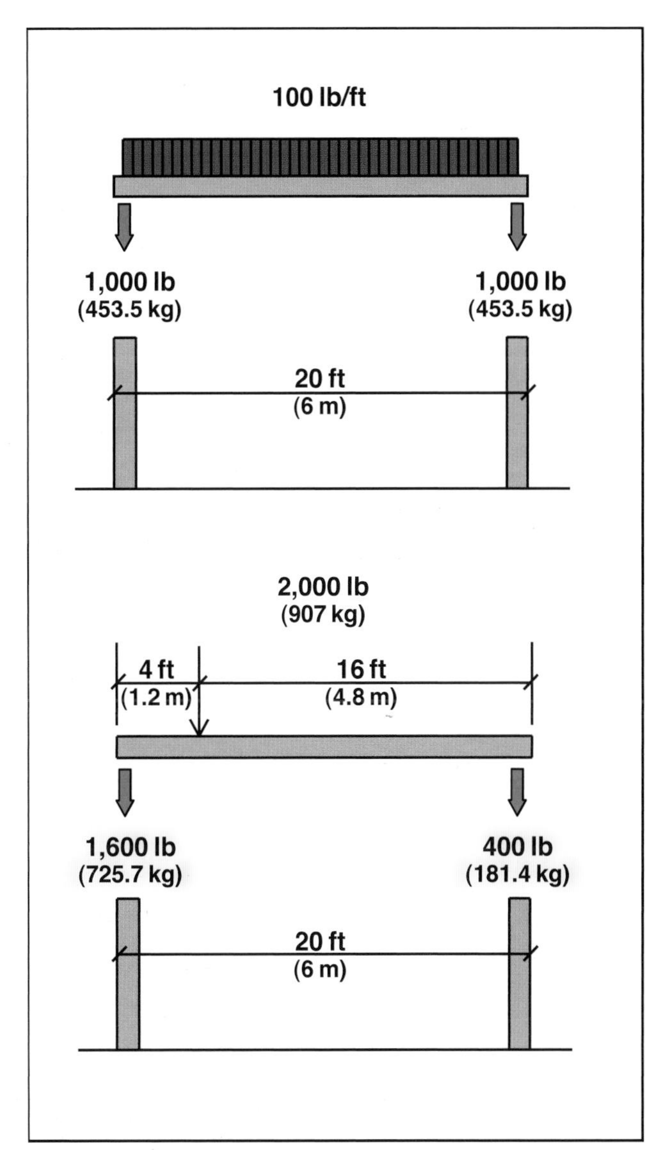

Figure 3.17 Unevenly distributed loads must have structural support to counter the different stresses.

Snow Loads

As noted previously, rain and snow can be considered live loads. Roofs are normally provided with drainage so rain water runs off, although some ponding can occur on large flat roofs if they are not uniformly level. Snow, however, does not drain off and can accumulate to a considerable depth **(Figure 3.18)**. The load exerted on a roof from the weight of snow, known as the *snow load,* can vary from virtually none in southern states to 60 pounds per square foot (291.6 kg/m2) in some areas such as northern Michigan. Even greater loads can occur in mountainous regions. The amount of snow that accumulates on a roof also depends on the slope or shape of the roof and the effect of adjacent structures.

Building codes contain requirements for snow loads depending on the particular region. The snow load expected on the ground is used as a starting point in calculating the snow load on a roof. Frequently, the snow load calculated for a roof may be on the order of

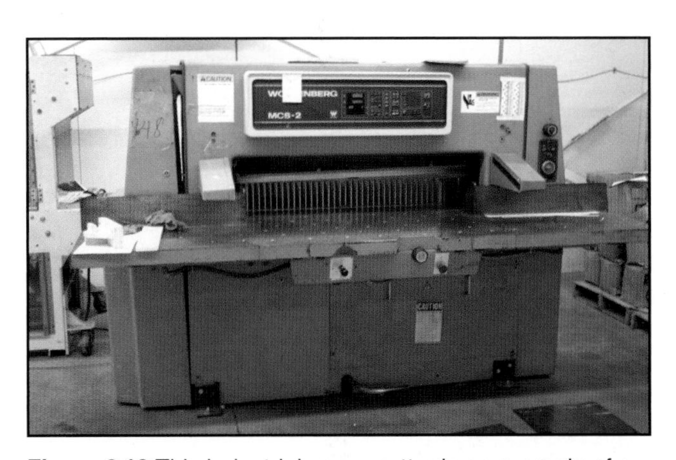

Figure 3.16 This industrial paper cutter is an example of a concentrated load.

Figure 3.18 Poorly maintained roofs are especially vulnerable to the accumulated weight of snow. *Courtesy of West Allis (WI) Fire Department.*

20 or 30 pounds per square foot (97 or 146 kg/m²). Loads of this magnitude are considerably *lower* than the live loads shown in **Table 3.1** for the floors. To the firefighter, this means that roofs in some areas are designed to support *lighter* loads than the interior floors of buildings.

Water Loads

Water from fire fighting operations can add an additional live load. A stream of water discharging 250 gpm (1 000 L/min) introduces 2,082 pounds of water per minute (944.3 kg/min). (In older multistory industrial buildings, scuppers were provided through the outside walls to provide for the drainage of water from the upper floors.) Most of the water will drain from upper floors by way of elevator shafts and stairwells. If the water begins to accumulate, however, its weight will impose an additional load on interior floors. A water depth of 3 inches (75 mm) will impose a load of 20.8 pounds per square foot (0.995 kPa).

Static and Dynamic Loads

Static Load(s) — Loads that are steady, motionless, constant, or applied gradually.

Dynamic Load — Loads that involve motion. They include the forces arising from wind, moving vehicles, earthquakes, vibration, falling objects, as well as the addition of a moving load force to an aerial device or structure. Also called Shock Loading.

In addition to their origin, the loads applied to buildings may be classified as static or dynamic. *Static loads* are loads that are steady or are applied gradually. The dead load of a building, the snow load, and many live loads are static loads. In evaluating static loads, the engineer can deal with a constant force usually equal to the weight of an object. Thus, if a structure is to support a water tank weighing 20,000 pounds (9 072 kg), calculations can proceed on the basis of a known force of 20,000 (9 072 kg).

Dynamic loads are loads that involve motion. They include the forces arising from wind, moving vehicles, earthquakes, vibration, firefighters, and falling objects. Dynamic loads differ from static loads in that they are capable of delivering energy to a structure in addition to the weight of an object. This concept can be illustrated with the example of the impact of a falling object:

A load of bricks that falls onto a roof has kinetic energy by virtue of its motion. As noted in the discussion of wind forces, the basic equation for kinetic energy is:

$$E = \frac{1}{2} mv^2$$

where m = the mass of an object

v = velocity

E = kinetic energy

If a falling load of bricks weighs 600 pounds (273 kg) and falls 10 feet (3.05 m) onto the roof, its velocity at impact would be 25.3 feet per second (7.7 m/sec). It would possess an energy of 192,027 foot-pounds (25 386 N-m).

For the bricks to come to a halt, the kinetic energy must be absorbed by the material of which the roof is constructed. Whether or not the roof can withstand the impact (load) depends on the design strength of the roof and the energy-absorbing properties of the roof material. Obviously, in many cases the roof would collapse under the impact.

Impacts of the type previously described can occur within a building involved in fire as individual components such as trusses or columns fail. The impact of falling contents and structural members subjects other portions of a building to impact, resulting in progressive failure.

Dynamic loads also have the ability to cause failure after repeated cycles. In some cases, a single peak load may not result in failure, but failure may occur after the repeated application of smaller loads because of the loss of some degree of resiliency. An example of this is the deterioration of a garage floor because of the repeated impact loads created by the movement of heavy vehicles.

Structural Equilibrium and Reactions

Both the individual structural members and the building as a whole must resist all the various applied loads. When the support provided by a structural system is equal to the applied loads, a condition known as *equilibrium* exists. A building collapses - or partially collapses — when the applied loads exceed the ability of the structural system to support them **(Figure 3.19)**. The collapse represents a loss of equilibrium. In a sense, when the building becomes a pile of debris on the ground, equilibrium is reestablished.

Equilibrium — Condition in which the support provided by a structural system is equal to the applied loads.

The simple beam shown in **Figure 3.20** is supporting three loads, each weighing 1,000 pounds (454 kg). The beam is supporting 3,000 pounds (1,361 kg) total. The beam supports at points A and B must support the three loads and the weight of the beam. If the beam weighs 500 pounds (227 kg), the total weight supported by A and B is 3,500 pounds (1 586 kg). In this simple example the three loads are uniformly spaced and the supporting forces A and B are equal; each exerts an upward force of 1,750 pounds (3 500/2). The upward forces equal the downward forces and equilibrium is established.

Figure 3.19 As a result of age and poor maintenance, this heavy timber beam collapsed onto a main gas line. When firefighters were finally called to the scene, the electrical panels in the building were still energized. *Courtesy of West Allis (WI) Fire Department.*

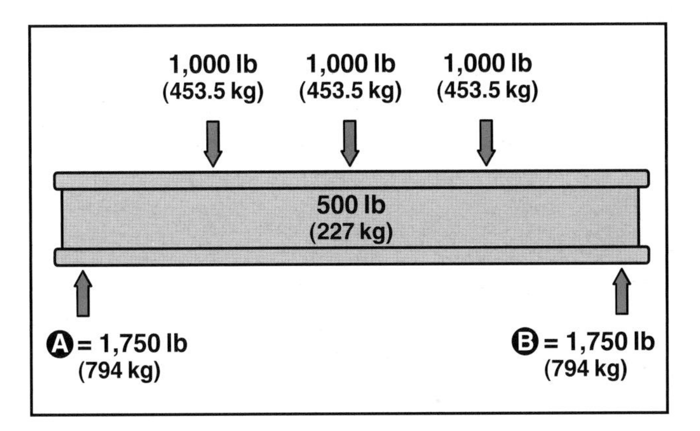

Figure 3.20 The beam is supporting three loads.

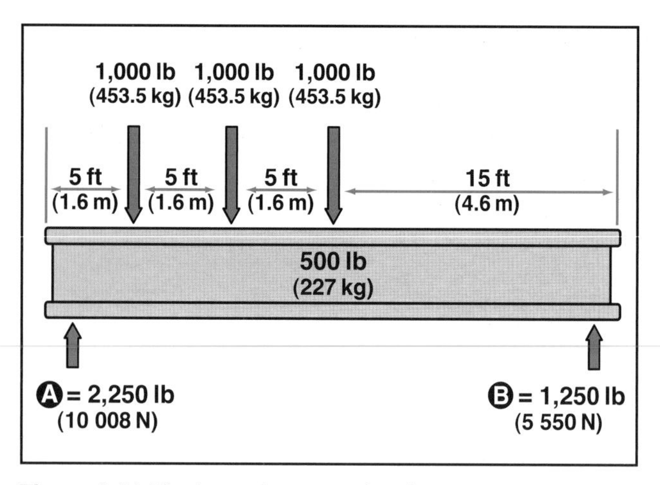

Figure 3.21 The beam is supporting three nonsymmetrical loads.

The forces that resist the applied loads are known as *reactions*. In Figure 3.20, the upward forces at A and B are reactions. If the loads were not uniformly distributed, the reactions would not be equal. In **Figure 3.21** with the loads repositioned, the reaction at A is 2,250 pounds (10 008 N) and at B is 1,250 pounds (5 560 N). **Figure 3.22** illustrates several different simple structures with applied loads designated as L. In every case, the reactions — designated R must equal the applied loads for structural equilibrium to exist.

Internal Forces — Stress

The loads and forces applied to a structural member create internal forces within the member. Proper design requires an evaluation of the internal forces to prevent failure in the form of cracking, crumbling, bending, or breaking.

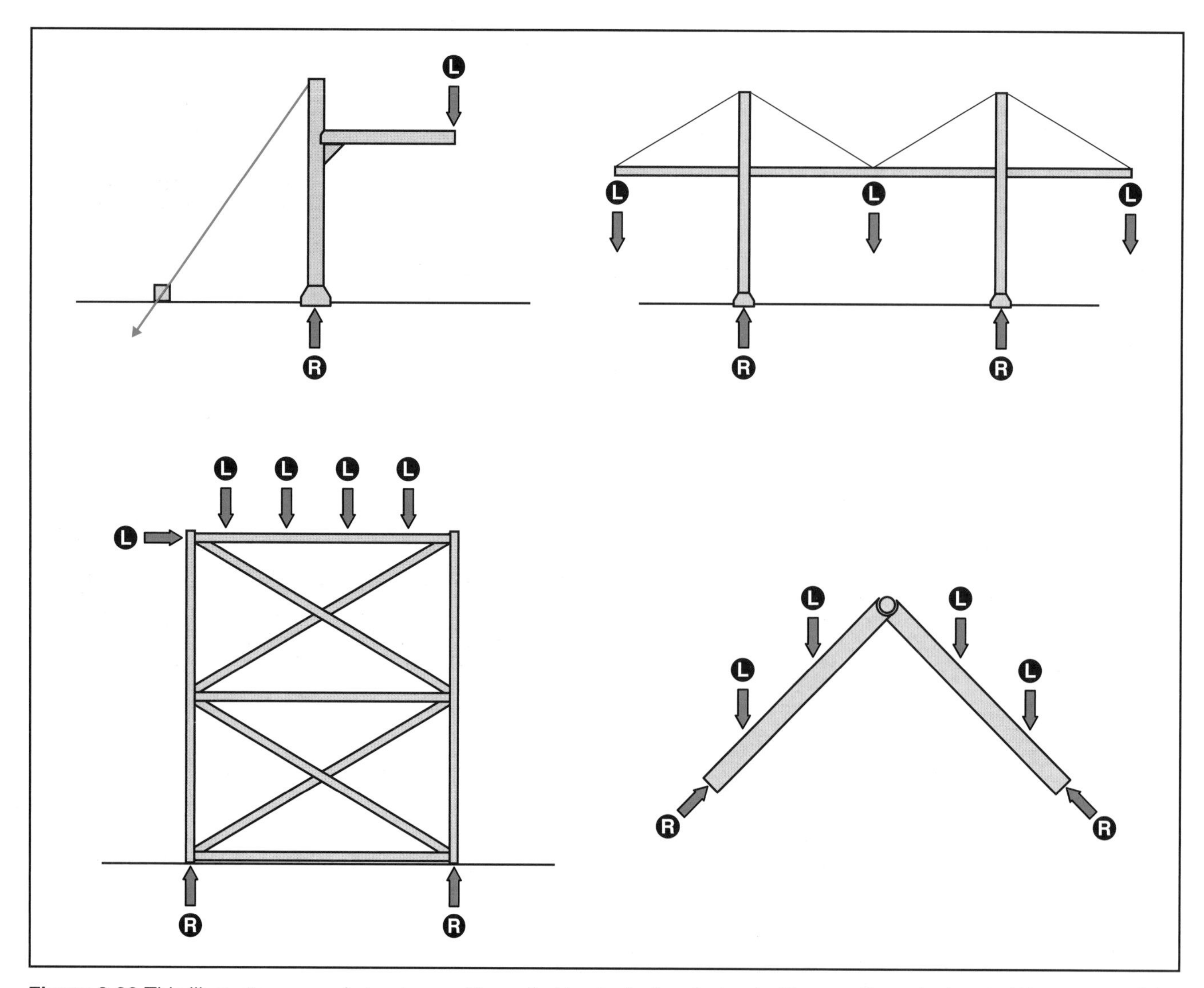

Figure 3.22 This illustrates several structures with applied loads designated as L. The reactions designated R must equal the applied loads to achieve structural equilibrium.

A beam that is supported at one end is known as a *cantilever beam* **(Figure 3.23)**. For equilibrium to exist in this case, the supporting bracket at A must be capable of supporting a vertical force equal to the load of 1,000 pounds (454 kg) plus the weight of the beam, 500 pounds (227 kg) **(Figure 3.24)**. The total downward force at A is 1,500 pounds (6,675 N). The support in turn must exert an equal upward force on the beam of 1,500 pounds (6 675 N). These external forces create a vertical force within the beam across the width and depth of the beam.

Cantilever —Projecting beam or slab supported at one end.

In addition to the vertical force at A, the support bracket must also resist a bending force. This bending force is known as a *bending moment* and is equal to the force multiplied by the distance at which the bending moment is applied. Looking at the beam in **Figure 3.24**, a bending moment exists at A equal to 10,000 foot-pounds (1,322 N/m) because of the load at B. In addition, a bending moment of 2,500 foot-pounds (331 N/m) due to the weight of the beam is also applied at A. Therefore, the support bracket at A must support two loads consisting of a vertical force of 1,500 pounds (6 672 N) and a total bending moment of 12,500 foot-pounds (1 652 N/m).

Exterior loads can create different kinds of interior forces in materials. The interior forces — tension, compression, or shear — are classified according to the direction in which they occur in the material **(Figure 3.25, p. 88)**:

- *Tension* tends to pull the material apart.
- *Compression* tends to squeeze the material.
- *Shear* tends to slide one plane of a material past an adjacent plane.

Tension — Those vertical or horizontal forces that tend to pull things apart; for example, the force exerted on the bottom chord of a truss.

The direction of the interior forces is important because the strength of materials varies with direction. For example, it is commonly recognized that wood has strength in the direction of its grain, which is different from its strength perpendicular to the grain. A material such as concrete has good compressive strength but little tensile strength.

The magnitude of the interior forces that occur in structural members is evaluated by a quantity known as *stress*. Stress is a measurement of force intensity and is expressed as force units divided by the area over which the force is applied [force/area (pounds per square inch or Newtons per square meter)].

Figure 3.23 Cantilever beams are often used to support balconies. *Courtesy of Ed Prendergast.*

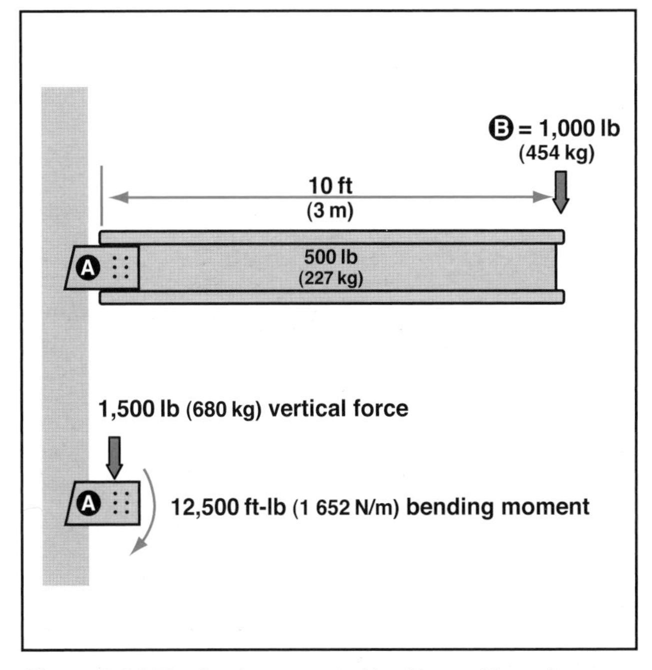

Figure 3.24 The loads supported by the cantilever beam create a combination of forces on the supporting bracket.

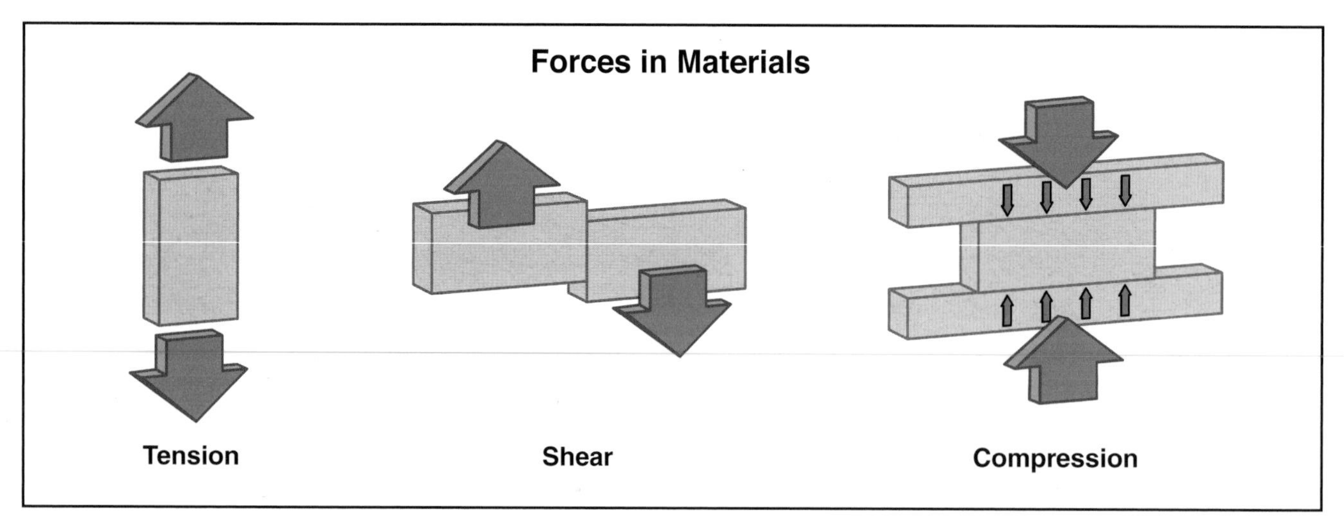

Figure 3.25 Loads are classified according to the direction in which they occur in the material.

In **Figure 3.26,** a steel member measuring 4 x 4 inches (100 mm x 100 mm) supports a load of 20,000 pounds (9 091) kg). The tensile stress in the member is:

20,000 /16 = 1,250 pounds per square inch (8,619 kPa)

In the design of structures, the stresses within structural members must be kept below the stresses at which the material being used would fail. As an example, with ordinary structural steel, the yield point stress, which is also known as the *failure point,* is about 36,000 psi (248 180 kPa). The failure point of a material is the stress at which it fails to perform satisfactorily. Because many factors used in design cannot be known precisely, a factor of safety is used. Factors of safety provide for variations in the properties of the construction materials, the workmanship, and the live and dead loads. A factor of safety is the ratio of the of failure point of the material to the maximum design stress. If a factor of safety of 2 were used in the above example, the maximum allowable design stress for the steel would be:

Failure Point — Point at which material ceases to perform satisfactorily. Depending on the application this can be breaking, permanent deformation, excessive deflection, or vibration.

36,000/2 = 18,000 psi (124 137 kPa)

Thus, the tensile stress induced in the steel support member is well below the maximum allowable stress.

It is typical for stresses to occur in combination within an individual member. In the beam shown in **Figure 3.27**, the applied loads create tension stresses in the bottom of the beam and compressive stresses in the top of the beam. A shear stress is also created across the vertical cross section of the beam. The shape and size of structural members are determined by the need to keep stresses within the allowable values for the particular material being used, such as wood, steel, or concrete.

Exterior loads can also be classified as axial, eccentric, or torsional according to the manner in which they are applied. These different loads create different stresses within the material **(Figure 3.28).**

Axial Load — Load applied to the center of the cross-section of a member and perpendicular to that cross section. It can be either tensile or compressive and creates uniform stresses across the cross-section of the material.

- An *axial load* is a load applied to the center of the cross-section of a structural member and perpendicular to that cross section. An axial load, which can be either tensile or compressive, creates uniform stresses across the cross-section of the material.

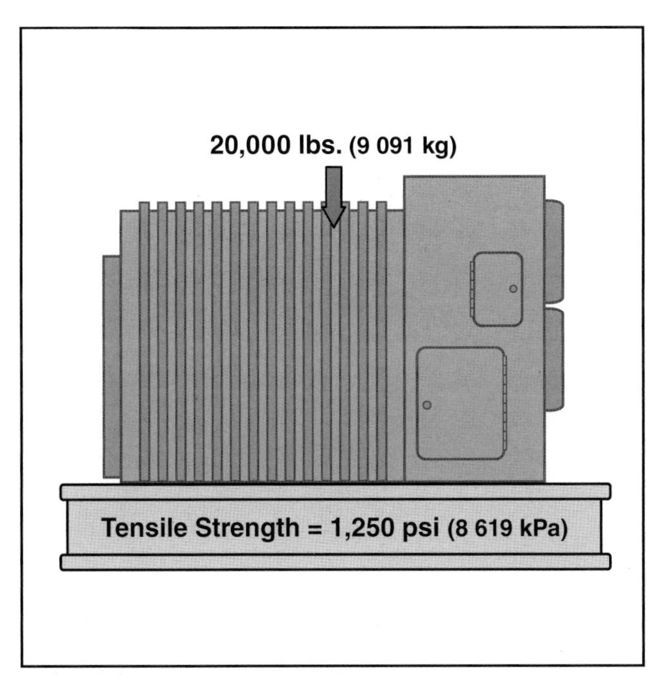

Figure 3.26 The stresses on structural materials must be kept below the point at which the materials would fail.

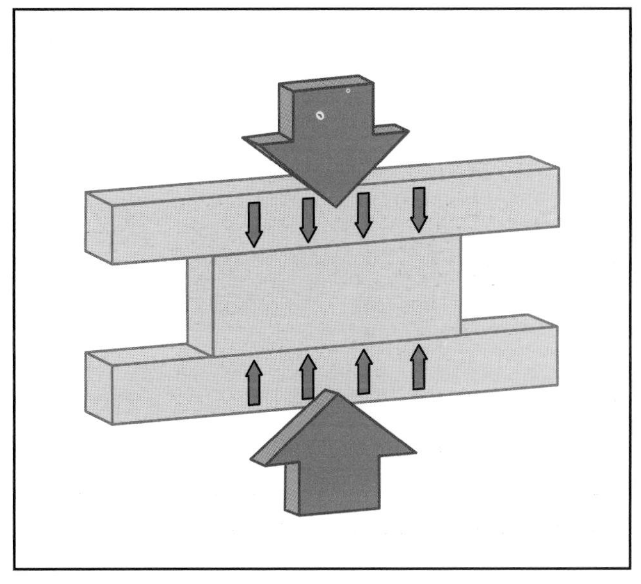

Figure 3.27 Compression forces cause a "squeeze" effect. The top of the beam is in compression and tension results from the stresses on the bottom of the beam.

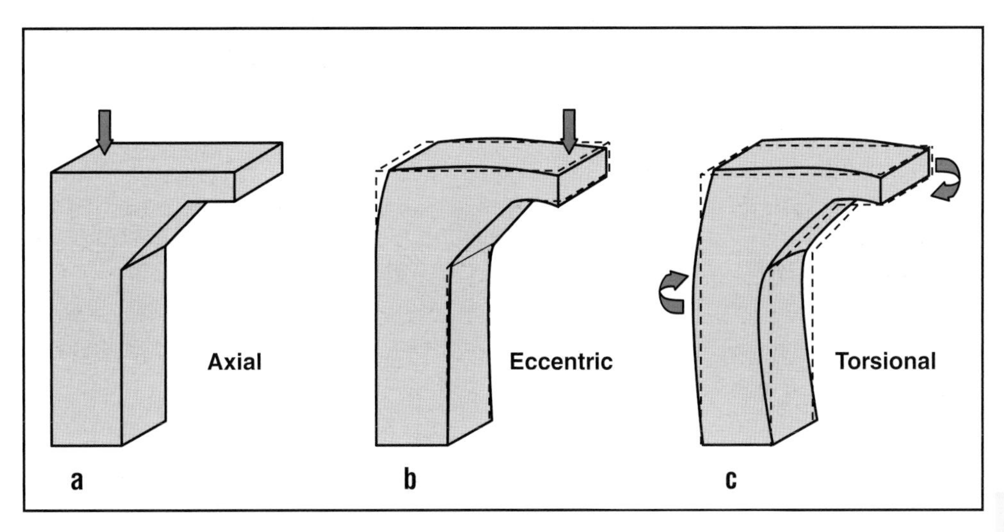

Figure 3.28 (a) Axial loads are applied along the member's axis. (b) Eccentric loads are applied to one side of the cross section, creating a bending tendency. (c) Torsional loads are applied at an angle to the cross section, creating a twisting tendency.

- An *eccentric load* is a load that is perpendicular to the cross section of the structural member but does not pass through the center of the cross section. An eccentric load creates stresses that vary across the cross section and may be both tensile and compressive.
- A *torsional load* is offset from the center of the cross section of the structural member and at an angle to or in the same plane as the cross section. A torsional load produces a twisting effect that creates shear stresses in a material.

Eccentric Load — Load perpendicular to the cross section of the structural member but does not pass through the center of the cross section. An eccentric load creates stresses that vary across the cross-section and may be both tensile and compressive.

Torsional Load — Load offset from the center of the cross section of the member and at an angle to or in the same plane as the cross section. A torsional load produces a twisting effect that creates shear stresses in a material.

As stated in the Introduction to this manual, making a detailed engineering analysis of buildings is not possible during the course of fire fighting. Nonetheless, firefighters must understand that structures and individual structural members are designed to support specific loads. Under fire conditions, the loads change because of the thermal energy released by the fire. Structural members can expand, sag, twist, or simply burn away. As the structural components become distorted, the loads within a building can shift and exert additional forces on adjacent members. Loads that were originally axial can become either eccentric or torsional. These shifting loads increase the probability of failure **(Figure 3.29)**.

Figure 3.29 When the steel column in the center of this photo buckled, the interior masonry wall collapsed. *Courtesy of Ed Prendergast.*

Structural Components

Larger structural systems can be constructed from several basic components, including some of the following:

- Beams
- Columns
- Arches
- Cables
- Trusses
- Space frames
- Connectors

Beams

A *beam* is structural member that can carry loads perpendicular to its longitudinal dimension. Beams can have several different designs **(Figure 3.30)**. A *simply supported* beam is supported at each end and is free to rotate at the ends. A wood joist resting on a masonry wall is an example of a simply supported beam. A cantilever beam is supported at one end; as illustrated in Figure 3.23, these types of beams are often used to support balconies.

Beam — Structural member subjected to loads, usually vertical, perpendicular to its length.

Beams can also be continuous across several supports or restrained at both ends **(Figure 3.31)**. Restrained beams are rigidly supported at each end. Under fire conditions, a rigidly supported beam will tend to retain its load-bearing ability longer than a simply supported beam. This is due to the fact that the end restraints provide more resistance to the applied bending moment of the beam.

Beams can be made of wood, steel, or reinforced concrete. The primary design consideration of beams is their ability to resist bending from the applied loads. The bending moment introduces stresses across the cross-section of a beam that are not uniform. If a beam supports loads as shown in **Figure 3.32,** the top of the beam is subjected to compressive stresses and the bottom of the beam is subjected to tension. It can be seen in **Figure 3.33, p. 92** that the maximum tension or compression stresses are in the top and bottom of the beam. At the middle of the beam, a point known as the neutral axis, the tension and compression stresses are actually zero.

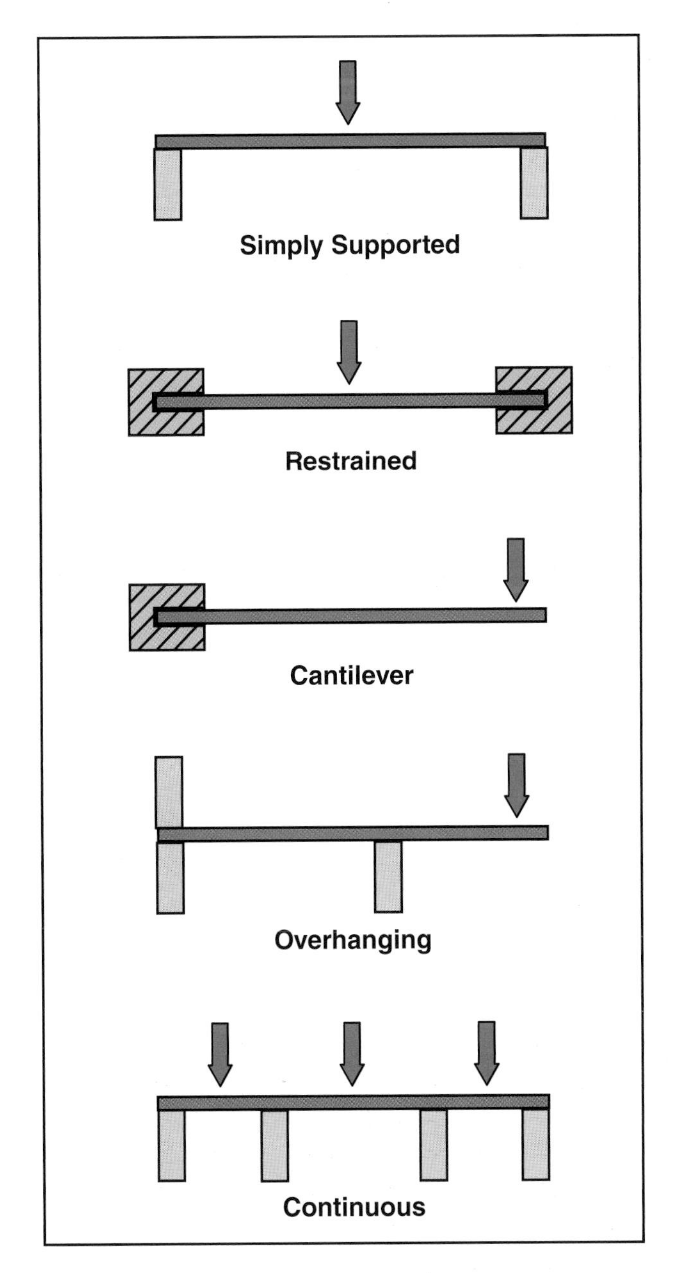

Figure 3.30 Beams are supported in a variety of configurations.

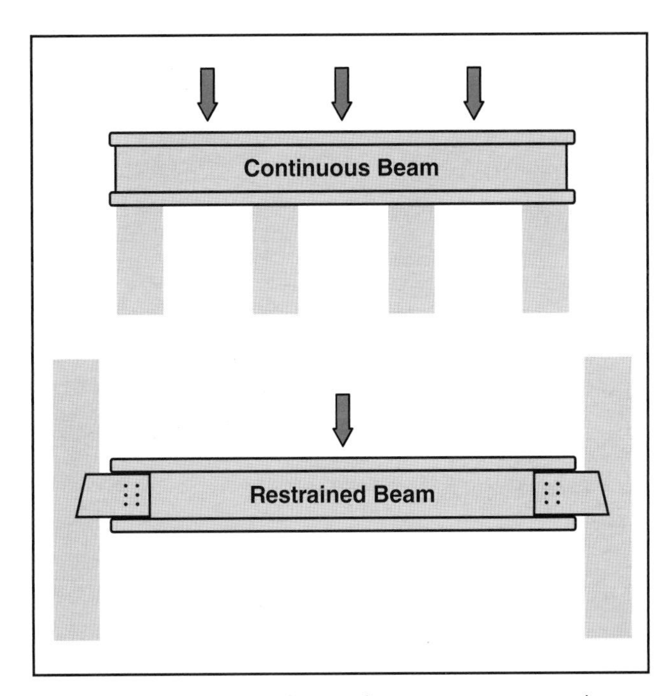

Figure 3.31 Beams can be continuous across several supports or restrained at both ends.

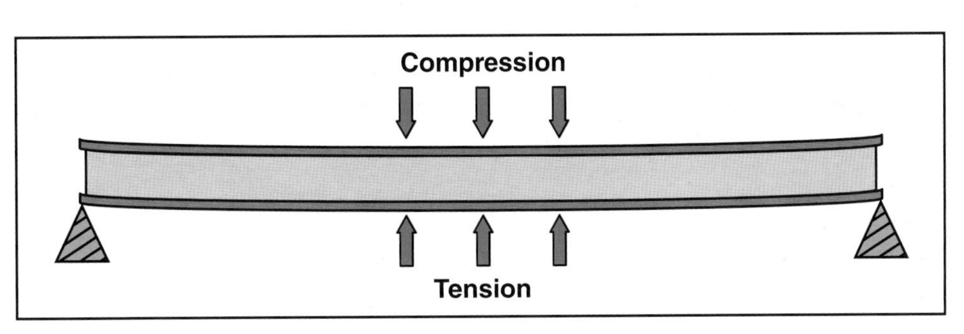

Figure 3.32 The top of the beam is in compression and the bottom of the beam is in tension.

It can also be seen from Figure 3.33 that the top and bottom portions of the beam do most of the work of resisting the bending moment. Because the middle portion of the beam does little actual work compared to the top and bottom, some of the material can be removed from the middle of the beam cross-section without greatly affecting the strength of the beam. This increases the efficient use of material and reduces the weight of the beam. It is this engineering principle that results in many beams being constructed in the shape of the letter I **(Figures 3.34 a and b)**.

In an I-beam, the top and bottom portions of the beam are known as the top and bottom flanges. Because the top and bottom flanges of an I-beam support most of the load, any alteration of the flanges, such as cutting the top flange of a wooden I-beam, can have the effect of greatly reducing its strength.

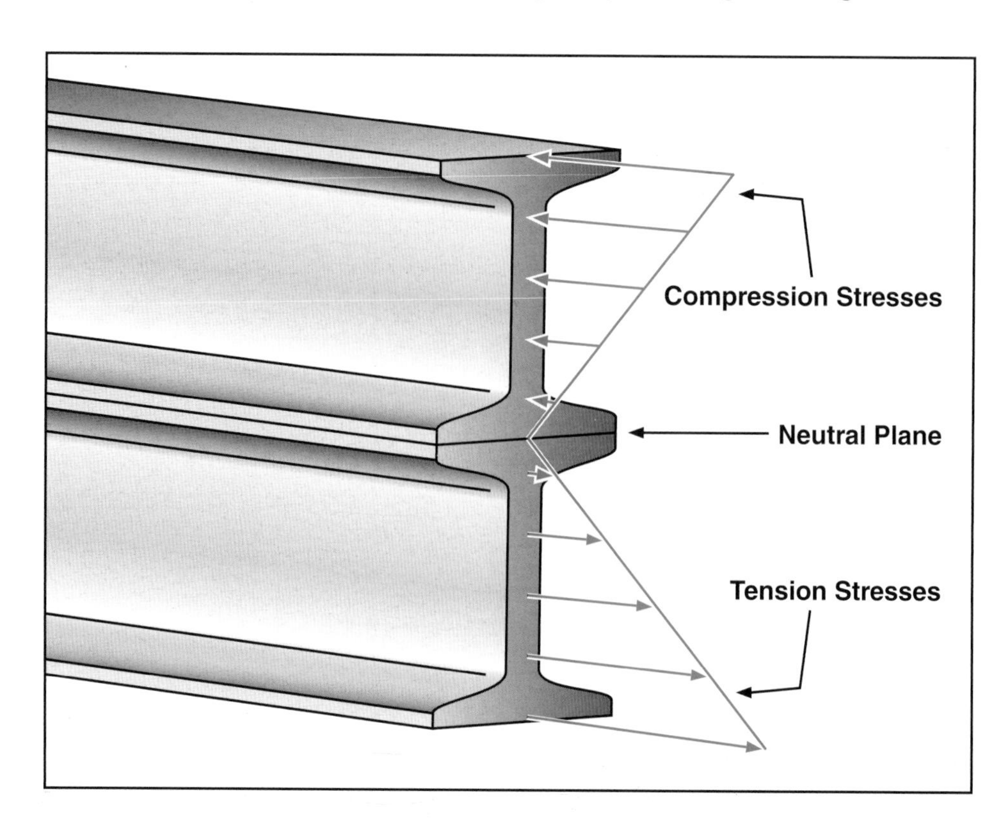

Figure 3.33 The maximum compression and tension stresses in a beam occur in the top and bottom of the beam.

Figures 3.34 a and b Engineered wood I-beam and steel I-beam. *Both photos Courtesy of Dave Coombs.*

The stresses in the flanges of a beam can be calculated mathematically. The stresses are a function of the cross-sectional area of the flanges and the vertical dimension of the beam. Tall beams are capable of supporting greater loads than short beams, even if they have the same cross-sectional area.

Columns

Columns are structural members designed to support an axial compressive load. The stresses created within a column are primarily compressive. Columns are not primarily designed to withstand stresses due to bending. However, it is not unusual for some bending stresses to occur because of shifting of the support beneath the column or shifting of beams attached to the column. Columns can be made of wood, steel, cast iron, concrete, or masonry **(Figure 3.35)**. In cases of failure, tall, thin columns fail by buckling and short, squatty columns fail by crushing **(Figure 3.36)**.

Column — Vertical supporting member.

Arches

An *arch* is a curved structural member in which the interior stresses are primarily compressive. Arches produce inclined forces at their end supports, which the supports must resist **(Figures 3.37 a and b, p. 94)**. Arches are used to carry loads across a distance and have application as support for roofs and entrances in masonry buildings.

Figure 3.35 A cast-iron column supporting a second floor.

Figure 3.36 This column was struck by an automobile. The resulting deformation has created bending stresses in the column, making its continued use highly questionable. *Courtesy of Ed Prendergast.*

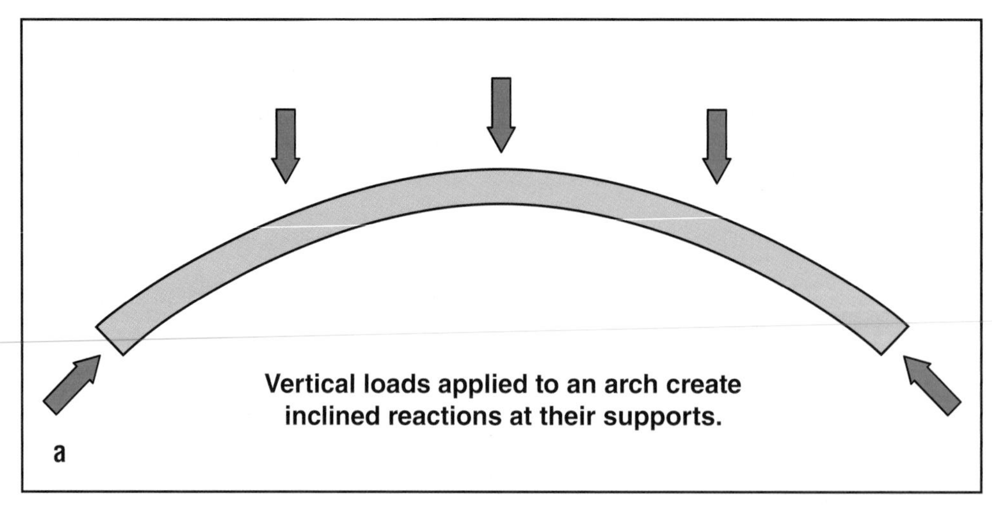

Figure 3.37 a and b An arch is a curved structural member in which the interior stresses are primarily compressive. *Photo b courtesy of Donny Howard.*

If the end supports of an arch are removed, the arch becomes unstable. **Figure 3.38** illustrates a situation in which half an arch was removed when a neighboring building was razed. This creates an inherently unstable situation in which the masonry wall supported by the arch could collapse. An effort was made to maintain stability by adding a wood column under the arch and a steel rod anchored in the masonry wall; however, this is a tenuous solution at best.

If the supports at the ends of the of the arches shift because of settling soil or thermal expansion, bending stresses may develop in the arch. To provide for minor adjustments, arches are sometimes designed with hinges **(Figure 3.39)**. If the end supports cannot support the arch, horizontal tie rods can be used to prevent the arch from spreading. Arches can be constructed of masonry, steel, concrete, or laminated wood.

Cables

Cables can be thought of as flexible structural members that can be used to support roofs, brace tents, and restrain pneumatic structures. Although cables can be used in applications where they are essentially straight, a cable used to support loads over a distance will assume the shape of a *parabola* **(Figure 3.40)**. The stresses in a cable are tension stresses. Cables are usually made of steel strands although aluminum may be used where weight is a critical factor.

Trusses

Trusses are framed structural units made up of a group of triangles in one plane. A true truss is made only of straight members. If loads are applied only at the point of intersection of the truss members, only compressive or tensile stresses will affect the members of the truss **(Figures 3.41 a and b)**.

Some types of roof trusses, such as the bowstring truss, have a curved top chord. These curved members are unavoidably subjected to bending forces. In addition, loads will be applied to the truss between the intersection points of the members, which will also create bending forces.

Truss — Structural member used to form a roof or floor framework. Trusses form triangles or combinations of triangles to provide maximum load-bearing capacity with a minimum amount of material; often rendered dangerous by exposure to intense heat, which weakens gusset plate attachment.

Figure 3.38 One half of this masonry arch was removed, rendering it unstable. Notice that a column has been added to add some stability. *Courtesy of Ed Prendergast.*

Figure 3.39 The hinge of an arch roof allows the roof to make minor adjustments while bending.

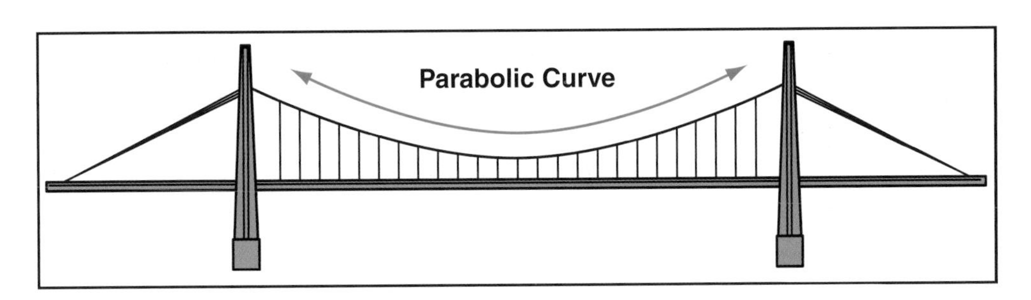

Figure 3.40 When a cable is used to support loads, it will assume the shape of a parabola.

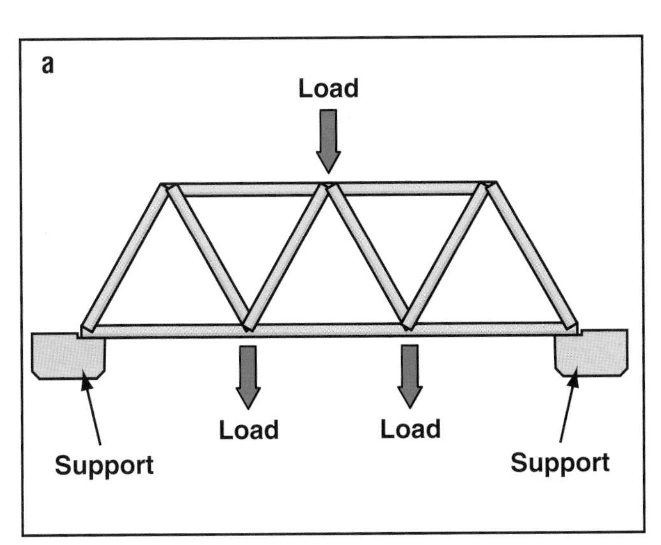

Figures 3.41 a and b (a) Forces in a truss under load. (b) Note the triangles that form a truss under construction. *Photo b courtesy of Steve Toth.*

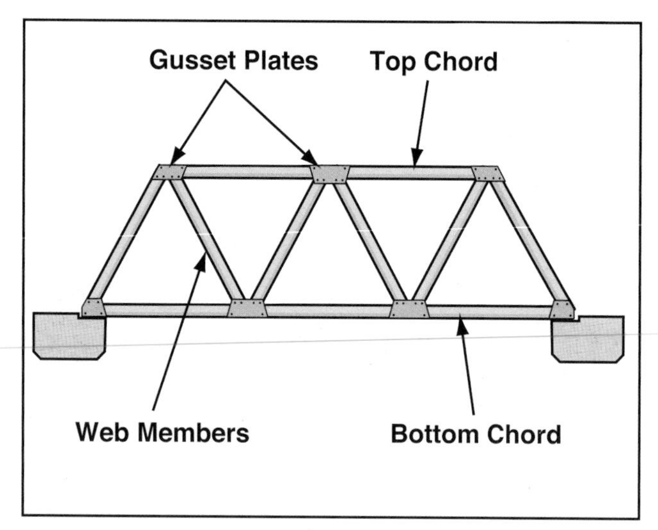

Figure 3.42 Truss components.

The top members of a truss are called the *top chords* and the bottom members are called the *bottom chords* **(Figure 3.42)**. The diagonal members are called either *diagonals* or *web members*. The joints may be formed by pin connections, welding, gusset plates, strap connectors, or structural adhesive. Trusses may be made of wood, steel, or a combination of wood and steel.

Geometrically speaking, the triangle provides an inherently rigid frame. If a diagonal brace is added to a framework, the resulting triangulation creates a stronger assembly. The basic triangles of which trusses are composed can be arranged in a large variety of styles. The most common truss configurations have names and are illustrated in **Figure 3.43**. Typical

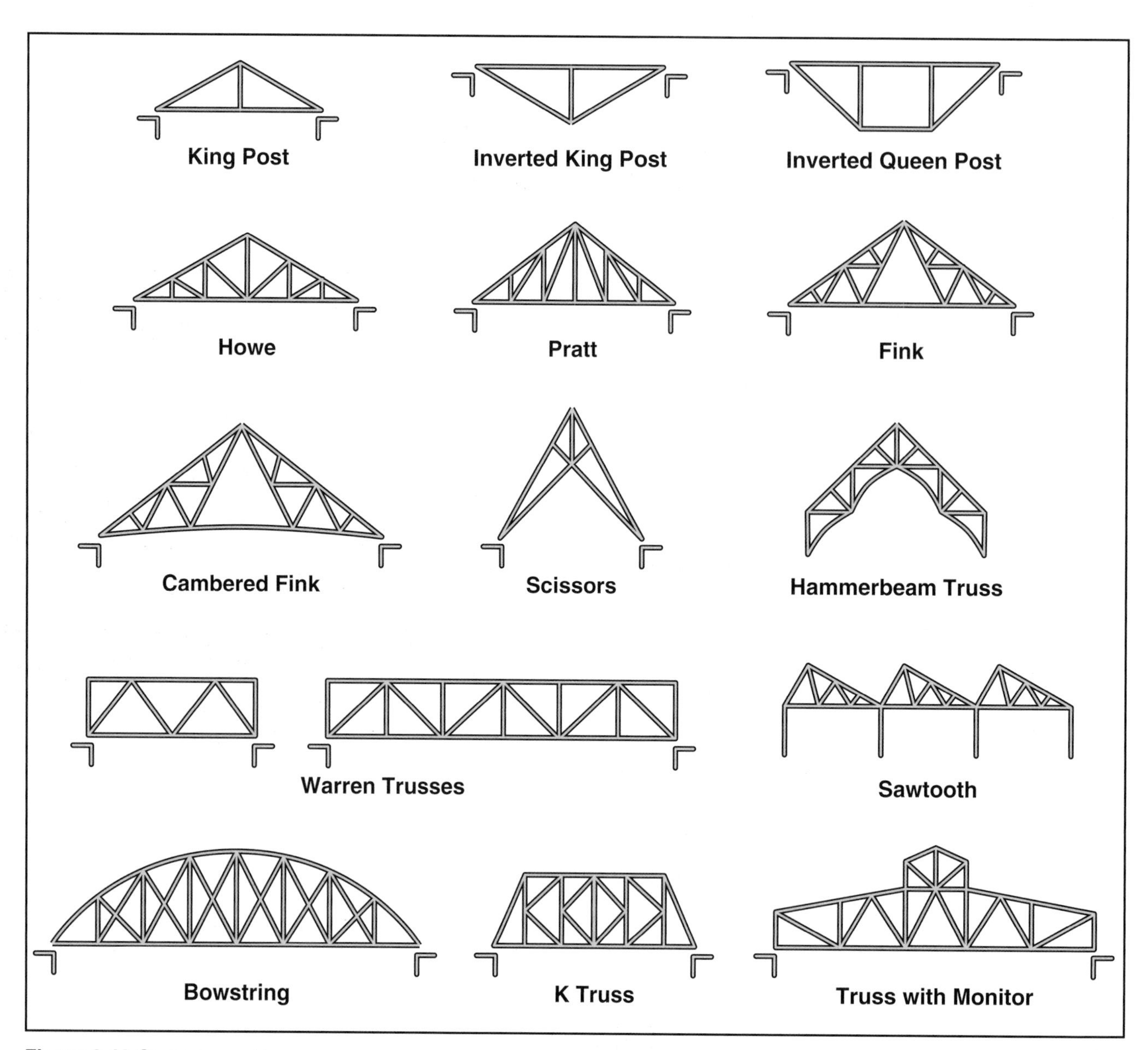

Figure 3.43 Common roof trusses.

truss shapes are available to span distances of 22 to 70 feet (6.7 m to 21.3 m), but in today's construction truss spans in excess of 100 feet (30 m) are not uncommon.

Trusses are a commonly used structural member. Most trusses are prefabricated. They also have the advantage of being able to span an equal distance using less material and being lighter than a comparable beam. Lightweight wood or metal trusses, known as *truss joists,* have become very common in floor construction, taking the place of solid joists **(Figure 3.44)**. Lightweight wood trusses are also commonly used in roof construction. From a fire fighting standpoint, trusses have the potential for early failure under adverse conditions. This is due to the fact that failure of any portion of the top or bottom chords results in failure of the truss. Lightweight steel trusses, known as bar joists, are also used for floor and roof construction in fire-resistive and noncombustible construction **(Figure 3.45)**.

NOTE: Roof trusses that are even lighter in weight than steel bar joists are available. In effect, these trusses are made of galvanized steel studs or channels similar to them, and assembled with self-drilling screws.

Space Frames

Space frames are truss structures that are developed in three dimension **(Figure 3.46)**. They offer many of the advantages of two-dimensional trusses in terms of economic use of material. Space frames are well-suited to support

Figure 3.44 Truss floor joists. *Courtesy of McKinney (TX) Fire Department.*

Figure 3.45 Lightweight truss used to support a roof. *Courtesy of McKinney (TX) Fire Department.*

Figure 3.46 A space frame used to support a roof. The black piping is part of the sprinkler system. *Courtesy of Ed Prendergast.*

uniformly distributed loads. The design of space frames is more complicated than with two-dimensional trusses because the forces must be analyzed in three dimensions.

Truss Connectors

Depending on the materials of which the truss is constructed, truss members can be connected by different means. Wood truss members can be connected by pins or bolts, gusset plates, adhesives, brackets, and metal straps. The members of steel trusses can be joined using steel gusset plates, rivets, or welds.

The connectors in truss assemblies are a critical part of the truss. Failure of a connector will result in failure of the truss. Quality control in the manufacture of trusses can affect the behavior of trusses under fire conditions. If a steel-toothed connector is not properly seated, the connector may work loose under fire conditions. In some wood truss assemblies a small gap may exist between the bends of the horizontal members at the joint.

NOTE: Various connectors are discussed in more details in later chapters.

Basic Structural Systems

Basic structural components are of little value unless they can be assembled into a composite system that will support a building. Just as architectural styles can vary widely, an almost infinite number of structural designs can be created. As with building types, practical necessity and economics result in a number of commonly encountered structural systems. Therefore it is possible to examine general types of structural systems, such as structural bearing walls and frame structural systems, that share fundamental characteristics. Each system has advantages or disadvantages related to material and cost limitations and applications to which it is most suited.

Bearing Wall — Wall that supports itself and the weight of the roof and/or other internal structural framing components such as the floor beams above it.

Structural Bearing Walls

A common method of construction uses the walls of a building to support spanning elements such as beams, trusses, and precast concrete slabs. These are appropriately known as *bearing wall* structures **(Figure 3.47)**. The bearing walls are usually the exterior walls with interior support system consisting of columns and beams; however, it is possible to use interior walls for structural support. Bearing walls provide lateral support to the structure along the direction of the wall.

In a bearing wall structure, the walls are subjected to compressive loads. The walls may be continuous or they may be interrupted for door and window openings. Materials used for bearing walls include concrete masonry units (commonly referred to as CMUs or concrete blocks), bricks, stone, and concrete panels. A log cabin is an example of the use of solid wood for a bearing wall.

Figure 3.47 These interior supports are part of a bearing wall structure. *Courtesy of McKinney (TX) Fire Department.*

Frame Structural Systems

In a frame structure, structural support is provided in a manner similar to the way the skeleton supports the human body. The walls act as the 'skin' to enclose the frame. The walls may also provide lateral stiffness but provide no structural support.

In the fire service, it is not uncommon to use the term *frame construction* to refer to a wood-frame building, but frame structural systems are also built using other materials. In addition to the framing associated with wood construction, other types of structural frame construction include the following:

- Steel stud wall framing
- Post and beam construction
- Rigid frames
- Truss frames
- Slab and column frames

Steel Stud Wall Framing

Steel stud wall construction uses relatively closely spaced vertical steel studs connected by top and bottom horizontal members. Historically, stud-wall frame construction has been associated with the use of 2 x 4 inch (50 mm x 100 mm) wood studs although the use of steel studs has become more common in recent years. A steel stud wall is frequently provided with diagonal bracing for stability **(Figure 3.48)**. When the exterior and interior of a stud wall are covered with paneling or sheathing, a rigid wall panel results.

Figure 3.48 Steel stud wall with diagonal bracing. *Courtesy of Ed Prendergast.*

Post and Beam Construction

Post and beam construction uses a series of vertical elements (the posts) to support horizontal elements (the beams) that are subject to transverse loads **(Figure 3.49)**. Historically, this system evolved from the use of tree trunks

Figure 3.49 The distinctive characteristic of post and beam framing is the spacing of the vertical posts and the cross-sectional dimension of the members.

for framing and is still commonly associated with wood beams and columns. Other materials can be used, however, including masonry for the posts and steel and precast concrete for the posts and beams.

The distinctive characteristic of post and beam framing is the spacing of the vertical posts and the cross-sectional dimension of the members. The vertical posts may be spaced up to 24 inches (600 mm) apart, unlike stud wall construction where the studs are 12 to 16 inches (300 mm to 400 mm) apart. The minimum dimensions used for the wood posts and beams are larger than the studs in stud wall construction. Typical dimensions for the posts are 6 x 8 inches (150 mm x 200 mm) when supporting roofs only. Post and beam construction requires the addition of other members such as diagonal braces to withstand lateral loads.

Figure 3.50 Rigid frame. *Courtesy of McKinney (TX) Fire Department.*

Figure 3.51 This home is a typical example of a rigid-frame building. *Courtesy of McKinney (TX) Fire Department.*

Rigid Frames

When the joints between a column and a beam are reinforced so bending stresses can be transmitted through the joints, the structural system is known as a *rigid frame* **(Figure 3.50)**. The most easily recognized rigid-frame structure is the single-story, gabled-roof and rigid-frame building. The peak of the roof is usually provided with a hinged connection to allow for slight movement between the two halves of the frame. This type of rigid frame can be constructed of steel, laminated wood, or reinforced concrete **(Figure 3.51)**.

Rigid frames are used in other types of structures including multistory and multispan designs. Because the joints are intended to transmit bending stresses, firefighters should pay particular attention to their design. The joints usually must be reinforced and will be the last portion of the assembly to fail under fire conditions **(Figure 3.52)**.

Truss Frames

The trusses previously discussed can be adapted to a variety of applications. It is possible to build components of a frame using a series of trusses, as with the arch and the rigid frame illustrated in **Figure 3.53**.

Figure 3.52 Collapse of an unprotected steel rigid frame. Notice that the rigid joint has remained in place. *Courtesy of Ed Prendergast.*

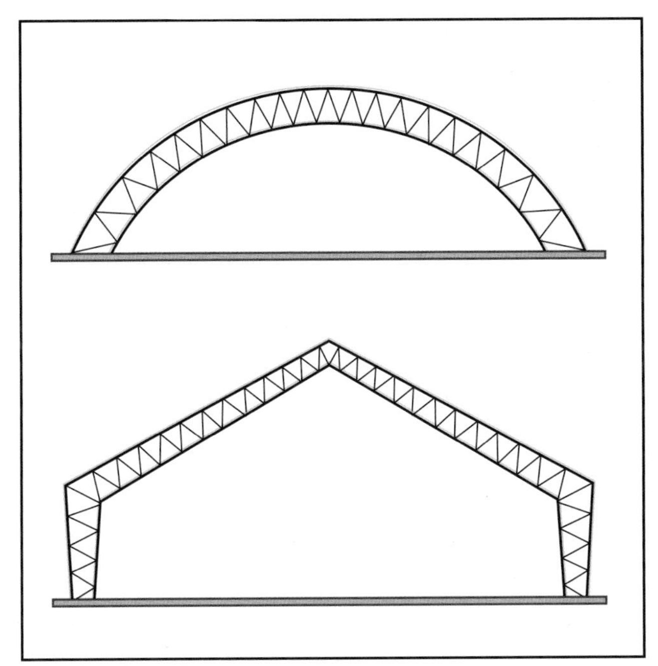

Figure 3.53 Frame structures consisting of truss systems.

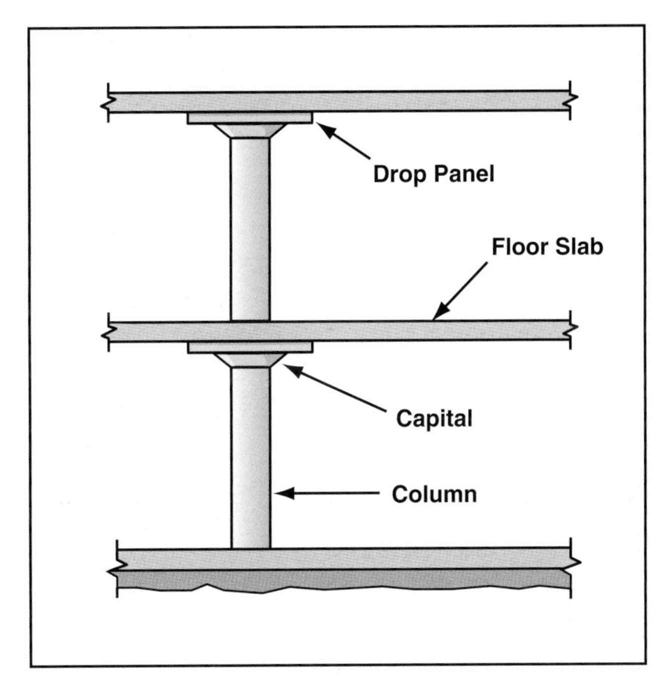

Figure 3.54 Reinforcement of concrete floor.

Slab and Column Frames

Slab and column frames are most frequently encountered in concrete structures. The floors of a multistory, reinforced-concrete building are concrete slabs, which can be designed by several methods, depending on the loads to be supported. These floor slabs are supported by concrete columns. The intersection between the slab and column is a region of high stress and usually is reinforced by additional material in the form of a capital or a drop panel **(Figure 3.54)**.

The concrete slab is one type of horizontal system that can be used to support floor loads. Others include wood decks and metal decks supported by beams and columns.

NOTE: Concrete framing is covered in Chapter 10, Concrete Construction.

Shell and Membrane Systems

It is possible to construct buildings that consist primarily of an enclosing surface and in which the stresses resulting from the applied loads occur within the surface. These structural systems are known as *surface systems.* When the enclosing surface is a thin stretched flexible material, the structure is known as a *membrane structure.* Examples of these structures are a simple tent or an air-supported structure.

Surface System — System of construction in which the building consists primarily of an enclosing surface and in which the stresses resulting from the applied loads occur within the surface bearing wall structures.

Membrane Structure — Structure with an enclosing surface of a thin stretched flexible material. Examples of these are a simple tent or an air-supported structure.

A membrane "structure" can be distinguished from a simple tent by its permanence. Tents are used for short periods; membrane structures are permanent. Building codes typically address membrane structures with a life of 180 days or more, while fire codes address those to be used for less than 180 days.

Membrane structures possess several design advantages. The fabrics used for roofs on membrane structures weigh less than other roof systems. Fabric roofs weigh only about 2 pounds per square foot (9.8 kg/m2). Membrane structures can usually be erected in less time than a rigid structural system. In addition, the fabrics used can flex and absorb some of the stresses caused by seismic and wind forces.

In early permanent membrane structures, polyvinyl-coated polyester fabric was used. More recent designs have used polytetraflouroethylene (PTFE) coated glass fiber. Fabrics cannot resist compressive forces; therefore, cables and masts or tubular or solid frameworks must support the fabrics **(Figure 3.55)**. Frameworks can be made of wood, concrete, or steel. In the U.S. and Canada, frames are usually steel and sometimes aluminum. Membrane structures can also be supported by internal air pressure **(Figure 3.56)**.

Shell Structure — Rigid, three-dimensional structure having an outer "skin" thickness that is small compared to other dimensions.

Shell structures are rigid three-dimensional structures having a thickness that is small compared to other structural material dimensions. Shell structures lend themselves to regular geometric shapes such as cones, domes, barrel vaults, and folded plates. Shell structures are most commonly constructed of concrete, although it is possible to construct them using plywood or fiberglass **(Figure 3.57)**.

NOTE: Additional information on air-supported structures can be found in Chapter 12.

Figure 3.55 Construction of membrane structure. *Courtesy of Ed Prendergast.*

Figure 3.56 Typical air-supported structure . *Courtesy of Ed Prendergast.*

Figure 3.57 A power station cooling tower is an example of a shell structure.

Summary

This chapter provided an introduction to the various forces to which a building is subjected. These forces must be analyzed and evaluated by the structural engineer. The forces on buildings arise from gravity, wind, soil pressure, snow, and other sources. The forces exerted on a building determine how a building's structural system is designed. Externally applied forces create different types of stresses within building materials.

It is important that firefighters have a basic understanding of the forces acting on structures so they can be aware of structural hazards and collapse dangers.

A variety of structural components are available to the engineer to support the applied loads. These components, which include beams, arches, trusses, and columns, support loads in different ways. It must be emphasized that under fire conditions, the loads and stresses exerted on a structural system are subject to change in magnitude and direction resulting in structural failure.

Review Questions

1. What is a load?
2. What factors determine the magnitude of the forces developed within a building during an earthquake?
3. What is a dead load?
4. How do dynamic loads differ from static loads?
5. What is a membrane structure?

References

1. Allen, Edward *Fundamentals of Building Construction, Materials and Methods 4th edition,* John Wiley and Sons.
2. Ambrose, James and Iano, Joseph *Building Structures, 2nd edition.* John Wiley and Sons.
3. Casey, Heather. *Vegas MGM Grand Fire 20 Years Ago Among the Worst.* Firehouse.com news.
4. Ishii, Kazuo *Membrane Design and Structures of The World,* Shinkenchiku-sha, Tokyo.
5. Saunders, Mark C. "Seismic Joints in Steel Frame Building Construction". *Modern Steel Construction.*
6. White, Richard N. and Salmon, Charles G. *Building Structural Design Handbook,* John Wiley and Sons.

Chapter Contents

chapter 4

Key Terms

FESHE Objectives

Fire and Emergency Services Higher Education (FESHE) Objectives: *Building Construction for Fire Protection*

3. Analyze the hazards and tactical considerations associated with the various types of building construction.
5. Identify the function of each principle structural component in typical building design.

Building Systems

Learning Objectives

After reading this chapter, students will be able to:

1. Discuss the various types of stairs and the structural requirements related to each.
2. Describe the various types of elevators and their safety features.
3. Discuss moving stairways, walkways, and conveyors as they relate to firefighting concerns.
4. Describe the uses of vertical shafts and utility chases and their impact on firefighting.
5. Describe the functions and components of HVAC systems and how they impact firefighting.
6. Distinguish between various smoke control methods.
7. Discuss the various types of electrical equipment found in building structures and the hazards posed by each.

Chapter 4
Building Systems

Case History

Event Description: Report of smoke on the 5th floor of a five-story, multi-residence building. The on-duty Assistant Chief arrived, established command, and observed light smoke rising from the side of the building. The first-arriving engine had an acting Captain with 28 years experience and a firefighter with 7 years experience. When they entered the lobby, a police officer reported the problem as smoke in apartment 518. The firefighter suggested taking the stairs but the acting Captain overrode this proper procedure and decided to take the elevator, along with forcible entry tools and a high-rise pack. The police officer said he would come along to assist.

When the elevator opened on the 5th floor the men were met with heavy smoke banked to the floor. The elevator door would not close and return to a lower floor, possibly because of equipment dropped in the doorway. The firefighters donned their facepieces but the police officer was in an IDLH situation. The firefighters left the police officer and located an apartment for shelter. They returned to assist him and the acting Captain shared his facepiece with the police officer as they crawled to the apartment. (**NOTE:** Sharing facepieces is contrary to standard procedure, but the Captain determined they were in a life-or-death situation.) An aerial ladder was raised to the apartment window and the police officer was removed with difficulty and transported to a hospital.

The firefighters returned to the hallway with the high-rise pack and crawled toward the apartment on fire; by this time it was fully involved and had burned through the entry door and into the hall. The fire was extinguished without any great difficulty from this point.

Lessons Learned:

— The building had a preincident plan in the first-arriving command vehicle with diagrams of all floors and information on stairs, standpipes, and apartment locations but it was not used until the fire had been knocked down.

— The Incident Commander and all fire crews cannot read smoke and fire conditions in this type and size of building from the exterior. Wired glass in parts of the structure prevented windows from breaking and concealed the size of the actual fire.

— Never take the elevator in a 5-story building when dispatched to a fire alarm; take the stairs.

— Never take police or civilians into an area that could be hazardous to them.

— Follow your training and use the preincident plan on all responses. When a working fire occurs it will come as second nature.

Source: National Fire Fighter Near-Miss Reporting System.

To be functional, buildings must contain a variety of basic systems to address the convenience, comfort, efficiency, and most important, life safety for occupants. Depending on the intended use of the building, special systems may be required to address operational requirements particular to that use. Examples of buildings with special requirements include hospitals, factories, warehouses, and detention centers.

All building systems, both basic and special, must be properly designed, installed, inspected, tested, and maintained to ensure the safety of building occupants and responding firefighters. A building is akin to a complex machine. Any defects in a system can have a negative impact on the building integrity.

NOTE: For more information on building systems, consult the IFSTA **Plans Examiner, Fire Detection and Suppression Systems**, and **Fire Inspection and Code Enforcement** manuals.

This chapter specifically discusses the basic vertical transportation systems, mechanical systems, electrical systems, and heating, ventilation, and air conditioning systems that provide the necessary usability and comfort levels for occupants. Of necessity, many building systems penetrate the walls and floors that provide this built-in fire resistance and compartmentation **(Figure 4.1)**. It is the responsibility of the building design team to ensure that

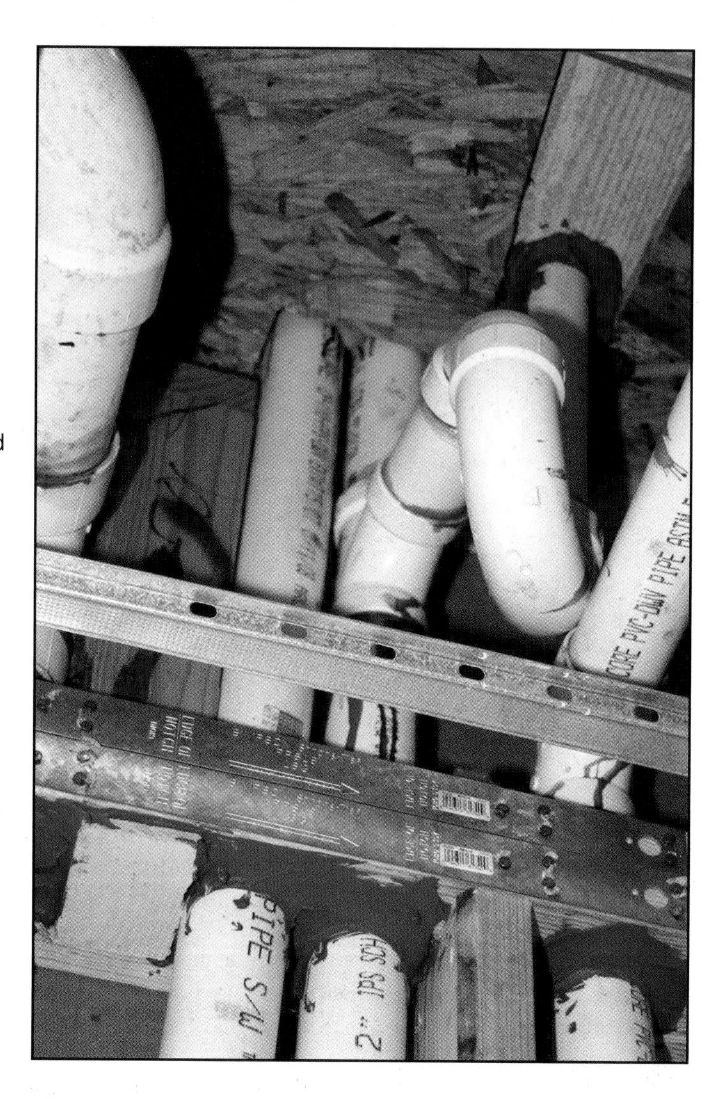

Figure 4.1 Unprotected openings are often found around pipes and cables. Lack of protection can allow fire to spread between floors and compartments.

the necessary building systems initially provide the intended level of fire and life safety. Subsequently, it is the responsibility of the building owner and/or management to maintain these systems over time.

NOTE: Systems specifically intended for life safety are addressed in Chapter 5, including fire resistance and compartmentation, which are critical fire and life/safety features of structures.

Stairs

Most stairs provide a dual role as a building system. First, they enable occupants to access various levels of the structure. Second, they serve as a basic component of building egress during an emergency. Stairs that are a part of the required means of egress must meet strict requirements of the applicable building code, and generally are either fully enclosed or protected open exterior stairs. In either case, the building code specifies the fire-resistance rating of the enclosure or separation when exterior stairs are utilized. The code also spells out other requirements to ensure safety during both nonemergency and emergency use. Exit stairs must resist fire and smoke to ensure safe passage during building evacuation. This level of protection also provides firefighters a safe route to access and attack a fire.

Means of Egress — Safe, continuous path of travel from any point in a structure to a public way; the means of egress is composed of three parts: the exit access, the exit, and the exit discharge.

Stairs that are not part of the means of egress are often referred to as convenience stairs. Typically, these stairs are open and limited by code to connecting only two levels, but the same general safety requirements are required as for exit stairs. It is not unusual to find "ship's ladders" provided to access mechanical spaces, roof hatches at the top of stairways, and between roof levels with portions at different elevations.

Convenience Stair — Stair that usually connects two floors in a multistory building.

In buildings four or more stories in height, one stairway is generally required to extend to the roof unless the slope of the roof is especially steep. This stair is required to be identified by signage in the stairway. Other stairways are often provided with a hatch.

Basic Components

All stair types have components in common as depicted in **Figure 4.2, p. 110**. Requirements for these components are specified in the applicable building code. A key component is the step itself, which consists of the tread and the riser, commonly referred to as the "run and rise" **(Figure 4.3, p. 110)**. For safety purposes, the code requires that the run and rise be consistent throughout the same stair. Other important features are the handrails and guards required when the stairs have open sides. The guards are intended to prevent objects from falling onto adjacent space.

Rise — Vertical distance between the treads of a stairway or the height of the entire stairway.

Run —The horizontal measurement of a stair tread or the distance of the entire stair length.

Types of Stairs

The design or layout of a set of stairs may take any of several different forms **(Figure 4.4, p. 111)**. The six basic types are described in this section.

Straight-Run Stairs

Straight-run stairs extend in a straight line for their entire length. Landings may be found, breaking up the stairs' vertical travel at intervals specified by codes.

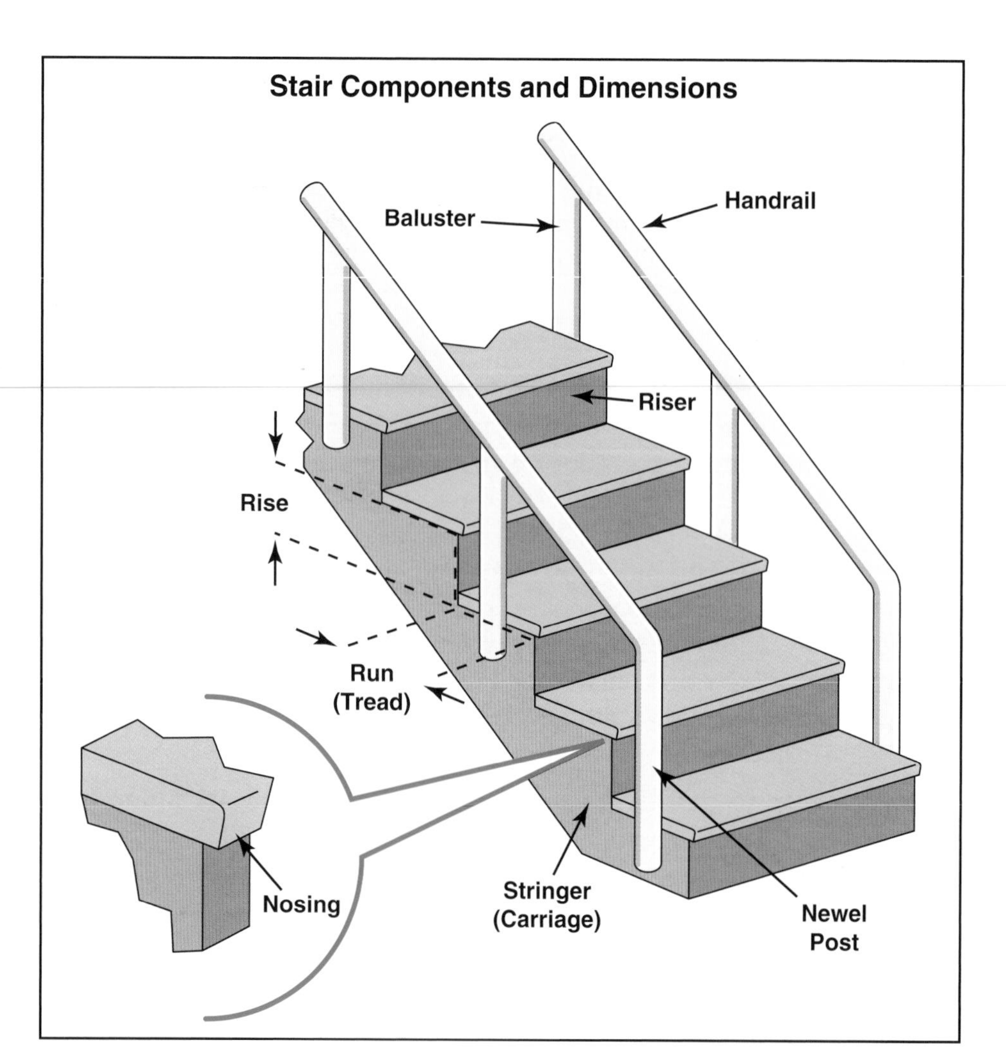

Figure 4.2 Building codes have specifications for stair components.

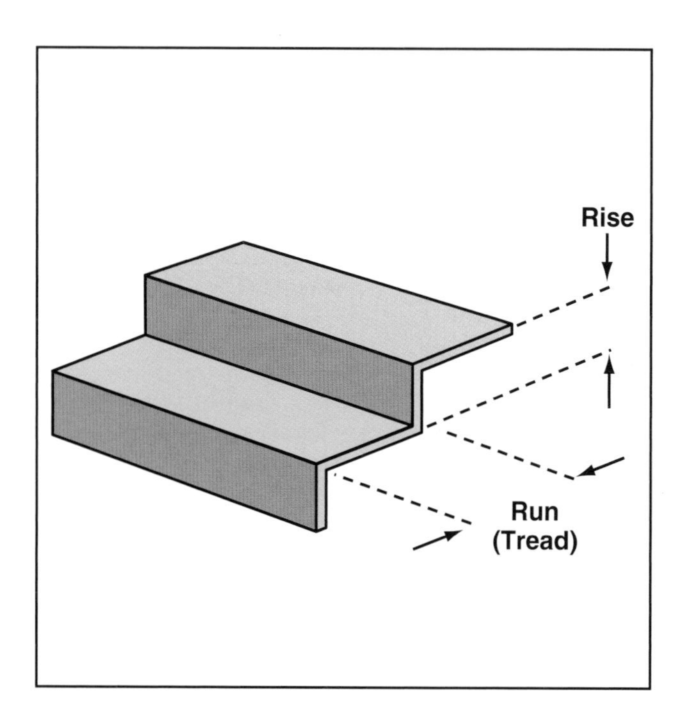

Figure 4.3 For safety, the measurements of the run and rise must be consistent within a set of stairs.

Return Stairs

Return stairs have an intermediate landing between floors and reverse direction at that point. Return stairs may have more than one landing where the height between floors is greater than normal. This type of stair design is common in modern construction.

Scissor Stairs

Scissor stairs are two separate sets of stairs constructed in a common shaft. They are less expensive to build than two separate stair enclosures and also use less floor space. Older scissor stair design consists of straight-run stairs that return to every other floor. This type of scissor stair is not found in recent design.

The newer design arrangement for scissor stairs allows for ingress and egress at each floor landing; this feature is not only less confusing but is also used to provide additional exit capacity. Sometimes the egress from the stairway is into one corridor on even-numbered floors and into the opposite corridor on odd-numbered floors **(Figure 4.5)**. This arrangement

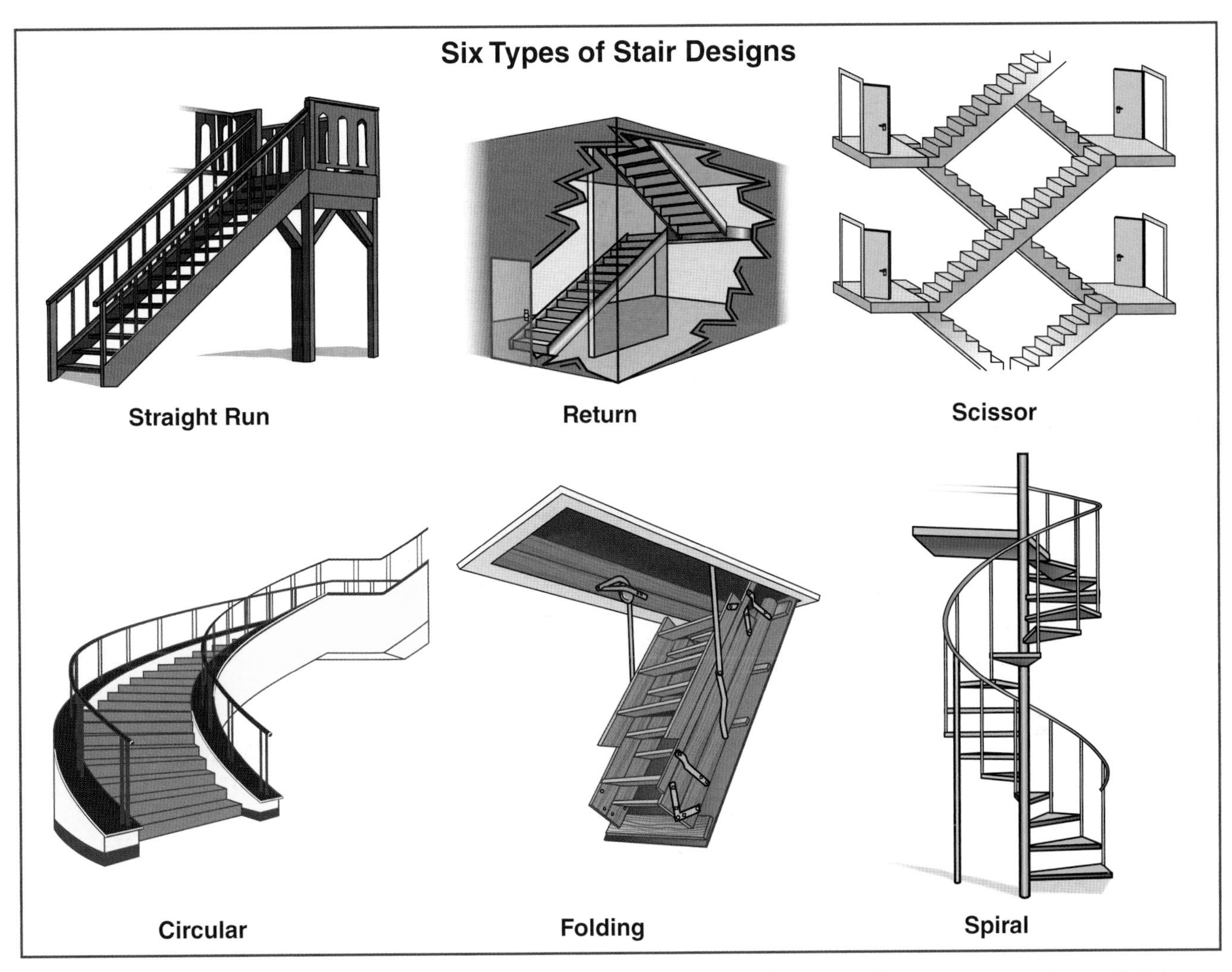

Figure 4.4 Shown are the most common types of stairs.

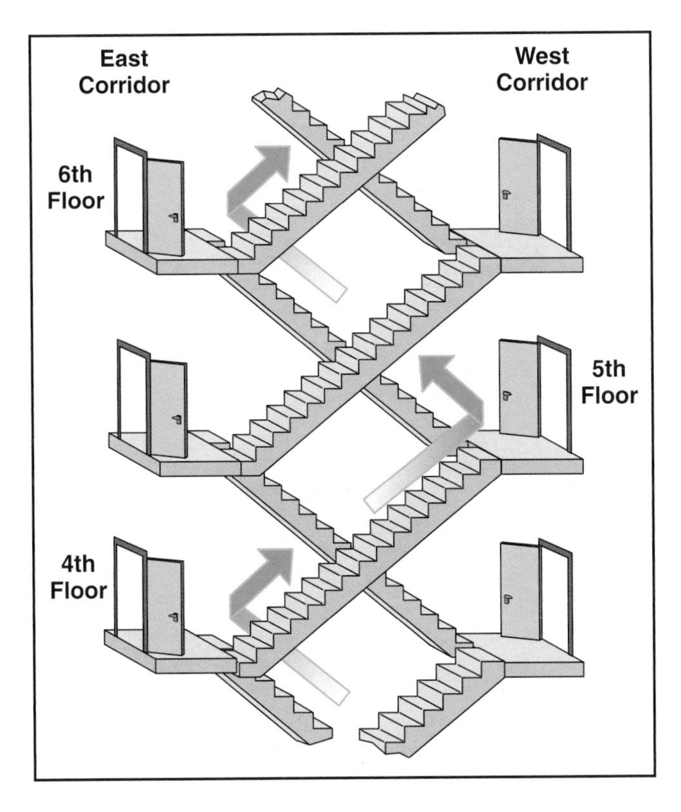

Figure 4.5 Scissor stairs can be confusing. Older designs exit on every other floor; newer designs have exits on each floor.

Scissor Stairs — Two sets of crisscrossing stairs in a common shaft; each set serves every floor but on alternately opposite sides of the stair shaft. For example, one set would serve the west wing on even-numbered floors and the east wing odd-numbered floors, while the other set would serve floors opposite to the first set.

can be confusing, particularly during a fire. Firefighters may need additional hose to connect to a standpipe on the floor below the fire. More often these scissor stairs use a pair of return stairs for assembly occupancies such as airport terminals, convention centers, malls, and cinema complexes that have high floor-to-ceiling heights. In these instances, the ingress/egress is from the same corridor, but several feet apart. This arrangement positions the standpipe on the same stair side at each floor.

Circular Stairs

Circular stairs are often found as grand stairs or convenience stairs serving only two levels. The minimum width of the run is usually 10 inches (250 mm). As shown in **Figure 4.6**, a special requirement for circular stairs is that the small radius (B) is not less than twice the width of the stairway (A).

Folding Stairs

Folding stairs are typically found in dwellings where they are used to provide access to an attic space that does not have a permanent access stair. The folding stair usually is made up of wooden sections: the main section that hinges from the frame and two articulating sections. After the lower stair sections are folded together, the stair swings up into the attic space and is held in place by either springs or counterbalances. This type of stair is most often located in a hallway. A light wooden panel, usually plywood, is attached to the main section and serves to conceal the stair when it is folded into the ceiling.

Although folding stairs can serve as a vertical path for fire and smoke spread, they can also serve as an access to the attic space for firefighters. Firefighters should be aware that springs in folding stairs lose tension rapidly when exposed to heat, and the stair assembly may swing down into the structure during a fire **(Figure 4.7)**. Firefighters must also be cautious when using folding stairs due to weight limitations.

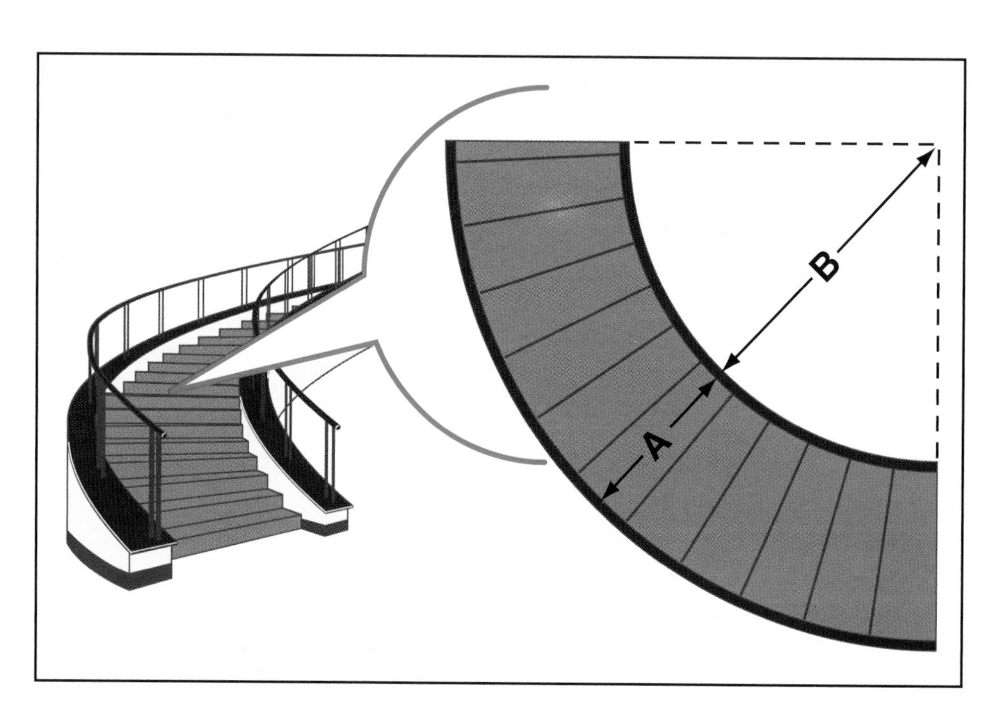

Figure 4.6 Circular stairs must remain a certain width.

Figure 4.7 Folding stairs are commonly placed for attic access. Firefighters must take proper precautions because these stairs, which can be very heavy, can collapse on them.

Spiral Stairs

This type of stair design allows stairs to be placed in a very small space. Spiral stairs consist of a series of steps spiraling around a single column. Each tread is tapered and connects to the column at the tread's narrow end. Typically custom-made, spiral stairs are not enclosed and are usually found in private homes; however, they may also be found in commercial occupancies for limited use. Steel spiral stairs are sometimes used for access to permit-required confined spaces, as a dry-well in a sewage pumping station, and in industrial applications. Because they can be difficult to traverse, spiral stairs are generally allowed as part of the means of egress only within residences.

Stairs as Part of the Means of Egress

Protected Stairs

Stairs can serve not only as a chimney to spread smoke and fire, but also as the lifeline to safety for occupants not on ground level. Most important, studies show that even a minor impairment in visibility significantly reduces occupants' ability to safely egress. Because protecting the stairwell from the products of combustion is extremely important, building codes require a high level of protection for most stairs used as a means of egress.

Protected Stair Enclosure — Stair with code required fire-rated enclosure construction. Intended to protect occupants as they make their way through the stair enclosure.

Generally, stair enclosures are required to be separated from the rest of the building. The only penetrations permitted in the enclosure are for light, fire protection, and environmental control. (Penetrations for services not required for the stair enclosure itself are generally prohibited.) Self- or automatic-closing fire-rated doors are required. These requirements indicate the high level of protection given to stair enclosures by the codes. Stair enclosures are considered to be a highly protected part of the means of egress because of their importance in overall building life safety.

NOTE: For details on stair enclosure constructions, consult the applicable building code.

Exterior Stairs

As defined by codes, exterior stairs are exterior to a building and are open to the air **(Figure 4.8)**. Open stairs are naturally ventilated but may be partially enclosed from the weather. They will have at least one side open to provide natural ventilation. The code will specify the minimum size of the opening. Enclosed stairs on the exterior of a building are considered the same as interior enclosed stairs **(Figure 4.9)**.

When provided as a part of the means of egress, open exterior stairs are generally protected by limiting or protecting the openings in the building's outside wall near the stairs. Thus, these stairs have some degree of protection from smoke and fire from inside the building that might impair the egress path on the stairs.

Figure 4.8 Exterior stairs provide a means of egress that is open to the elements.

Figure 4.9 Exterior enclosed stairs are a good means of egress because they can be protected from the fire and smoke that develop inside a building.

Fire Escapes

Fire escapes are open metal stairs and landings attached to the outside of a building. The lowest flight may consist of a swinging stair section to limit unwanted access. Stairs that have been exposed to weather must be continually inspected and maintained. Open exterior stairs that are not supported at the grade level but are supported only from the side of the building must be used with extra caution **(Figure 4.10)**.

Codes have not permitted fire escapes in new construction for many decades, so the fire escapes that do exist have been in place for many years **(Figure 4.11, p. 116)**. Many older fire escapes have failed when loaded with people during an emergency. Fire service personnel attempting to use fire escapes present a significant higher load with tools, breathing apparatus, and other gear compared to occupants using a fire escape during evacuation.

Fire escapes are usually anchored to the building and are not supported at ground level. These anchor points are subject to the freeze-thaw cycle, corrosion from pollution and weather, and temperature changes. The mortar in which the anchors are set may have deteriorated or may have originally been inadequate for their load potential **(Figure 4.12, p. 116)**.

Fire Escape — (1) Means of escaping from a building in case of fire; usually an interior or exterior stairway or slide independently supported and made of fire-resistive material. (2) Traditional term for an exterior stair, frequently incorporating a movable section, usually of noncombustible construction that is intended as an emergency exit. It is usually supported by hangers installed in the exterior wall of the building.

WARNING!

Because the structural soundness of a fire escape may not be apparent, firefighters must use extreme caution during an emergency response. The fire escape structure itself may be severely weakened due to constant exposure and lack of maintenance over a period of years. Firefighters should keep these factors in mind when deciding whether a fire escape is usable.

Figure 4.10 Fire escape stairs typically found with older buildings.

Figure 4.12 Note that the mortar supporting the base of this fire escape has weakened with age. The access ladder has been bent to prevent access.

Figure 4.11 This very old fire escape cannot be considered safe for use today.

Smokeproof Enclosures — Stairways that are designed to limit the penetration of smoke, heat, and toxic gases from a fire on a floor of a building into the stairway and that serve as part of a means of egress.

Smokeproof Stair Enclosures

Building codes have traditionally required a minimum of one smokeproof stair enclosure for stairs serving buildings five stories or higher, and more recently, stairs serving floor levels more than 30 feet (10 m) below the level of exit discharge. Stair enclosures may be classified as *smokeproof* using either active or passive smoke control. Typically, smokeproof stair enclosures are on the exterior perimeter of the building and are entered through ventilated vestibules or open exterior balconies. Refer to the applicable code for specific design details.

Active Smokeproof Enclosures

A common alternative approach over the past 25 years to provide a smokeproof enclosure is to pressurize stairwells when the building is in a fire mode **(Figure 4.13)**. Activated by automatic fire/smoke detection equipment, a dedicated mechanical air-handling system is designed to keep smoke out of the stair enclosure by pressurizing the shaft. Each stairwell needs to be designed for the particular installation, including addressing the stack effect created by the height of the stair shaft. A properly designed, installed, and maintained system should allow firefighters to begin suppression operations while occupants are still using the stairs for escape. The mechanical pressurization system keeps the stair enclosure free of smoke, even when a door is open to the fire floor.

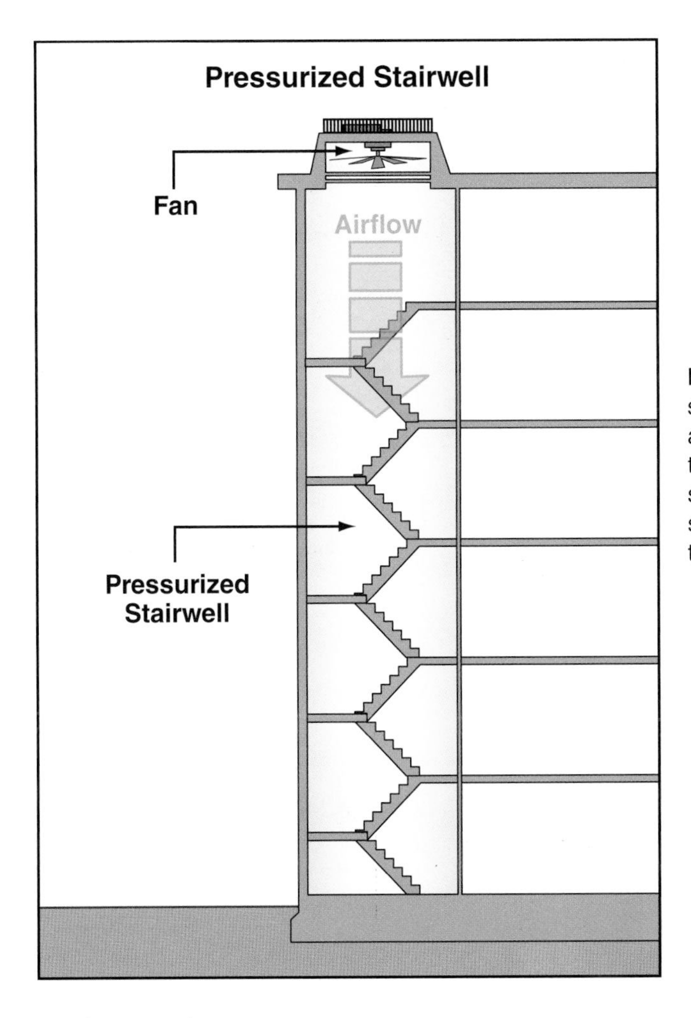

Figure 4.13 A pressurized stairwell incorporates a ventilation system that pushes air into the stairwell, which helps keep smoke from a fire out of the stairwell.

The building codes specify maximum and minimum allowable pressure differentials between the stair enclosure and the building to allow the doors in the enclosure to be opened with a reasonable amount of force. Earlier versions of codes allowed the stair door to open directly from a corridor; newer versions typically require a vestibule between the corridor and stairway. These vestibules are also pressurized to assist in keeping smoke out of the stairway. Refer to the applicable code for specific mechanical system design guidance.

NOTE: Additional information about smokeproof stairwells, as well as smokeproof towers, is provided later in this chapter.

Passive Smokeproof Enclosures

This type of stair enclosure is accessed by using a vestibule or an exterior balcony (**Figure 4.14, p. 118**). Regardless, the design protects the stairway enclosure from smoke by providing a means for the smoke to be vented to the outside before it enters the stair enclosure.

Open Stairs

Because they are not enclosed with fire-rated construction, open stairs will likely serve as a path for spread of fire and smoke and will not protect anyone using them from exposure to the products of combustion. Building codes typically allow the use of open stairs in buildings only when they connect no more than two adjacent floors above the basement level. These stairs are sometimes referred to as "convenience" stairs and can be used as part of an exit system in a two-story building **(Figures 4.15 a and b)**. Refer to the applicable code for specific requirements.

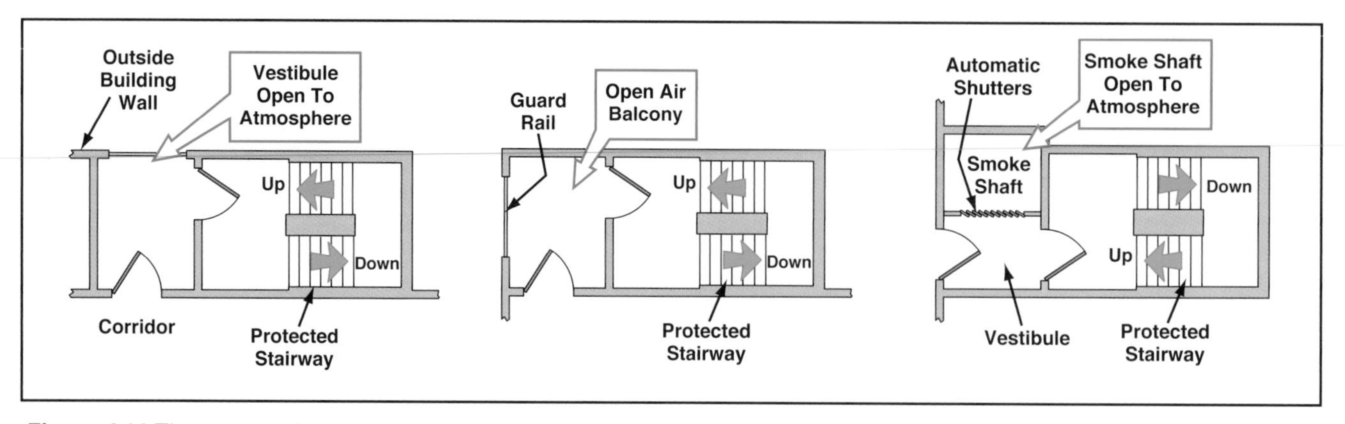

Figure 4.14 Three methods of passive smoke containment: a vestibule, a balcony, and a smoke shaft. All can be used to prevent smoke from traveling into a building and/or contaminating a stairwell.

Figures 4.15 a and b Convenience stairs connect different areas of a building but do not lead to an exit.

Elevators

Elevators are a key building system for providing access to above- or below-grade stories. They can be a positive or negative factor for firefighters during a fire response. Frequently, firefighters must utilize elevators to access the upper floors of high-rise buildings. Responders also use elevators during such non-fire situations as medical emergencies. Problems with the elevators themselves or loss of the electrical power system can create potentially dangerous rescue situations.

Because of their importance and the obvious need for safety and reliability, elevators have developed into safe and reliable modes of transportation. Elevator design, construction, and operation are stringently controlled and monitored by all levels of government. Most elevator regulations are based on ASME/ANSI A17.1, *Safety Code for Elevators,* published by the American Society of Mechanical Engineers.

In recent years, elevators have taken on added importance as a means of egress during emergencies. The Americans with Disabilities Act (ADA) has mandated that public buildings be made accessible to individuals with disabilities. Multistory buildings, therefore, require either a ramp or an elevator. The elevator is far more common due to the large floor areas required for ramps. If a fire occurs, the elevator becomes an important element of the evacuation program for individuals with disabilities.

Americans with Disabilities Act (ADA) of 1990 - Public Law 101-336 — A federal statute intended to remove barriers — physical and otherwise — that limit access by individuals with disabilities.

This section discusses how elevators impact the fire service. It describes how elevators function, their control and safety systems, their power supplies, and the fire-resistance requirements of the shafts and hoistway doors. This manual does not cover emergency elevator rescue techniques such as poling.

NOTE: For more information on these topics, see the IFSTA **Fire Service Rescue** manual.

NOTE: This manual does not intend to set any particular standard operating procedures for the use of elevators during fire operations; rather, it describes *some* basic safety precautions. Each jurisdiction is urged to establish its own set of policies and procedures on the use of elevators during fires and to practice these procedures during drills.

Types of Elevators

An elevator is a hoisting and lowering mechanism equipped with a car or platform that moves in guide rails and serves two or more levels or landings. Elevators can be classified according to their use as either passenger or freight **(Figure 4.16)**. *Service elevator* is another term used to describe elevators and often is defined as a passenger elevator that has been designed to carry freight. The various types of elevators use some form of power to perform the hoisting and lowering operations. The two most common types of power are hydraulic and electric.

Figure 4.16 Typical passenger and freight elevators. *Photo a courtesy of McKinney (TX) Fire Department.*

Hydraulic Elevators

The operating principle of hydraulic elevators involves a fluid being forced under pressure into a cylinder containing a piston or ram. As the fluid is pumped in, the ram rises and the attached elevator car moves upward. As the fluid drains out, the car is lowered by gravity. Hydraulic elevators do not have brakes; cars are slowed and stopped by controlling the flow of hydraulic fluid back into the reservoir.

Because the elevator car is attached to the top of the ram, the ram must be long enough to reach the highest floor served by that elevator. For many years this meant that the ram had to extend an equal distance into the ground. This put a practical upper limit for hydraulic elevators at about six stories; consequently, hydraulic elevators were not usually found in high-rise buildings. However, some hydraulic elevators installed in taller buildings now use a multi-stage hydraulic cylinder rather than a single-stage ram. This reduces the overall length of the cylinder as well as the depth of the well that needs to be provided.

Electric Elevators

Electric elevators are subdivided into either drum or traction devices.

Drum Elevators. Older style elevators employ a drum on which the hoisting cable is wound. The drum is located in a motor room directly over the hoistway. The car is connected to a set of moving counterweights to reduce the effort the motor must produce to raise the car. Like hydraulic elevators, drum-type elevators have practical height limitations because of the size of the drum required in a tall building. This type of elevator is obsolete and is found only in very old structures. However, drum elevators still may be found in use as freight elevators.

Figure 4.17 Looking up into a typical elevator cable hoistway.

Traction Elevators. The most common type of elevator in buildings over six stories is the traction elevator because it is very fast and does not have the height limitations of either hydraulic or drum-type elevators. Traction elevators, similar to drum elevators, use counterweights to reduce the amount of energy needed to raise the elevator. Hoist cables attached to the elevator car run up and over the drive sheave at the top of the hoistway and then down the back wall of the hoistway to connect to the movable counterweights. The hoist cables do not wind around the drive sheave — they merely pass over it **(Figure 4.17)**. Friction between the cable and the sheave is created by the weight of the car on one side and the counterweights on the other, as well as the weight of the cable. The drive equipment for traction elevators is contained in a machine room that is usually located directly over the hoistway.

NOTE: Historically, elevator cables have utilized conventional wire cables. However, a recent development has been flat polyethylene-coated steel belts that increase energy efficiency.

Even though counterweights reduce the amount of energy needed to raise the elevators, the heights to which they operate may require them to have as much as a 500-volt power supply. The drive motors may be either direct current (DC) or alternating current (AC) types. Obviously, firefighters must be extremely careful when conditions require them to work in the vicinity of such high-voltage equipment.

Figure 4.18 The brake on a typical traction elevator.

Traction motors have a braking system that operates during both normal operation and malfunctions. The system employs a brake drum located on the shaft of the drive motor **(Figure 4.18)**. Under normal conditions, the spring-operated brake shoes are held away from the drum by electromagnets. In the event of power failure, the electromagnets release and the brake shoes are forced against the drum. This results in the car being stopped wherever it was when the power failed.

During normal operation, the brakes on traction elevators with AC motors aid directly in stopping the car at the correct floor. On those with DC motors, the brakes do not play any part in actually stopping the elevator car. The motor stops the car and then the brakes are applied to hold the car in place.

Safety Features

The excellent safety record of elevators can be attributed to strict regulation, rigorous engineering to reduce the likelihood of failure, and numerous safety devices designed to limit the effects of any failures that do occur. Maximum passenger protection is maintained by the use of equipment that can safely stop a car in the event of a malfunction. These safety devices include terminal switches, buffers, speed-reducing and overspeed switches, and car safeties.

NOTE: Some of these devices may not be found on every type of elevator.

- A *terminal device* is an electric switch designed to stop the car by removing power before it reaches the upper or lower limits of the hoistway.
- *Buffers* are large springs or hydraulic cylinders and pistons located at the bottom of the pit that act as shock absorbers should the terminal switch fail. Buffers cannot safely stop a free-falling car. They only stop one traveling at its normal rate of speed.
- A *speed-reducing switch* is also known as the speed governor. This switch slows the drive motor when an elevator starts to exceed a safe speed. If the car continues to accelerate, it applies the car safeties and trips the overspeed switch.
- An *overspeed switch* is also connected to the speed governor. This switch is activated if the speed-reducing switch fails to slow the car sufficiently.
- *Car safeties* are tapered sets of steel jaws that wedge against the guide rails and bring the elevator to a stop. Elevator safeties are designed to stop a free-falling car.

Elevator Hoistways

An *elevator hoistway* is the vertical shaft in which the elevator car travels and includes the elevator pit **(Figure 4.19)**. The pit extends down from the lowest floor landing to the bottom of the hoistway. Hoistways are required to be constructed of fire-resistive materials and are equipped with fire-rated door assemblies. However, some hoistways, such as those located in an atrium, are not required to be enclosed.

Elevator hoistway enclosures usually are required to be a fire-rated assembly with a 1- or 2-hour rating, depending on the particular situation. Therefore, the integrity of the rated hoistway assembly must be maintained. Penetrations through the hoistway walls are protected by the installation of an appropriately rated assembly such as a listed and labeled door and frame. No wiring, ductwork, or piping should be run within the hoistway unless it is required for the elevator itself.

In low-rise buildings, the entire hoistway enclosure may consist of gypsum, cement block, or other easily penetrated material. In tall buildings of reinforced concrete construction, the elevator hoistway may be enclosed on three sides with poured concrete, leaving only the wall that faces the elevator car doors to be built of block. While this construction serves to stiffen the entire building against the wind load, it makes the hoistway difficult to breech. Firefighters must keep this fact in mind when considering making openings in elevator shafts during emergency operations.

Mushrooming — Tendency of heat, smoke, and other products of combustion to rise until they encounter a horizontal obstruction. At this point they will spread laterally until they encounter vertical obstructions and begin to bank downward.

Hoistways present a potential for acting as a vertical chimney to spread fire and smoke throughout a building. If the hoistway is not vented at the top, the accumulated hot gases and smoke may tend to *mushroom* or spread horizontally into the upper floors. To prevent mushrooming, building codes

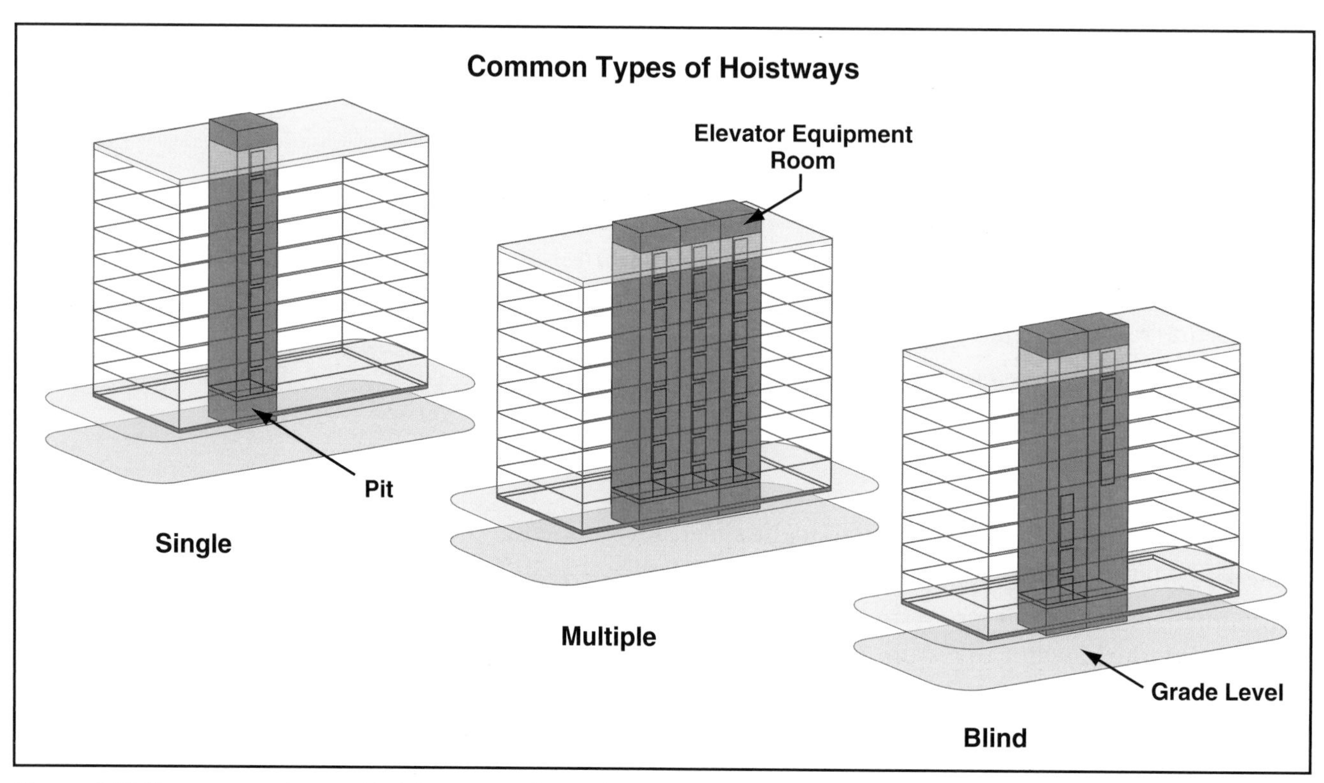

Figure 4.19 The common types of hoistways are chosen based on the size of the structure.

require venting at the top of practically every hoistway built today. The codes also require fire-rated vestibules, or equivalent, at each floor to prevent smoke and hot gases from moving throughout the building.

If a building contains three or fewer elevators, the codes permit them to be placed in one hoistway. When four or more elevators are provided, there must be a minimum of two separate hoistways. This requirement minimizes the possibility that all elevator services will be eliminated by a fire.

When more than one hoistway is provided, up to four elevators may be located in one hoistway. Elevators within a given hoistway usually are not separated by any sort of partition. Hoistways containing express elevators serving only upper floors of tall buildings have no entrances to the hoistway shaft between the main entrance and the lowest floor served. Single-elevator hoistways that only serve upper levels have access doors for rescue purposes every three floors or so in that portion without normal hoistway doors.

In very tall buildings, elevators are divided into zones, with one zone serving the lower floors and another zone serving the upper floors **(Figure 4.20)**. A zone usually serves 15 to 20 floors. The upper zone cars operate express from the first floor to the lowest floor of the upper zone. In some buildings, there may be more than two zones.

Blind hoistways are used for express elevators that serve the upper elevator zones in tall buildings. There will be no entrances to the hoistway on floors between the main floor and the lowest floor served. If a single-car hoistway is used, however, access doors will be provided for rescue purposes. Generally, these are placed every three floors.

Blind Hoistway — Used for express elevators that serve only upper floors of tall buildings. There are no entrances to the shaft on floors between the main entrance and the lowest floor served.

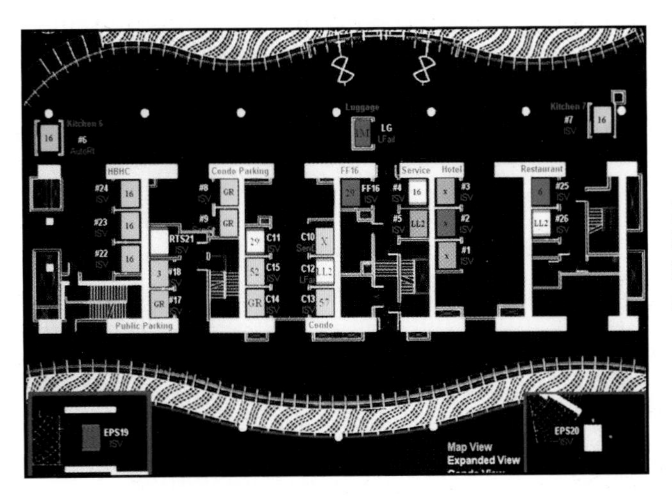

Figure 4.20 Elevators in large buildings are divided into zones. *Courtesy of Ed Prendergast.*

Elevator Doors

Doors in elevator installations include both car doors and hoistway doors. The two are usually designed to open in the same direction and to open and close together.

Passenger elevator car doors are powered by an electric motor mounted on top of the elevator car. The car door does not have locks and can be pushed open at any time. However, electric interlocks will not allow a car to move when the car doors are open and a moving car will immediately stop if the doors are pushed open. When the doors are closed again, most elevator cars will start to move again. Some types will not start moving again until they have been reset.

Elevator doors are designed to open and close automatically when the car stops at the floor to which it has been summoned. When the elevator stops at the correct level, the hoistway doors are held open by a driving vane attached to the car door. As the car door opens, the vane strikes a roller that releases the hoistway door lock. The car doors then push the hoistway doors completely open. When the controller signals the doors to close, a weight forces the hoistway doors closed, the driving vane moves away from the roller, and the hoistway doors are relocked.

Access Panels

During some emergencies it may be necessary to remove occupants from a disabled elevator. It must be noted that only trained personnel should attempt an evacuation of passengers from a stalled elevator car. ASME Standard A17.4, *Guide for Emergency Personnel,* contains procedures to follow in performing evacuations.

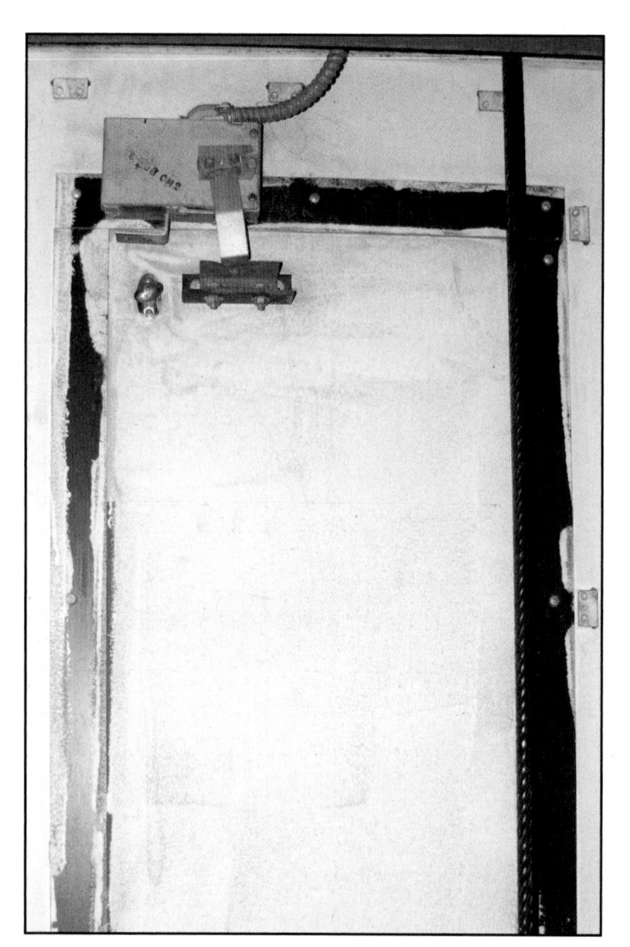

Figure 4.21 A typical side door exit viewed from the hoistway.

If the car is stopped in a blind hoistway, the emergency exits from the car must be used. These exits consist of either a hinged access hatch through the top of the car or hinged or removable panels on the sides of the car. Using either of these exits is time-consuming and involves some added risk to passengers, so should be done only as a last resort.

A top exit is provided on all electric traction elevators. On hydraulic elevators, a top exit may or not be provided depending on whether the system is equipped with a manual lowering valve. This valve permits the lowering of the car in the event of malfunction. A top exit is optional otherwise.

Some top panels are designed to be opened from inside the car, but all can be opened from outside and all open outward. Some cars are also provided with electrical interlocks that prevent movement of the car while the panel is open. This feature is not required and is not found on all models, especially freight elevators.

In multiple-elevator hoistways, most elevator cars are equipped with side exits to allow passengers to be transferred laterally from a stalled car to a functioning car next to it. Side exits can be opened from the outside where a permanent handle is provided **(Figure 4.21)**. Some panels are locked from the inside and cannot be opened without a special key or handle. Side exits are required to have electrical interlocks to prevent car movement when the panels are open. Side exits may not be provided on cars in hydraulic elevator systems where a manual lowering valve is provided.

NOTE: Emergency use of elevators is covered in Chapter 12.

Moving Stairways and Walkways

Moving stairs, commonly called escalators, are stairways with electrically powered steps that move continuously in one direction. Older escalators usually operate at speeds of either 90 or 120 feet per minute (27.4 m/min or 36.5 m/min); however, speeds now have been standardized at 100 fpm. Each individual step rides a track. The steps are linked and move around the escalator frame by a chain called the *step chain*. The driving machinery is located under an access plate at the upper landing. Continuous handrails also move at the same speed as the steps.

Escalators are commonly found in such buildings as hotels with conference facilities, retail stores, malls, transportation terminals, convention centers, sports areas, and other facilities that contain large numbers of people. Escalators require periodic maintenance and may be out of service for lengthy periods.

Typically, the vertical openings created by escalators need to be protected when serving more than two floors. The most common method of protecting the vertical opening is to use closely spaced sprinklers in conjunction with draft stops around the opening **(Figure 4.22)**. This approach consists of an 18-inch (450 mm) deep draft stop with a row of automatic sprinklers on all sides outside the draft stop.

Vertical opening protection also can be provided by a rolling shutter at the top of the escalator **(Figure 4.23)**. A partial enclosure uses separate fire-rated enclosures for the up escalator and the down escalator.

Escalators can be used while stopped as if they were simply fixed stairs. Moving escalators should not be used during emergency operations but should be stopped to be available as fixed stairs. An emergency stop switch is available that will stop the escalator and set the brake **(Figure 4.24)**.

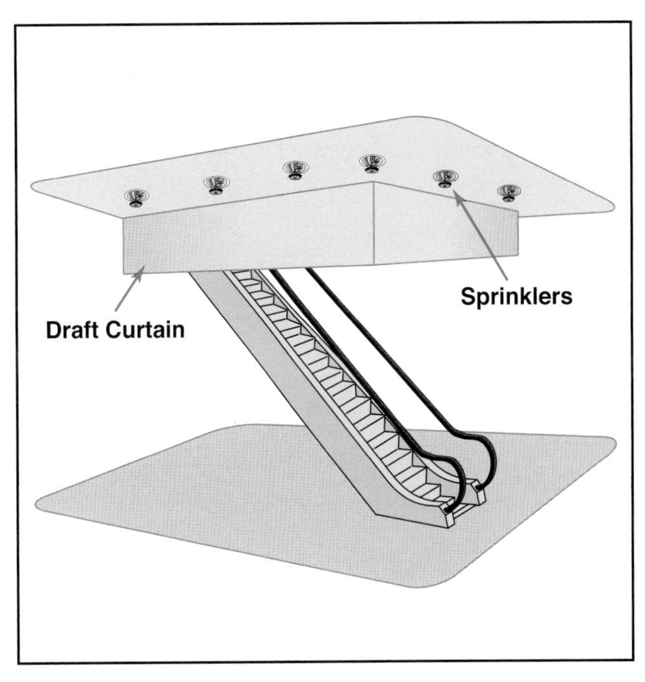

Figure 4.22 Sprinklers protection escalators must be well supplied with water.

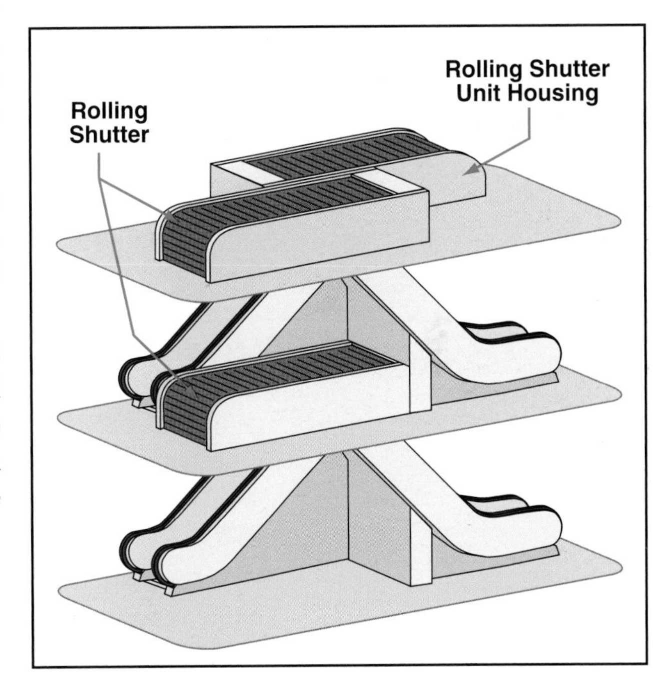

Figure 4.23 Some escalators are equipped with automatic fire shutters.

Figure 4.24 Firefighters should be aware of the location of emergency shutoffs for escalators.

Moving walkways are similar to escalators but are only used to move people horizontally or up slight inclines. Typically, they use attached metal plates in a continuous pathway equipped with moving handrails. Some installations utilize moving rubber-like belts over metal rollers. Moving walkways are often found in airports to move occupants long distances between the main terminal and remote concourses.

Conveyor Systems

A conveyor system is typically found in manufacturing or storage occupancies and is used to transport items and materials. Conveyors are also found in airport baggage handling facilities. Types of conveyors include belt, roller, chain, screw, bucket, and pneumatic systems. Some conveyors, such as screw or pneumatic systems, are enclosed. Dust is often produced in materials-handling processes and this is usually controlled by using enclosed conveyor systems.

Conveyor systems often pass through fire barriers. These penetrations are usually protected by either providing a fire door or shutter or by utilizing a water-spray method. A primary concern for conveyor penetrations during a fire is incomplete door or shutter closure. Several methods are used to prevent incomplete closure, including automatic stop controls, breaks in the conveyor, and multiple layers of doors or shutters. Regardless of the specific operation method, conveyors should be routinely inspected and tested.

Vertical Shafts and Utility Chases

Vertical shafts also include stairways and elevator and dumbwaiter hoistways, which have been previously discussed. This section concentrates on those vertical shafts associated with utility and other building services, particularly pipe chases, refuse and laundry chutes, and grease ducts.

Utility chase is a term generally applied to the vertical pathway (shaft) in a building that contains utility services. These services include plumbing, electrical raceways, telecommunications, data cables, and ductwork for heating, ventilation, and air conditioning (HVAC). Vertical shafts are also provided for refuse chutes, laundry chutes, and light shafts. Knowledge of chases and shafts is critical because they can provide a vertical path for smoke and fire as well as serve as the area of origin for fires.

Vertical shaft enclosures are built using fire-rated construction methods but may contain combustible materials, such as electrical wiring, trash, and laundry within the shaft. Vertical openings such as rubbish chutes, laundry chutes, and air shafts present special problems for firefighters because of their function. It is usually not practical to provide any horizontal fire barriers along the length of the shaft.

Pipe Chases

A *pipe chase* is a type of utility chase used to contain piping needed for building services such as hot and cold potable water, drain lines, steam, hot and chilled water for heating and air-conditioning, and sprinkler piping. One or more pipe chases may be provided in a building depending on building size and design.

Pipe Chase — Concealed vertical channel in which pipes and other utility conduits are housed. Pipe chases that are not properly protected can be major contributors to the vertical spread of smoke and fire in a building. Also called *Chase*.

As with any vertical opening in a building, pipe chases can spread smoke and fire to other floors of the building if not properly protected. Building codes specify shaft enclosure protection, typically requiring fire-resistive construction **(Figure 4.25)**. The access openings must also be rated accordingly.

Figure 4.25 Pipe chases must be surrounded by fire-resistive construction to ensure fire safety.

Occasionally buildings do not have pipe chases but instead use stacked mechanical equipment rooms. Pipes, electrical raceways, and other services pass through these rooms, which are stacked one above the other on each floor. The walls enclosing these rooms are then treated as shaft walls to separate the utility space from the rest of the building. There may or may not be fire-rated horizontal separations between the mechanical rooms.

Although they are not required to be installed in a rated chase (shaft), plumbing pipes in one- and two-story residential and small commercial buildings of wood-frame construction typically form pathways in walls that are capable of spreading fire and smoke. Plumbing fixtures drain into a vertical pipe connecting to the underground sewer pipe, which also extends above the roof to ventilate the system **(Figure 4.26)**. This pipe typically travels through walls and horizontal layers and can serve as a pathway for fire and smoke if it does not fit tightly or is not firestopped at the penetrations.

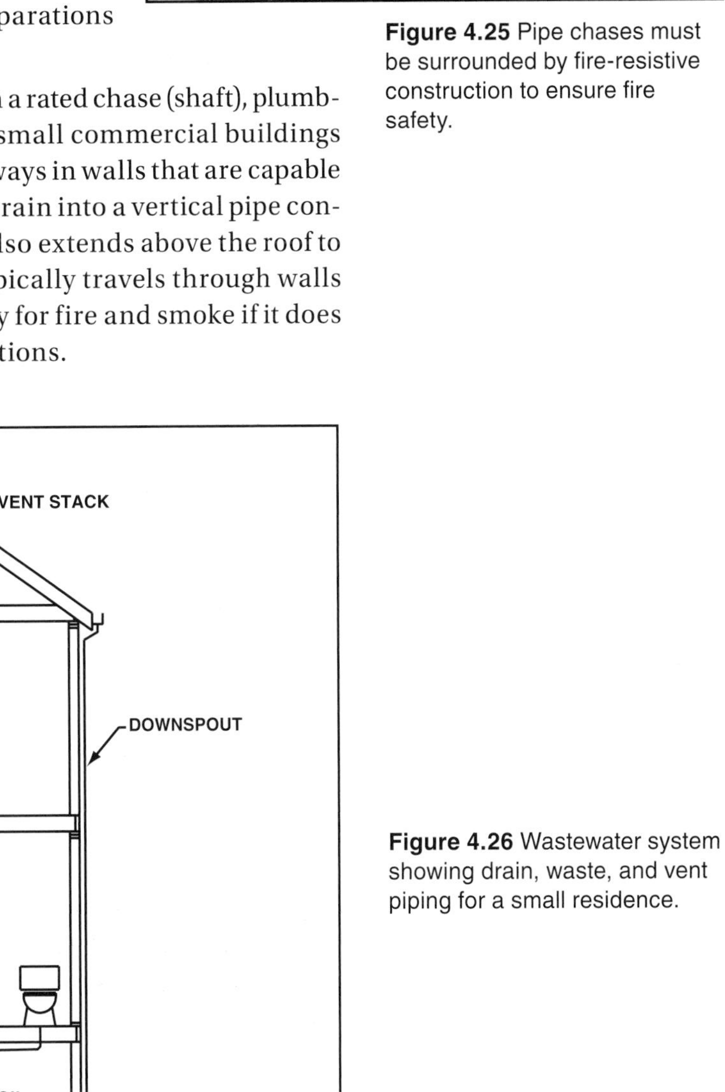

Figure 4.26 Wastewater system showing drain, waste, and vent piping for a small residence.

Refuse Chute — Vertical shaft with a self-closing access door on every floor; usually extending from the basement or ground floor to the top floor of multistory buildings.

Refuse and Laundry Chutes

A *refuse chute* provides for the removal of trash and garbage from upper floors of buildings such as residential properties. These chutes extend through the building and have openings on each floor for depositing trash or linen **(Figure 4.27)**. The chute often terminates in a room at grade level or in the basement where the refuse or laundry is collected. These chutes and rooms create a frequent fire response.

An improperly constructed or maintained refuse or laundry chute creates a potentially severe fire problem. The trash or laundry deposited in the chute is mostly combustible; it is easy to inadvertently or intentionally drop an ignition source such as a cigarette, down the chute. A refuse or laundry chute can also become jammed, resulting in a large quantity of combustible material becoming lodged in the chute.

Refuse chutes are required to be constructed of noncombustible material with rated doors **(Figure 4.28)**. The chute typically is surrounded by a fire-rated shaft enclosure. Sprinklers are required at the top of the chute and in its termination room. Current codes require access openings for chutes to be in a separate room from the corridor, but older buildings often have the openings directly on the corridor.

Figure 4.27 Trash chutes can pose a significant fire hazard because of combustibles placed in them.

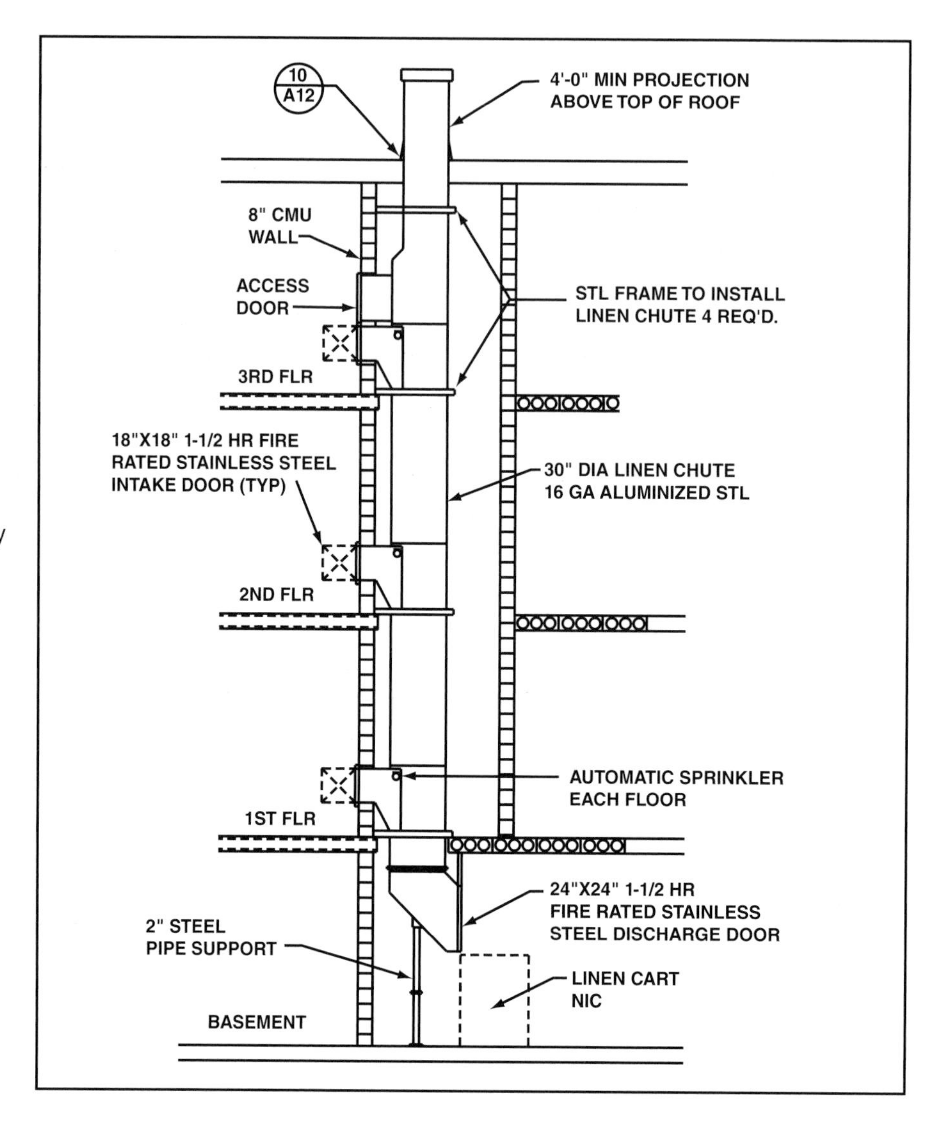

Figure 4.28 A section view of a laundry chute shows details of its enclosure.

A fire in a properly designed, installed, and maintained refuse or laundry chute should be contained within the chute. It is not uncommon, however, that some smoke will leak out of the chute through the access doors in older installations because of poor maintenance or loss of operational integrity. Smoke can then be transferred throughout multiple floors and there may be heavy smoke in upper floors.

Grease Ducts

A *grease duct* is installed as part of an exhaust system for commercial cooking appliances that produce grease-laden vapors. Typically, grease ducts are installed over deep-fat fryers and grills. A grease duct travels vertically and carries grease vapors to the outside of the building, often utilizing in-line fans or fans on the roof. A proper installation has no areas, such as dips or horizontal runs, where grease may become trapped.

Some design applications include horizontal ducts. In these cases, a grease removal system is provided that minimizes the likelihood of grease-laden waste material collecting in the horizontal sections. The application, design, and protection required for grease ducts are specified in codes. Codes also require that the grease duct be enclosed in fire-resistive construction, similar to the requirements for a shaft when penetrating rated floors or rated roofs.

Heating, Ventilating, and Air Conditioning (HVAC) Systems

HVAC systems are provided in buildings primarily to maintain a comfortable environment for occupants. The functions provided by an HVAC system include heating, cooling, filtering, humidifying, and dehumidifying. The system will also provide for exhaust and intake (or make-up) air. In addition to maintaining a comfortable temperature, HVAC systems also regulate the intake of outdoor air and the recirculation of indoor air to provide acceptable air quality.

As is the case with all building systems, HVAC systems have the potential to significantly impact any fire event. (See Chapter 5 for additional information.) With the trend toward green building design, HVAC systems of the future may incorporate such natural ventilation features as openable windows and vents in roofs. These elements could then be used when outdoor conditions were favorable.

Heating, Ventilating, and Air Conditioning (HVAC) System — Heating, ventilating, and air-conditioning system within a building and the equipment necessary to make it function; usually a single, integrated unit with a complex system of ducts throughout the building. Also called Air-Handling System.

Depending on a building's age and complexity, an HVAC system can be very simple or very sophisticated. In a one-story convenience store, heating and air conditioning might be provided by a single rooftop unit **(Figure 4.29, p. 130)**. In a high-rise office building, the total HVAC system could include hundreds of feet of ductwork, heating equipment, refrigeration equipment, motors, and blowers — all monitored and controlled by computers.

The early forced-air systems provided heat only through a ducted system. Cooling and/or ventilation was provided by such building features as openable windows and rooftop cupolas that permitted a flow of air through a building **(Figure 4.30, p. 130)**. Buildings were not airtight and infiltration of air around cracks facilitated ventilation.

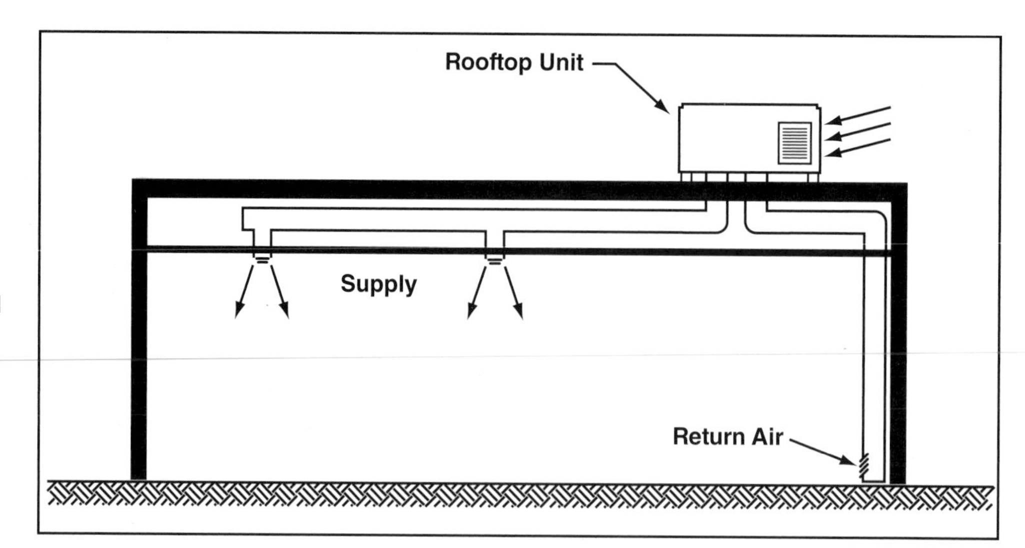

Figure 4.29 A small building supplied by a rooftop-mounted heating unit.

Figure 4.30 Many years ago, cupolas were a common way to facilitate ventilation. A modern one is shown here. *Courtesy of McKinney (TX) Fire Department.*

With the advent of economical cooling systems it became possible to both heat and cool the interior environment. The evolution of this technology has had a major impact on overall building design. Today, many buildings have windows that do not open or no windows at all. The consequence has been near-total dependence on the HVAC system to provide not only a comfortable environment but also a livable environment. In many complex buildings the HVAC system has become, in effect, a life-support system. Any interruption of this system, such as occurs with a fire, is potentially life-threatening.

A distinction should be made between an HVAC system and a simple ventilation or exhaust system. An HVAC system provides *conditioned* air to building occupants. A ventilation or exhaust system may only provide for the removal of contaminated air, for example, exhaust from a paint spray booth. Toilet facilities are vented separately and therefore are exhaust systems.

It is possible to provide heating and/or cooling in a building without the use of ducted systems. A building that relies on the transfer of heat by means of hot water and radiators would be an example. Systems that make use of water as the heat transfer medium are known as *hydronic* systems. However, many HVAC systems involve the distribution of conditioned air through a building from one or more mechanical equipment rooms. Such systems are generally known as *forced-air* systems.

The distribution of the conditioned air through a forced-air system involves ductwork **(Figure 4.31)**. In fact, this is one of the disadvantages of a forced-air system because the ducts are bulky and take up space.

The ductwork associated with a forced-air HVAC system can provide a path for communication of heat and smoke through a building. In addition, the ducts must frequently penetrate fire-rated assemblies where they can create an opening that can destroy the integrity of the assembly. Therefore the systems that make use of extensive ducting are of greatest interest to fire protection engineers and firefighters.

NFPA® 90A, *Standard for the Installation of Air-Conditioning and Ventilation Systems,* which is widely used, contains requirements for protection when the HVAC ducts (both horizontal and vertical) penetrate a fire-rated assembly or a smoke barrier. In general, a fire and/or smoke damper may be required wherever a duct penetrates a fire-rated assembly. **Figure 4.32** illustrates the typical points where fire dampers would be provided. The local mechanical code may also contain requirements for the operation of the dampers.

HVAC ducts often travel vertically up through a multistory building. With a few minor exceptions, the codes require that HVAC ducts be enclosed in a fire-rated shaft enclosure.

Figure 4.31 HVAC ductwork being installed in a building. *Courtesy of Ed Prendergast.*

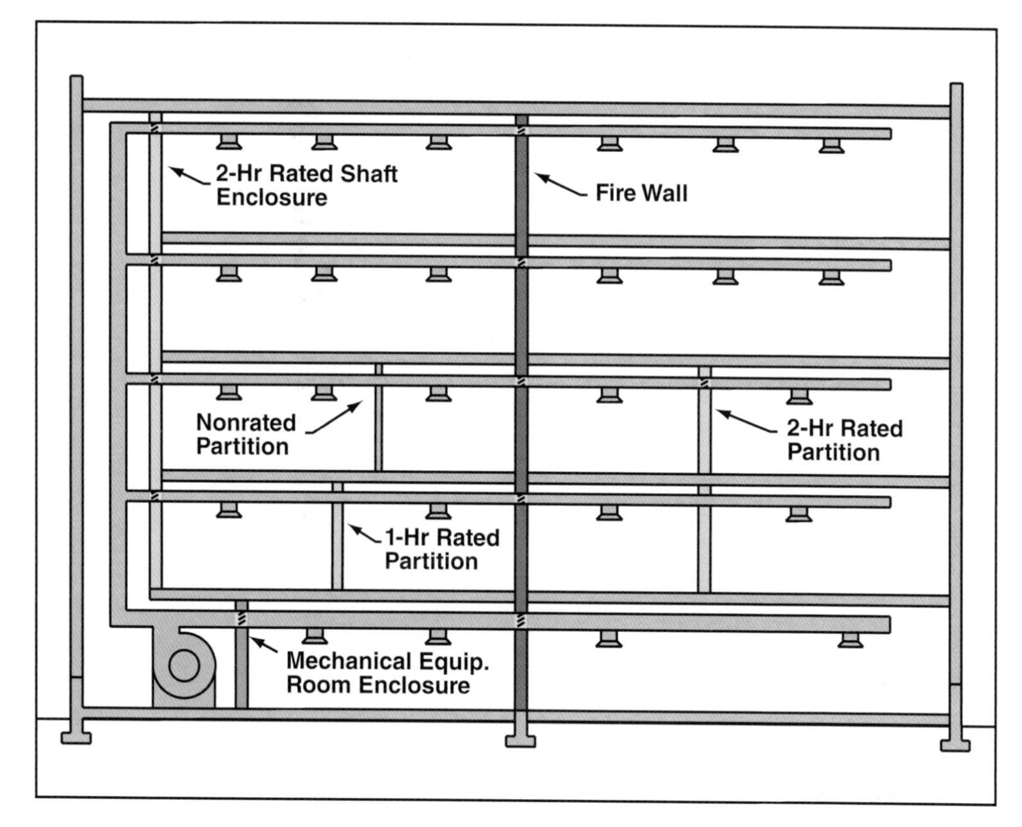

Figure 4.32 Typical locations for dampers in HVAC ductwork. Dampers are not usually required for nonrated partitions or for 1-hour rated walls.

For example, NFPA® 90A requires that the enclosure be 1-hour rated for buildings less than four stories in height and 2-hour fire rated for buildings four stories or greater.

Figure 4.33 illustrates the basic functional components of a forced-air system for a multistory building. The mechanical equipment room is shown located at the top of the buil ding. The advantage in placing an HVAC equipment room on the top floor is that it is easy to take in makeup air and eject exhaust air to the atmosphere. In addition, some large systems make use of cooling towers. In a congested urban setting it is easier to locate a cooling tower at the roof than at grade level.

There are some advantages to locating an HVAC equipment room in the basement of a building and some buildings are equipped that way. It is also possible in tall buildings to have more than one equipment room, with one room in the basement and the other at the top of the building. In very tall buildings HVAC equipment rooms may also be located at intermediate levels. In these cases the HVAC systems are divided into zones of several floors with each zone supplied from a separate equipment room.

Figure 4.33 shows that most of the air flowing through the HVAC system can be recirculated through the building space. Recirculating air increases system efficiency by reducing the amount of air that must be either heated or cooled to the desired building temperature. The control dampers are used to vary the airflow depending on the building needs. In large buildings the HVAC system can be controlled through the use of temperature sensors connected

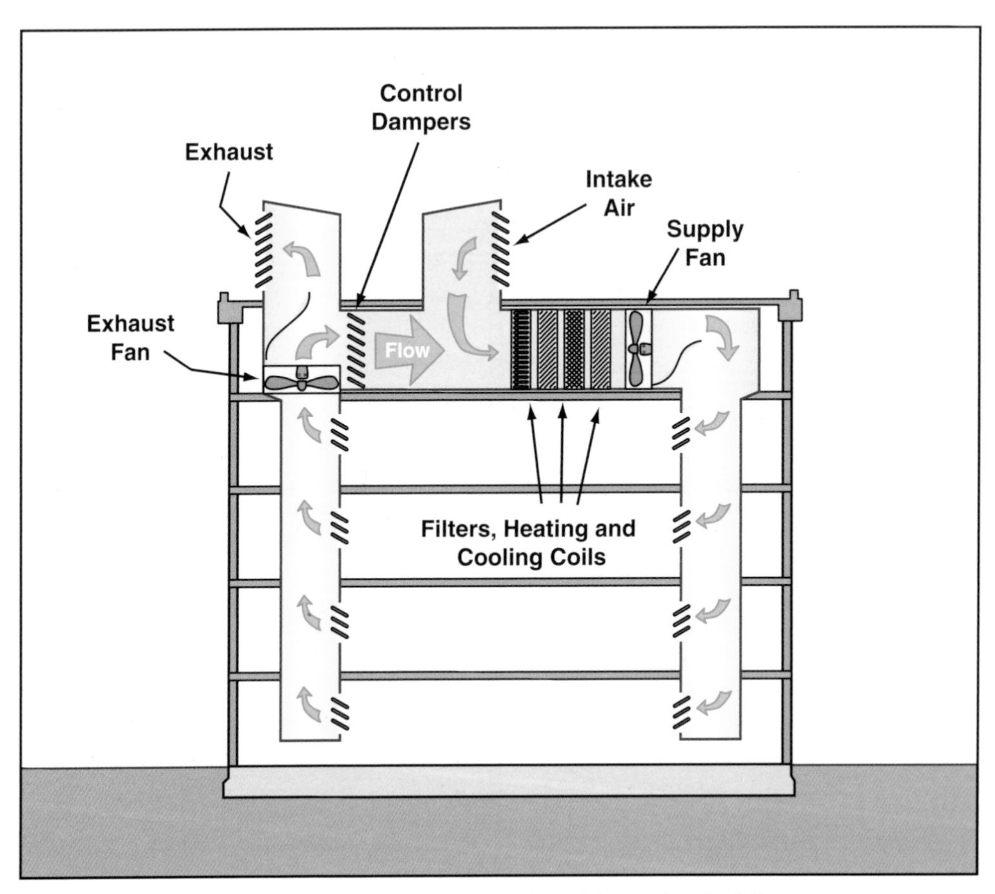

Figure 4.33 An HVAC system recirculating a portion of the air in a building.

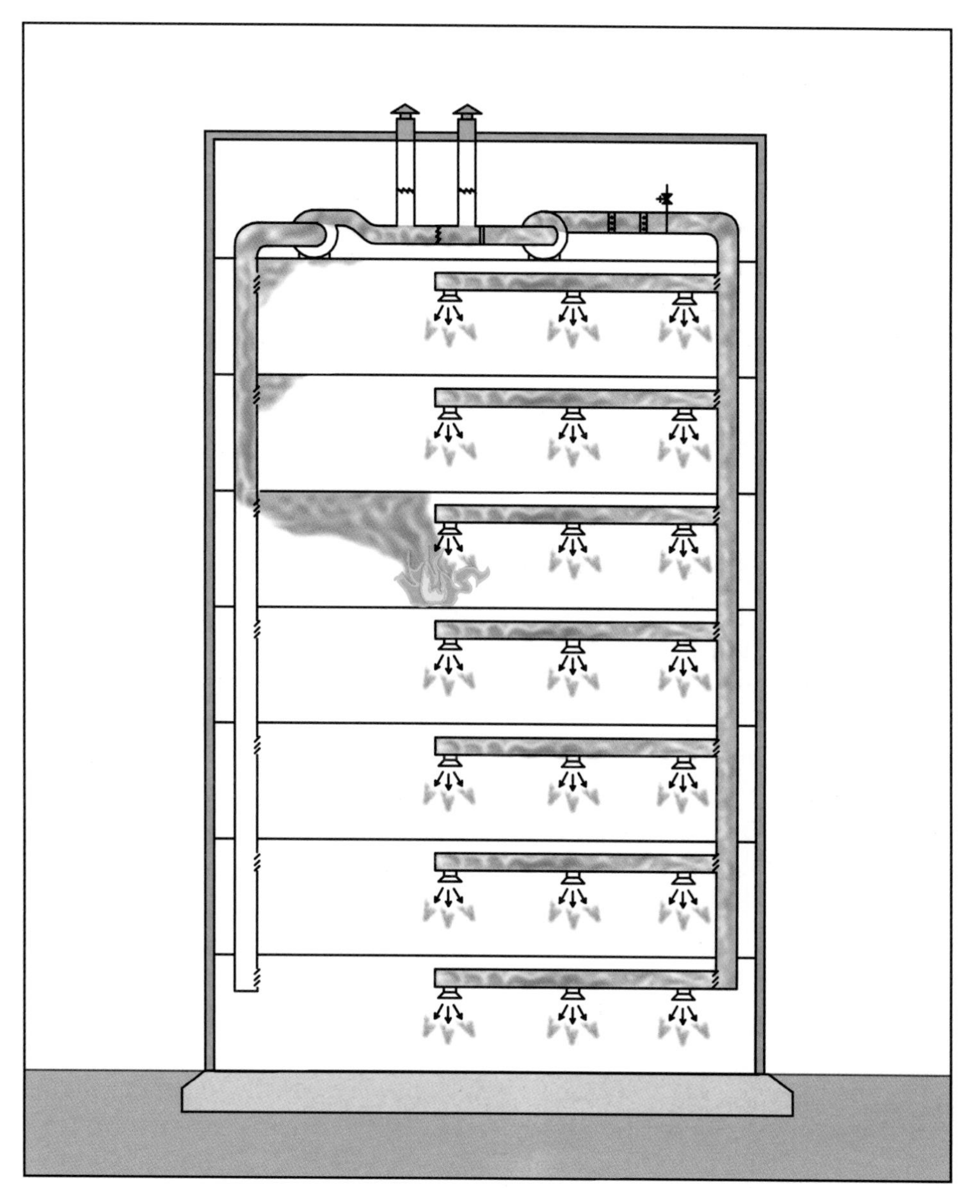

Figure 4.34 The HVAC system can draw the products of combustion into the ducts and transport them throughout the building.

to computers. The computers can adjust dampers, blower motors, and heating units to achieve the desired rate of airflow and temperature. It can also be seen that if a fire occurs in the building, the products of combustion can be drawn through the HVAC system and distributed throughout the building **(Figure 4.34)**.

System Components

Outside Air Intakes

An outside air intake draws outside air into the system. Codes regulate the location of outside air intakes to minimize drawing in combustible, flammable, or toxic substances, vehicle exhaust, or smoke from fires in nearby structures.

Fans

Separate fans are generally used for supply air and for exhaust air, but some fans contain dual units or are reversible, and can be used for both purposes. Fans move the air throughout the system, but will also move smoke and heat if it is introduced into the system. Duct detectors are typically provided for fans with capacities exceeding 2,000 cubic feet per minute, to shut down these systems under fire conditions and minimize unwanted smoke movement.

Air Filtration

Air can be cleaned with filters, with electrostatic equipment, or both. Filters should be made of approved materials to minimize their fire potential. A filter using liquid adhesives may present a combustible liquid hazard; the adhesive should be appropriately stored. Electrostatic equipment can present a significant electrical equipment hazard.

Air Heating and Cooling Equipment

Many types of equipment are used to heat and cool the air circulated in buildings. Heating equipment can be fuel-fired, such as a natural gas or oil burner, or the heat may be produced using electricity or steam. Each method has particular hazards associated with the fuel. Natural gas, for example, is lighter than air and rises to upper levels of a containment, but LPG (propane) is heavier than air and tends to collect in low areas.

Figure 4.35 An air-cooled chiller on a roof.

The cooling equipment hazards are limited mainly to hazards associated with the refrigerant and the electrical equipment. With the prohibition of some halogenated refrigerants due to environmental concerns, replacements that may be used, such as butane and propane, pose a greater hazard for firefighters. The use of flammable refrigerants warrants special awareness on the part of emergency responders. Air-cooled chillers are often found on roofs **(Figure 4.35)**. These devices remove heat from a piped water system in the building that passes through fan-coil or other air-handling equipment. Cooling towers may also be located on roof tops.

Duct — (1) Channel or enclosure, usually of sheet metal, used to move heating and cooling air through a building. (2) Hollow pathways used to move air from one area to another in ventilation systems.

Interstitial Space — In building construction, refers to generally inaccessible spaces between layers of building materials. May be large enough to provide a potential space for fire to spread unseen to other parts of the building.

Air Ducts

An *air duct* is the air distribution component of the HVAC system. Ducts also provide a direct means of spreading fire, heat, and smoke from one area of the building to another. HVAC ducts frequently penetrate fire-rated assemblies. Building, mechanical, and fire codes contain many requirements for ducts, including allowable materials and the requirement for smoke and fire dampers to maintain the integrity of the fire-rated assemblies **(Figure 4.36)**.

Sometimes *interstitial spaces* (the space between a suspended ceiling and the roof deck) are used as a return air plenum. This is less expensive than installing ducts to carry return air. This design technique is limited by the codes and, if improperly done, can result in very dangerous exit conditions during a fire. It should also be noted that one of the early proposals for addressing green or sustainable building design objectives is to develop under-floor spaces to provide supply air. It is likely that building codes will be amended to control this approach.

Figure 4.36 A partly-installed fire damper. This unit is large enough that it will contain four dampers, each separately operated by a fusible link and closing springs. *Courtesy of Gregory Havel, Burlington, WI.*

Smoke Control Systems

Experienced firefighters know that when a fire occurs in any building, fire control and occupant safety are enhanced by venting the products of combustion to the outside by the shortest route. The venting of buildings involved in a fire is an age-old fire fighting tactic. The design of many modern buildings, however, can make traditional ventilation methods difficult or impossible. Simply breaking out windows to vent products of combustion must be carried out very carefully when the window is 30 stories above grade! In addition, some newer glazing, such as hurricane glazing, may be very difficult to remove.

Smoke-Control System — Engineered system designed to control smoke by the using mechanical fans to produce airflows and pressure differences across smoke barriers to limit and direct smoke movement.

Newer building design plus the complexity of contemporary forced air HVAC systems has given rise to the concept of smoke control. *Smoke control* means the use of mechanical equipment to produce pressure differences across smoke barriers to inhibit smoke movement.

To prevent the recirculation of smoke through the HVAC system and to facilitate removal of the smoke, an HVAC system of the type shown in **Figure 4.37, p. 136,** can be switched from its normal operating mode to fire mode. The transfer can be accomplished either automatically or manually.

Automatic Smoke Control

Automatic transfer of an HVAC system to fire operation can be accomplished by smoke detectors serving the various floors of the building. When the smoke from a fire is detected, the HVAC controls are signaled to switch to the fire mode. The automatic transfer of the system to fire mode can also be initiated by sprinkler waterflow switches or heat detectors. When the system goes into fire operation, dampers can be opened or closed, depending on the location of the fire, to redirect the flow of air and to exhaust the smoke.

Figure 4.37 illustrates an HVAC system operating in a fire mode. In the illustration, the system exhaust fan is discharging smoke from the fire floor to the outside without returning air to the supply fan. The supply of air to the fire floor has been stopped by closing a damper while air continues to be supplied to adjacent floors. Continuing the supply of air to the non-fire floors creates a "pressure sandwich" of higher air pressure on floors above and below the fire floors. This reduces the migration of smoke into those areas.

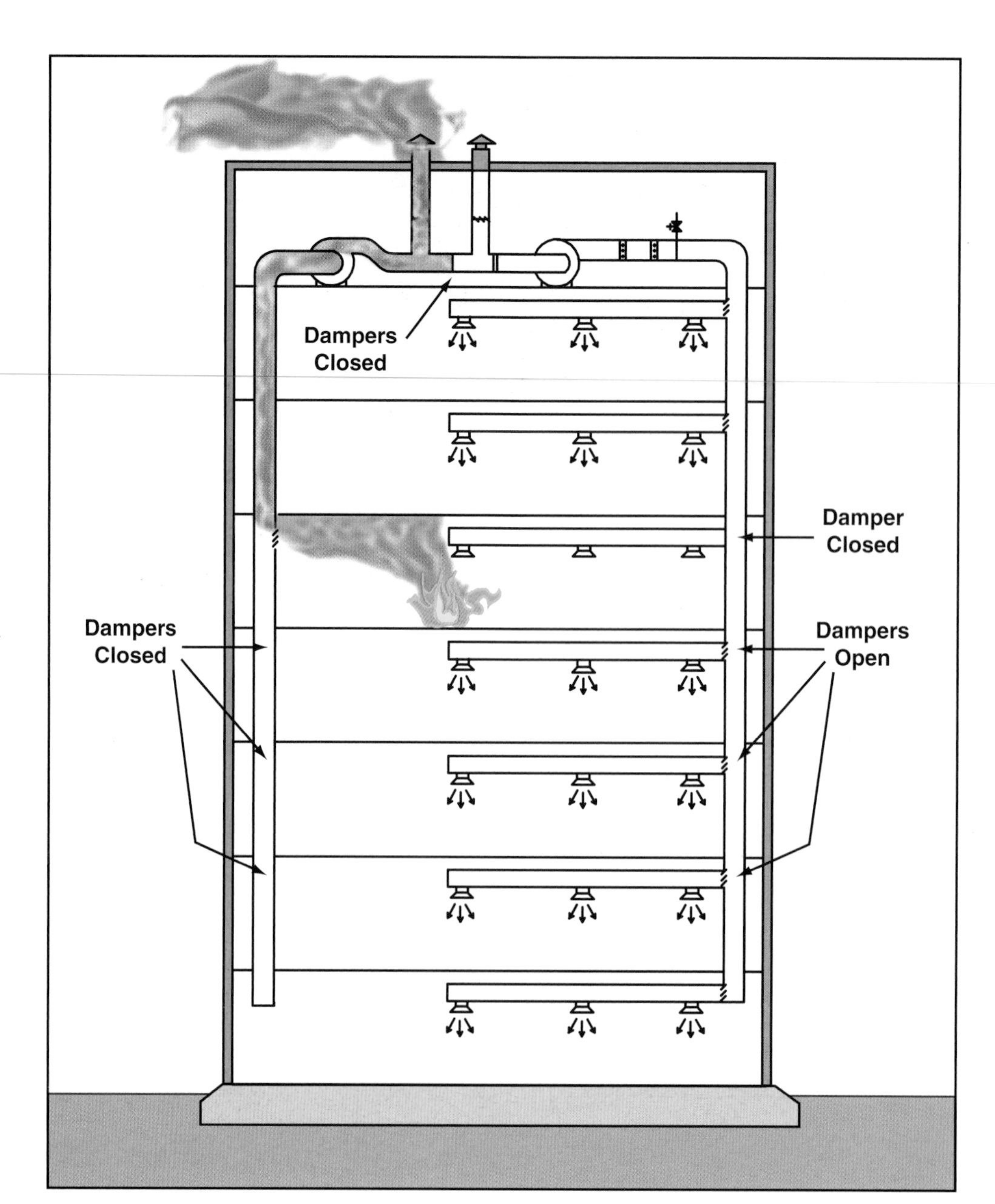

Figure 4.37 An HVAC system operating in the fire mode will exhaust smoke from the fire zone and supply fresh air to adjacent zones.

Automatic transfer of an HVAC system to the smoke control mode has advantages and disadvantages. Automatic operation is relatively fast and will be activated during nighttime hours when occupants may be asleep. However, coordination of fire alarm, automatic sprinkler, and HVAC zones is critical to maximize the benefits of a smoke control system. The system detectors must be carefully designed to eliminate the possibility of a detector outside the fire area being activated first. This would result in the misdirection of the system with the wrong dampers being operated. Provision must also be made for a situation in which smoke detectors operate in more than one zone. In any case, automatic initiation of fire operation in an HVAC system must always take priority over the normal system functions.

The design of a smoke control system as illustrated in **Figure 4.38** requires extensive engineering analysis. In the design of the system, the designer must take into account such factors as the anticipated fire size, outside weather

Figure 4.38 Portion of a firefighter's smoke control station (FSCS) showing the fan controls and HVAC diagram.

conditions, the volume of a fire zone, and maximum pressure differences across barriers such as stairwell doors. In addition, any equipment such as fans and ducts used to exhaust the products of combustion must be capable of withstanding the anticipated temperatures.

Manual Smoke Control

The transfer of an HVAC system to fire mode can be accomplished manually through controls provided for that purpose. The advantages to manual control are elimination of system disruption due to false alarms and more specific system control. The system can be controlled from a dedicated control panel, the building's main control room, or a firefighter's smoke control station (FSCS). When a system has both automatic and manual capability, the manual shall take priority over the automatic control.

An obvious disadvantage to manual operation is that it is slower than automatic operation. Manual activation usually occurs after arrival of the fire department and may not occur until late in the fire development when a danger to life may have already developed.

A smoke control system should not be activated by using the manual pull stations on the building fire alarm system. This is to avoid the possibility of occupants seeing a fire but not operating a pull station until they have fled to another area of the building. Special situations may occur in some occupancies, such as jails, where the fire alarm system is under control of the guard staff.

When a firefighter's smoke control station is provided (such as the one shown in Figure 4.38) it should have complete system monitoring capability. Codes require status indicators and switches for all fans and dampers serving a smoke control function. The FCSC should contain a diagram of the building that indicates the type and location of system components such as fans and dampers.

Smoke Control in Stairwells

As was discussed earlier in the chapter, stairwells must be kept clear of smoke if they are to function as a safe means of escape. If a stairwell door is opened on the floor of the fire, as in the case of an occupant fleeing from the fire, smoke will flow into the stairwell. Even if the door returns to the closed position, enough smoke can flow into a stairwell to render the stairwell unusable. Smoke can also seep into the stairwell around door cracks.

There are two methods that can be used to protect an enclosed stairwell from smoke. One method is to make use of a *smoke-proof tower.* The other is to provide for stairwell pressurization.

Smokeproof tower. A smokeproof tower is a type of smokeproof enclosure that makes use of a vestibule between the corridor and the stairwell that is open to the atmosphere **(Figure 4.39)**. Smoke that enters the vestibule from the corridor is exhausted to the atmosphere and the stairwell remains free of smoke.

Smokeproof towers are effective and simple but they take up otherwise usable floor space. The vestibule takes up space and the smokeproof tower usually has to be located on the periphery of the floor plan. Floor space must be sacrificed to provide corridor access to the stairwell.

Pressurized stairwell. A pressurized stairwell utilizes a blower or fan to provide a slightly greater pressure in the stairwell than the corridor to prevent the infiltration of smoke from the corridor into the stairwell.

Stairwell pressurization systems require careful engineering analysis. The pressure in the stairwell must be high enough that it will prevent the flow of smoke into a stairwell, but not so high that people cannot open the door into the stairwell. As an example, consider a stairwell door with dimensions of 36 in. x 78 in. (91.4 x 198.1 cm). The area is 2,808 sq. in. (18,106 sq. cm.). If a pressure of only 0.01 psi (68.9 Pa) is exerted against the door, a force of 28 pounds would result. This force could prevent some people from opening the door and could exceed the maximum opening force permitted by the codes for exit doors.

NFPA® 92A, *Standard for Smoke-Control Systems Utilizing Barriers and Pressure Differences*, requires a minimum pressure difference of 0.05 inches, water gauge across a smoke barrier in a *sprinklered* building. This corresponds to a pressure of 0.0018 pounds per square inch (12.5 Pa).

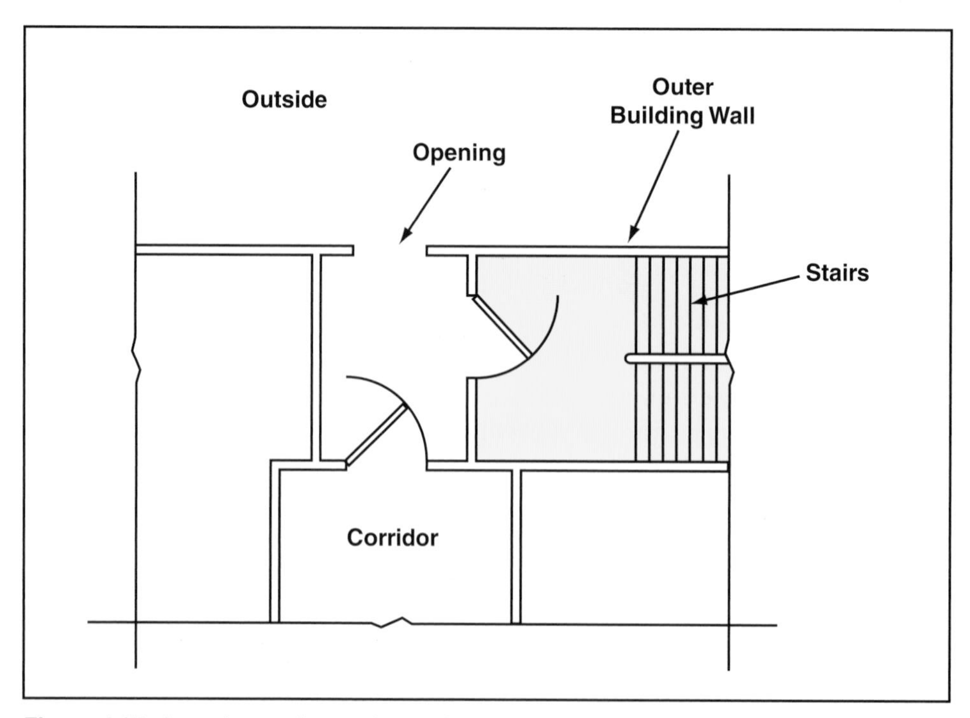

Figure 4.39 A smokeproof tower located at periphery of building.

One limitation to the effectiveness of pressurized stairwells is the loss of pressure that can occur when doors are opened to the stairwell. A pressure difference greater than the 0.05 inch water gauge stated in the previous paragraph may be necessary to compensate for the opening of doors. Therefore the designer of a stairwell system must work between narrow pressure limits to achieve an effective system.

There are two general design methods used for pressurized stairwells. One method is to have a fan supply air to a stairwell at a single point **(Figure 4.40)**. This is known as the *single-injection method.* A limitation of the single-injection system is that if stairwell doors are opened at a point close to the air supply, all of the air can flow out of the stairwell and the pressurization can be lost.

To overcome the possible limitations of a single-injection system, a *multiple-injection system* can be used **(Figure 4.41, p. 140)**. In a multiple-injection system an air supply shaft is run parallel to the stairwell from the supply fan. Air discharge points into the stairwell are located at several points along the stairwell. This method provides a more uniform flow of air into the stairwell.

A problem confronting the designers of stairwell pressurization systems is in attempting to determine the number of doors that might be opened into a stairwell during building evacuation. If it is anticipated that several doors

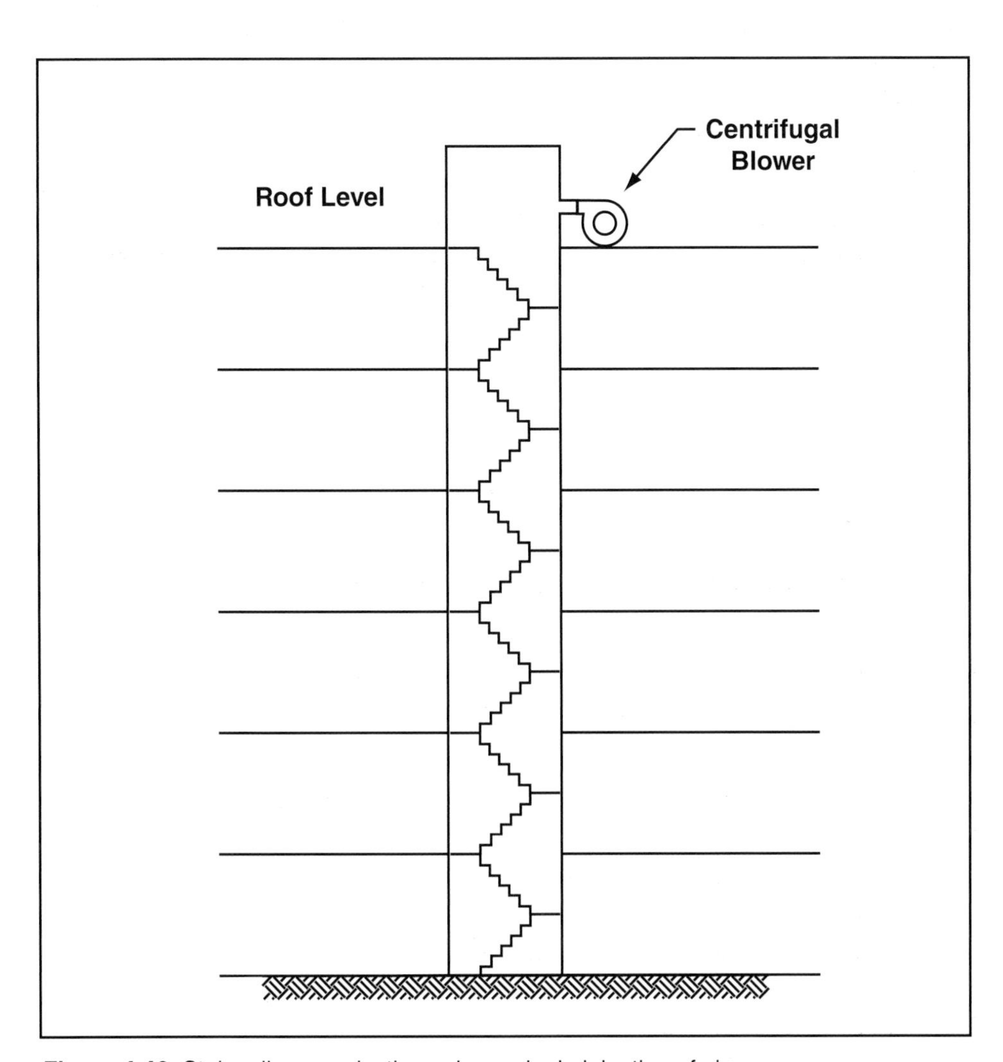

Figure 4.40 Stairwell pressurization using a single injection of air.

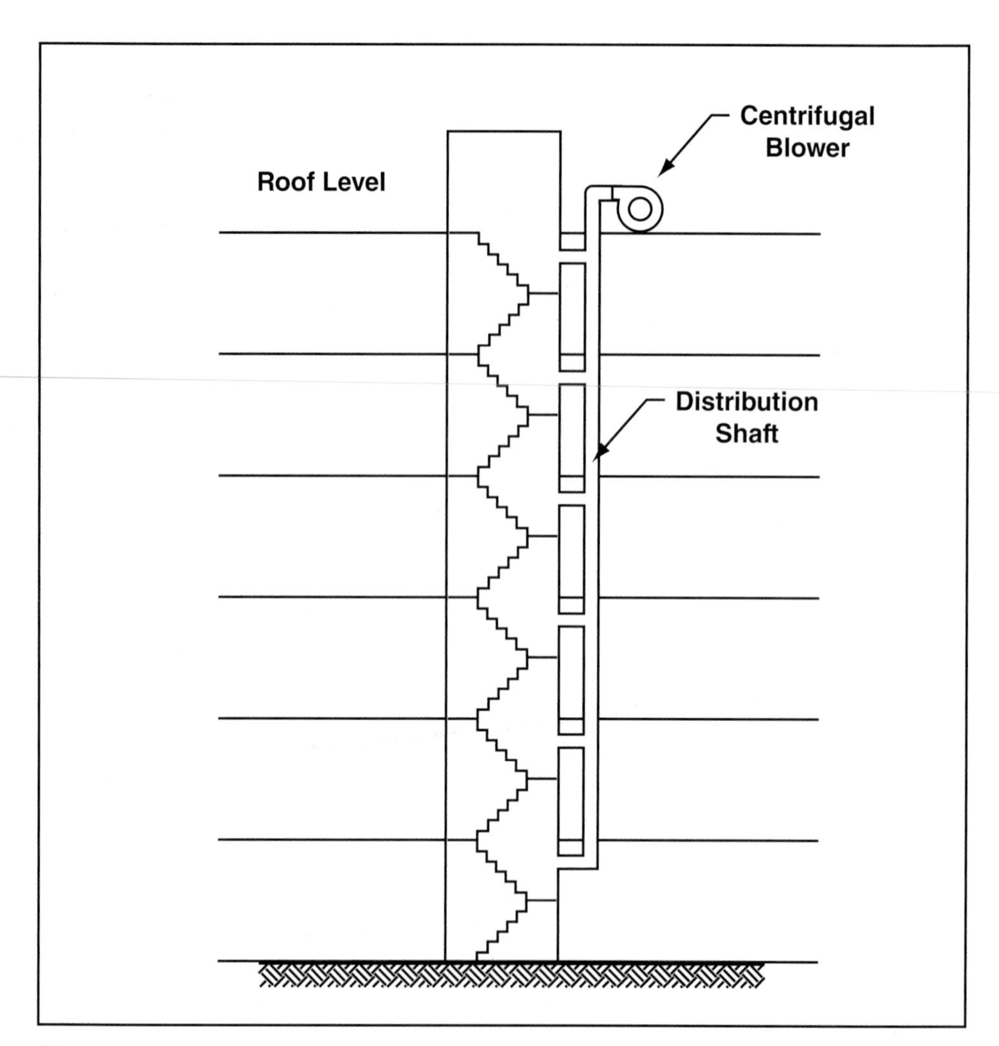

Figure 4.41 Stairwell pressurization using multiple injection points for air.

might be open, a higher pressure would be necessary to maintain a smoke-free atmosphere. If the pressure is initially too high, though, occupants may not be able to open the stairwell doors. This problem can be overcome through the use of a compensated system.

In a compensated system, the airflow into the stairwell can be adjusted depending on the number of doors that may be open. A compensating system can adjust the pressure in a stairwell either by modulating the air supply or through overpressure relief. Modulating the air supply into the stairwell can be accomplished with pressure sensors that sense the pressure difference between the stairwell and the interior of the building. The air supply modulation can be accomplished by varying the speed of the fan, inlet vanes, variable pitch fan blades, or varying the number of fans operating.

Overpressure relief in a stairwell is accomplished through the use of dampers that open to the outside, thus relieving a pressure buildup in the stairwell. The simplest means of controlling the exhaust dampers is to use dampers with adjustable counterweights. The dampers open when an overpressure condition in the stairwell is reached. A second method is to use motor-operated dampers controlled by pressure-sensing switches. This method is somewhat more complicated and therefore more costly.

Smoke and Heat Vents

Vents can be provided in the roofs of buildings to permit the release of smoke and heat from a fire. This venting is especially useful in large-area buildings and in buildings with few windows. Venting the heat and smoke enables firefighters to make a faster and safer interior attack and dissipates some of the thermal energy of a fire. Without a means of venting the heat and smoke, firefighters could be forced to withdraw to a peripheral attack while the fire continued to burn in the interior.

In modern practice, rooftop smoke and heat vents are typically required on the roofs of large industrial and storage buildings. Incorporating smoke and heat vents in building design is not a new concept; for example, theaters have been required to have smoke vents over stages since the early part of the twentieth century **(Figure 4.42)**.

Figure 4.42 Theaters are required to have smoke vents above the stage. *Courtesy of Ed Prendergast.*

The type of heat and smoke vents typically encountered are individual small-area hatchways (a minimum of 4 feet [1.2 m] in either direction is typical) with single- or double-leaf metal lids or plastic domes designed to open automatically or manually **(Figure 4.43)**.

Vents with metal lids have spring-operated opening mechanisms that can be released by a fusible link or other automatic detectors, including smoke detectors. Plastic vents make use of a thermoplastic dome that softens from the heat of a fire and falls out.

Figure 4.43 Firefighters must be sure they are not drawing smoke into uninvolved areas when they open smoke vents.

The use of smoke detectors may pose a problem of sensitivity, resulting in annoying opening of the vents under nonfire conditions. The vents can also be released manually or by fusible links. Firefighters manually activating these vents must ensure that they are operating the appropriate vents to prevent drawing the fire to other uninvolved areas.

Curtain boards are used in conjunction with smoke vents to increase their effectiveness. The curtain boards reduce the dissipation of the heated air currents from a fire and increase the speed of operation of the vents **(Figure 4.44, p. 142)**. The depth of a curtain board will vary depending on the nature of the hazards within an occupancy, but should be not less than 20 percent of the ceiling height. Curtain boards should be spaced so that they are not farther apart than eight times the ceiling height.

Curtain Boards — Vertical boards, fire-resistive half-walls, that extend down from the underside of the roof of some commercial buildings and are intended to limit the spread of fire, heat, smoke and fire gases.

The size and spacing of smoke and heat vents depend on the floor-to-ceiling height of a building and the nature of the contents. NFPA® 204, *Standard for Smoke and Heat Venting*, contains the design methodology for determining the required vent area. The method used to determine the vent area requires an analysis of the rate of heat release of the fuel, the ceiling height, and the depth of the curtain boards.

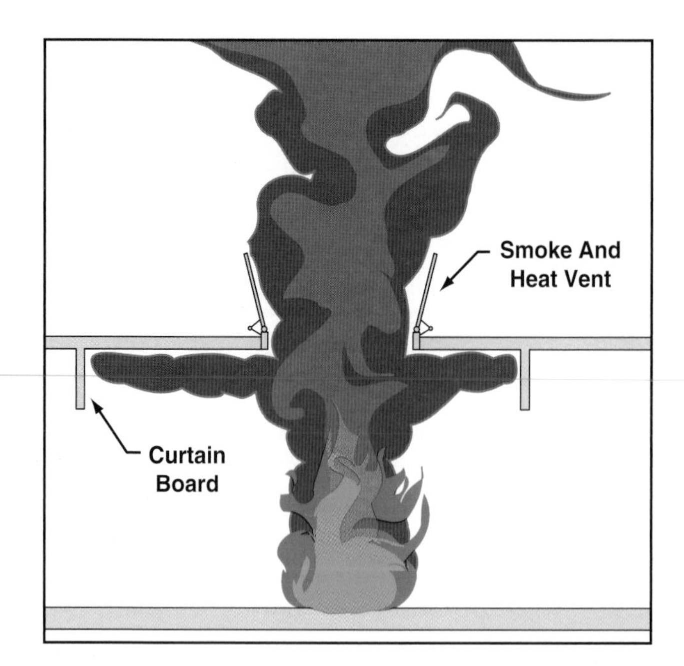

Figure 4.44 Curtain boards direct products of combustion to a vent.

Some controversy exists with respect to the effectiveness of smoke and heat vents in sprinklered buildings. When sprinklers operate, the discharge of the sprinklers cools the products of combustion (smoke and gases) and they lose their buoyancy. This loss of buoyancy reduces the natural tendency of the products of combustion to rise and escape through the roof vents. Thus, the roof vents in sprinklered buildings are not as effective as they are in non-sprinklered buildings. Ultimately, however, the removal of the products of combustion from the building must still be accomplished and the existence of roof vents facilitates this process. Roof vents in a sprinklered building may be of maximum value in removing products of combustion during the final or overhaul stages of a fire.

Voltage — The electrical force that causes a charge (electrons) to move through a conductor. Sometimes called the electromotive force (EMF). Measured in volts (V).

Electrical Equipment

Electrical equipment may be installed in separate rooms or vaults. Electrical switch gears, transformers, and panels may be required by codes to be separated from the rest of the building by fire-rated construction. High-voltage equipment is defined as operating at 600 volts or higher; low-voltage equipment operates at less than 600 volts.

Transformers

Transformers are used to convert high-voltage electricity, as supplied by the electric utility service, to an appropriate voltage for use in the building. Some dedicated transformers are used to supply special systems and equipment in industrial and commercial buildings. Transformers generate heat, and the method of cooling the unit directly affects the hazard presented to emergency response personnel. The two most common cooling methods use air and oil. Transformers may be high or low voltage.

Air-cooled transformers, also called "dry" transformers, use the surrounding air to cool the unit through fins and heat sinks installed on the body of the transformer **(Figure 4.45)**. The hazards they present are those of energized electrical equipment.

With the increasing cost of building materials, many buildings are now supplied by the electric utility with 480/277-volt services, which can carry the same amount of energy in smaller wire and conduit when compared to traditional 120/240-volt services. These include schools, medical facilities, strip malls, factories, and "big box" stores. The electricity is used at 480 volts by HVAC equipment, refrigeration, and other motors; and at 277 volts by light fixture ballasts.

Figure 4.45 Electrical transformers located outside buildings generate heat and must have a cooling solution in place to function safely and efficiently.

Dry transformers are located throughout the building to step down the voltage to the 120-volts required by office equipment, table lamps, and household appliances. Large transformers are usually set on the floor, while smaller units may be mounted on wall brackets or suspended from the roof or floor above. Although they are air-cooled and have ventilation openings, most modern dry transformers are designed for use inside sprinklered spaces.

Oil-cooled or oil-filled transformers contain oil to conduct heat away from the core and also to electrically insulate internal components from arcing. In addition to the hazard of energized electrical equipment, they also have the potential to be the source of a combustible liquid leak. Because the oil also provides electrical insulation, it must have *dielectric* (nonconducting) properties. Some older transformer cooling oils contained highly hazardous PCBs because they have excellent dielectric properties. Transformers containing PCBs are required to be labeled. Newer oils are much less toxic and may contain a less flammable-type oil.

Dielectric — Nonconductor of direct electric current. Term usually applied to tools that are used to handle energized electrical wires or equipment.

Transformers are located in rooms or vaults that may be inside or outside the building. When transformers are inside the building, they are generally located at or near grade. However, they may also be located on upper levels in high-rise buildings. When transformers are inside, the transformer rooms or vaults are required by code to be enclosed in 3-hour fire-rated construction if not protected by automatic sprinklers, or 1-hour fire-rated construction if they are protected by sprinklers.

Rooms or vaults that contain electrical gear or transformers should be protected with sprinklers if the building has an automatic sprinkler system. Some power utility companies may not allow sprinkler protection for their equipment. In these instances, the utilities may permit carbon dioxide systems. Fires involving electrical equipment usually de-energize the equipment early in the event. Fire operations involving electrical rooms or vaults should use the same safety guidelines for operations involving any electrical source. Preincident planning can provide the information needed in an emergency to protect both fire personnel and to preserve property.

Emergency and Standby Power Supplies

When required by codes or operational needs, backup power supplies for building systems dependent on electric power may consist of generators, batteries, or a combination of both. Some occupancies, such as correctional, detention, and health-care facilities, are often required by codes to have emergency power systems. Other types of occupancies may have an emergency power supply because of the facility's function. Examples of these types of facilities include data processing, telecommunications, semiconductor fabrication, and electrical utility facilities. Buildings that are required to have smoke management systems, such as high-rises, covered malls, and buildings with atriums, also must have emergency backup generator systems.

Generator — Auxiliary electrical power generating device. Portable generators are powered by small gasoline or diesel engines and generally have 110- and/or 220-volt capacities.

Generators

Standby generators may be either permanent or portable, with the latter typically being used for small commercial or single-family residential properties. The following discussion will focus on permanent generators installed to meet code or major operational requirements.

Generators are typically engine-driven, using a gasoline, diesel, or natural gas internal combustion engine. A diesel fuel engine is the most common type because gasoline is more hazardous and natural gas may be subject to outages at the same time as the electrical power distribution system **(Figure 4.46)**. The size of the generator depends on the amount of power needed to provide the desired services when the normal power fails. At a minimum, the generator will need to supply the required life safety systems plus critical needs of the occupancy. In a hospital, for example, power is required at all times for life-support systems and monitoring equipment.

Codes specify the minimum required fuel storage for diesel- or gasoline-driven generators. The amounts are stated in terms of the expected duration of operation which may vary from two to eight hours. Operational requirements for hospitals and other critical facilities, however, may be up to 48 hours or longer. Thus, for many facilities, operational needs will determine the fuel storage requirements. Generators and their fuel storage may be located either outside or inside the building.

Often generators are located on the roof, in an open parking garage, or outside the building to easily provide ventilation and cooling. When generators are located inside buildings, current codes require that the generator and the main fuel storage be installed in separate fire-rated rooms. The fuel storage rooms must be provided with fuel spill control and containment systems. In older buildings, the fuel storage could be in the same room with no spill control or containment provisions.

An exception to the fuel storage separation is a day-tank located in the same room or mounted directly on the generator. This tank is typically limited by fire codes to 60 gallons (240 L) for diesel fuel. Day tanks are usually kept full by pumping fuel from the main tank.

Other components of the emergency power supply system include transfer switches. Codes often require the transfer switch to be located in a protected room separate from the main electrical panel room for the building. In buildings with fire control rooms, there should be a status panel and sometimes a remote start/stop switch for the generator **(Figure 4.47)**.

Lead-Acid Batteries

Emergency power supplies that require batteries commonly use lead-acid type storage batteries. Because of the materials they contain and their potential electrochemical reactions, these batteries present a significant potential hazard. Lead-acid batteries contain sulfuric acid and lead. The sulfuric acid is hazardous to humans: skin contact causes injury and inhalation of the acid vapors can cause serious injury or death.

Figure 4.46 The size of a generator depends on the amount of power needed. *Courtesy of McKinney (TX) Fire Department.*

Figure 4.47 Transfer switch for a generator, located in a separate area from the generator. *Courtesy of McKinney (TX) Fire Department.*

Sulfuric acid reacts with other materials and can cause a fire through chemical reaction. The sulfuric acid in lead-acid batteries also releases flammable hydrogen gas during battery charging and also during contact with some metals. The batteries can undergo unusual electrochemical reactions such as "thermal runaway" or a battery fire, which may require an emergency response.

The metallic lead in the batteries is a toxic heavy metal. Both long-term and short-term exposure to the lead can cause heavy metal poisoning with potentially severe health effects.

Lead-acid batteries have many different names that include the following:

- Wet cell
- Gel cell
- Starved electrolyte cell
- Sealed cell
- Maintenance free cell
- Flooded cell

Large numbers of these batteries may be found in buildings and they often are overlooked when assessing the hazards of the facility. Entire rooms or even entire floors of a building may be found to contain lead-acid batteries. The areas containing the batteries are not usually diked or sealed to contain a liquid acid spill. Small uninterruptible power supplies (UPS) containing lead-acid batteries are found near fire alarm system control panels, under desks, or next to computers in many offices. Some occupancies using battery-powered vehicles, such as storage, manufacturing, and golf clubhouses, may contain a battery charging room in the building or on the property. When located inside, these rooms should be separated by fire-rated construction and well ventilated.

The presence of lead-acid batteries should be documented during pre-fire incident planning. Cooperation with the building owner/operator will help ensure the safety of emergency operations in facilities with emergency battery power supplies.

WARNING!

Standard fire fighting personal protective gear is not designed to protect the wearer against acid exposure. Walking, crawling, or falling into an acid pool can result in serious injury.

Summary

A building consists of the exterior shell and the interior core. This inner portion contains the usable space to satisfy the occupant use intended for the building. Building systems provide the ability for the occupants to efficiently and safely use this space in a comfortable environment. These same systems, however, create fire protection concerns. Many building systems, such as stairs, elevators, conveyors, vertical shafts, and utility chases, of necessity, must penetrate both vertical and horizontal fire-rated components.

Modern buildings rely extensively on HVAC systems to provide interior comfort. In large buildings these systems can become a means for communicating products of combustion through a building. Therefore these systems must be designed so they do not facilitate such spread. This can be accomplished either automatically or manually.

These penetrations provide the opportunity for fire and smoke to spread throughout a building if the proper code provisions are not installed and maintained. The firefighter needs to be aware of the potential for fire and smoke spread due to building systems and not take anything for granted. Likewise, it is important for the firefighter to note any possible loss of the integrity of vertical building elements during company inspections.

Review Questions

1. What are the basic components common to all stair types?
2. Why are fire escapes no longer permitted in new construction?
3. Where are conveyor systems typically found?
4. In what ways do HVAC systems potentially affect fire events?
5. What are the potential hazards encountered with lead-acid batteries?

References

1. International Building Code 2006, International Code Council®, Washington, D.C.
2. NFPA® 5000, *Building Construction and Safety Code*, 2003 ed. National Fire Protection Association, Quincy, MA.
3. Strakosch, George R., The Vertical Transportation Handbook, 3rd ed. John Wiley and Sons.
4. The American Society of Mechanical Engineers, ASME A17.1 - 2007/CSA B44-07, Safety Code for Elevators and Escalators.

Fire Behavior and Building Construction

Chapter Contents

Divider page photo courtesy of Gregory Havel, Burlington, WI.

Key Terms

FESHE Objectives

Fire and Emergency Services Higher Education (FESHE) Objectives: *Building Construction for Fire Protection*

2. Classify major types of building construction in accordance with a local/model building code.
3. Analyze the hazards and tactical considerations associated with the various types of building construction.
5. Differentiate between fire resistance, flame spread, and describe the testing procedures used to establish ratings for each.

Fire Behavior and Building Construction

Learning Objectives

After reading this chapter, students will be able to:

1. Discuss the factors affecting combustibility of various interior finishes and their effects on fire behavior.
2. Explain the methods used to evaluate the surface burning characteristics of interior finish material.
3. Discuss compartmentation as it relates to fire and smoke containment.
4. Describe the types of walls used to prevent fire spread and their effectiveness in providing fire and smoke containment.
5. Describe the requirements for fire doors and their contribution to fire and smoke containment.

Chapter 5
Fire Behavior and Building Construction

Case History

The Station Nightclub

Highly Flammable Interior Finish Materials:
On the night of February 20, 2003 a fire at The Station nightclub in West Warwick, Rhode Island killed 100 people and injured 200 more. Several factors contributed to this disaster; however, the interior finish used on part of the walls and ceiling played a major role.

This case is unusual in the history of fire protection because a video camera recorded the start of the fire and subsequent events. Furthermore, the attention of the audience was focused on the point of origin of the fire and people realized the danger within a few seconds of its start. The photographer survived the fire.

The Station nightclub was housed in a one-story wood frame structure originally built in 1946. It had a total area of approximately 4,484 sq. ft. (412 sq. m). **(Figure 5.1, p. 152)**. A raised platform that served as a bandstand was situated at one end of the building. The walls and part of the ceiling in the vicinity of the bandstand had been covered with a polyurethane foam for sound attenuation purposes **(Figure 5.2, p. 152)**. In addition, a portion of the interior walls in the area of the bandstand and dance floor consisted of wood paneling.

The band began its performance with fireworks, which consisted of four fountain-type devices (gerbs) emitting sprays of sparks. When the gerbs were ignited, the sparks impacted on the polyurethane foam, igniting it. The foam began to burn rapidly with flames quickly spreading to and along the ceiling of the nightclub.

Lessons Learned: The videotape permitted a very precise reconstruction of events – a rare occurrence in fire investigation. The reconstruction indicated that flames reached the club's ceiling within 25 seconds. Within 90 seconds a thick black smoke layer appeared to have dropped to within 1 foot (0.3 m) of the floor of the club. Within five minutes flames were photographed emitting from the front door and windows of the club.

Subsequent tests and fire modeling by fire protection engineers at the National Institute for Standards and Technology (NIST) indicated that conditions inside most of the club would have become untenable on the basis of temperature alone within 100 seconds of ignition of the polyurethane. The rapid combustion of the polyurethane used for the interior finish had resulted in a deterioration of conditions within the nightclub before the crowd could exit.

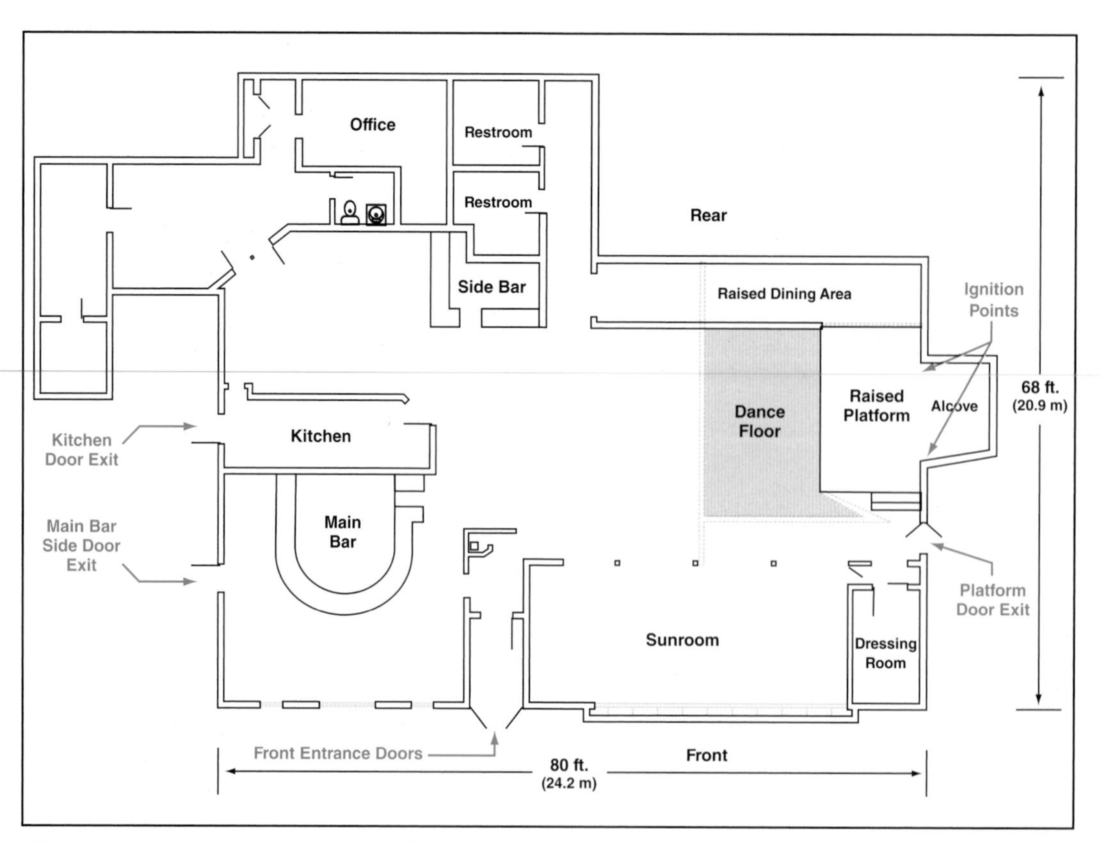

Figure 5.1 Floor plan of The Station nightclub. *Source: NIST (National Institute for Standards and Technology).*

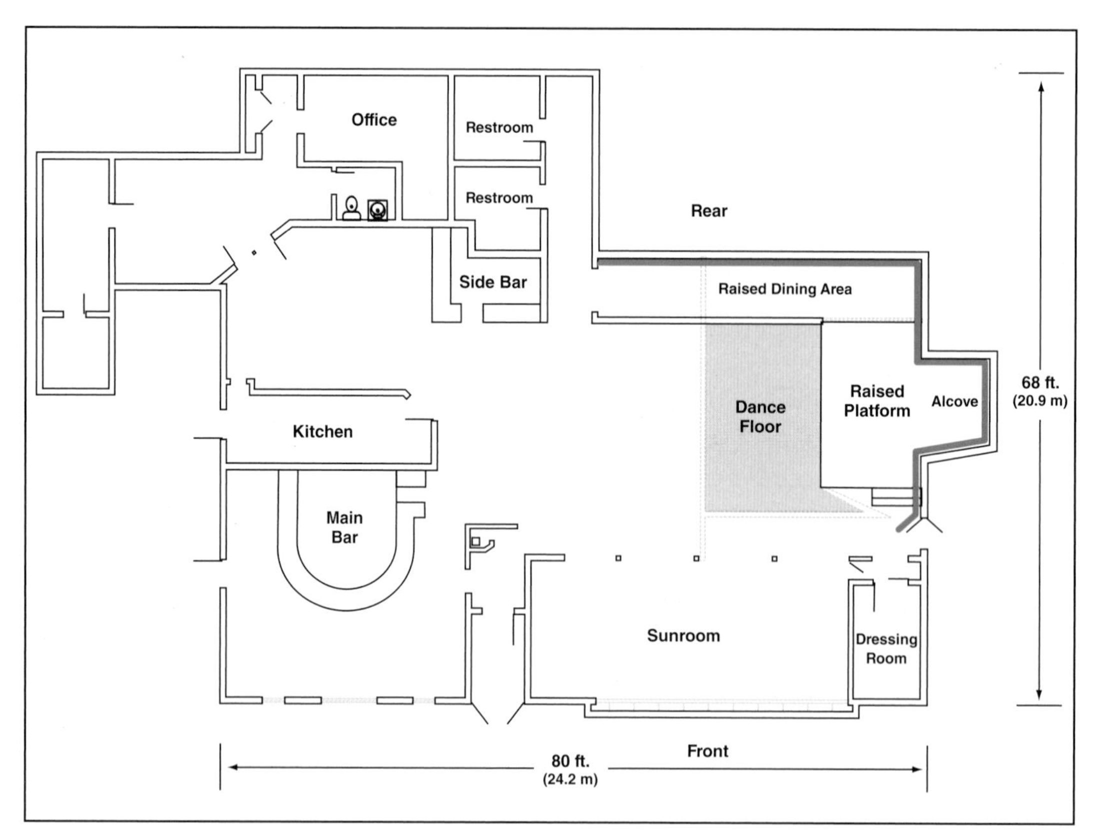

Figure 5.2 The shaded area shows the portion of the nightclub that had polyurethane mounted on the wall. *Source: NIST.*

The fire behavior of buildings is determined by a variety of factors, some of which are not directly related to the basic construction or occupancy of a building. Interior features of a building such as wall coverings, compartmentation, and fire protection systems have a cumulative effect on the outcome of a fire event.

Several disastrous fires have illustrated the fact that combustible interior finishes can contribute greatly to loss of life even if the basic building is fire-resistive or noncombustible. Noteworthy examples of these fires include the LaSalle Hotel fire in Chicago in 1946 (61 lives) and the Beverly Hills Supper Club in Southgate, Kentucky in 1978 (165 lives).

The combustibility of interior finish affects the behavior of fire in several ways:

- It can contribute to the fire extension by flame spread over the surface of walls and ceilings.
- It affects the rate of fire growth (which can lead to flashover).
- It adds to the intensity of a fire because it contributes fuel.
- It can produce smoke and toxic gases that contribute to the life hazard.

As building codes developed in the early twentieth century, their primary focus was on the structure of the building, namely the construction materials and the structural system. As time passed, however, people realized it was also important to consider the fire behavior characteristics of materials used for the interior finish in evaluating the overall behavior of a building under fire conditions.

Other chapters in this manual cover the specifics of building construction types and fire behavior. This chapter discusses the classification and testing of interior finishes, heat release rates, and compartmentation and fire doors.

Interior Finishes

The term *interior finish* is generally applied to the materials used for the exposed face of the walls and ceilings of a building. Interior finishes can include such materials as plaster, gypsum wallboard, wood paneling, ceiling tiles, plastic, fiberboard, and other wall coverings. Interior finish may or may not be structural materials. For example, if wood paneling were to be applied over a concrete block wall, the wood paneling would be considered the interior finish. On the other hand, if a masonry bearing wall is left exposed as part of the decor of a restaurant, the surface of the masonry would be part of the interior finish.

NOTE: In the international community the term *interior lining* is used instead of interior finish. The term interior lining may ultimately replace interior finish in North America.

Interior Finish — Exposed interior surfaces of buildings including, but not limited to, fixed or movable walls and partitions, columns, and ceilings. Commonly refers to finish on walls and ceilings and not floor coverings.

Building codes may exclude from the definition of interior finish such surfaces as countertops, doors, and window frames. Incidental trim, such as chair railing, is considered interior finish but is less rigidly controlled because it does not present a continuous surface. Such material is generally considered trim as long as it does not exceed ten percent of the wall and ceiling area and is distributed throughout the space so that it does not constitute a large continuous surface. Movable partitions, such as those often used to subdivide a banquet facility, are treated as interior finish.

Building codes usually exclude surface treatments such as paint and wallpaper that are no thicker than 1/28 inch (0.9 mm). During a fire, thin materials tend to behave in a manner similar to the material to which they are attached because the heat is transmitted to the material beneath the surface material. If there are multiple layers of surface material, however, such as several layers of vinyl, the surface material acts as an insulator and will contribute to the spread of fire.

NOTE: At one time, floor coverings were not considered to contribute significantly to the development of fires because heat naturally rises. The introduction of various deep-pile floor carpets increased awareness of a need to regulate floor coverings. The building codes treat floor coverings separately from wall and ceiling finishes.

Combustible Interior Finishes

The first efforts at evaluating and controlling the combustibility of interior finish materials began after several disastrous hotel fires in the 1940's. The potential danger created by the materials used for interior finishes stems from the fact that interior finish materials typically cover a large area and are relatively thin compared to structural components. Thus, if wood paneling is used over a concrete block wall, fire can spread rapidly over the surface of the paneling even though the quantity of wood is small compared to the concrete block **(Figure 5.3)**. The early investigations of the role of interior finish materials, therefore, logically concentrated on evaluating the speed with which flame can spread over the surface of a material. The degree to which fire can spread over the surface of a material is technically referred to as the *surface burning characteristics* of the material.

In order to control the potential hazard represented by interior finish materials, it is necessary to evaluate their surface burning characteristics. Fire behavior is highly dynamic and is influenced by several thermal variables. The speed of flame spread over an interior finish is influenced by such factors as the following:

- The composition of the material
- Ventilation
- The shape of the space in which the material is installed
- Whether the finish material is applied to the ceiling or wall

NOTE: Standardized fire test methods are an inexact science and cannot precisely duplicate the wide variety of real-life situations in which interior finish materials are found. The test results have to be taken as general guidelines.

Figure 5.3 Interior finishes like wood paneling can contribute to the spread of a fire in a compartment or structure.

The manner in which an interior finish is mounted can greatly affect the material's burning characteristic. A thin product, ¼-inch (6.4 mm) or less, will propagate flame more quickly when it is attached to studs with an air space behind the material than when it is attached directly to a more solid material such as gypsum board. Building codes therefore may require that materials with greater surface burning rates be installed over a noncombustible material.

Classification of Interior Finishes

Building codes make use of the flame spread ratings of materials to establish some control over interior finishes. The codes establish three classifications of interior finishes using a letter designation as shown in **Table 5.1**. These classifications are used to restrict the materials in vertical exits and exit corridors to those with low flame spreads and to permit materials with a higher flame spread rating in other areas:

Table 5.1

Class of Material	Flame Spread
A	0-25
B	26-75
C	76-200

- Materials with a Class A (0-25) rating are required in the vertical exits of most occupancies.
- Materials with a Class B (26-75) rating are required in corridors that provide exit access.
- Class A, B, or C (76-200) materials may be required in other rooms and spaces depending on the occupancy. The rooms of health care and assembly occupancies, for example, will require either Class A or B interior finish materials.

Rooms in other occupancies would be permitted to have Class C materials. Building codes generally allow an increase in the flame spread rating of interior finish materials in buildings equipped with an automatic sprinkler system. The maximum flame spread rating allowed, however, is 200.

Heat Release Rate

Heat Release Rate (HRR) — Total amount of heat produced or released to the atmosphere from the convective-lift fire phase of a fire per unit mass of fuel consumed per unit time.

In Chapter 2 the concept of a fire load was introduced. The fire load is a measure of the *total* fuel available to a fire and, therefore, the total heat that can be released in a fire. The fire load is not by itself a measure of the severity or rate of fire development of a fire. The severity of a fire is determined by the fire load *plus* the rate at which the fuel burns. The faster the available fuel burns the greater will be the heat release rate (HRR). A greater heat release rate results in a faster developing fire.

The rate at which a fuel burns is determined by several factors. The most significant is the combination of the fuel and the available oxygen. Materials that have a high exposure surface to the surrounding oxygen will burn more rapidly with a correspondingly higher heat release rate. (See also Chapter 7).

In the case of interior finishes used on walls and ceilings, the combustible material always has a large area exposed to the surrounding oxygen. This fundamental concept explains why the combustibility of interior finish materials is a significant factor in the development of fires in buildings.

Testing Interior Finishes: Steiner Tunnel Test

Because different materials burn at different rates, it is important to evaluate them by means of a standard test. Interior finishes are tested to derive several measures of a material's flammability: the flame spread rating and the smoke developed rating. The Steiner Tunnel Test is the most commonly used method for evaluating the surface burning characteristics of materials.

Flame Spread Rating

Flame Spread Rating — Numerical rating assigned to a material based on the speed and extent to which flame travels over its surface.

Steiner Tunnel Test — Unofficial name for the test used to determine the flame spread ratings of various materials.

The Steiner test derives its name from A.J. Steiner, an engineer at Underwriters Laboratories, Inc. who developed the test in the late 1940's. The test is frequently referred to as the *tunnel test,* but is formally identified as ASTM Standard E84 and UL 753. It is also NFPA® 255, *Standard Method of Test of Surface Burning Characteristics of Building Materials.* The tunnel test produces a numerical evaluation of the flammability of interior materials, which is known as the *flame spread rating.*

Examination of the role of interior finishes is directed at materials applied to the ceiling of a room or the upper portions of a room. Because the heat of a fire rises, it is assumed that the most critical application of a material would be on the ceiling **(Figure 5.4)**. The "tunnel" used in the tunnel test consists of a horizontal furnace 25 feet (7.6 m) long with a removable top. The interior of the furnace is 17⅝ inches (448 mm) wide and 12 inches (300 mm) high.

Assembly — All component or manufactured parts necessary for and fitted together to form a complete machine, structure, unit, or system.

To conduct the test, the sample material to be tested is attached to the underside of the top of the furnace and the assembly is lowered into place. A gas burner located at one end of the tunnel produces a 4½-foot (1.37 m) flame that is projected against the test material. The flame is adjusted to produce approximately 5,000 BTU's (5,270 kj) per minute. The test is continued for ten minutes, during which time the travel of the flame along the test sample is observed **(Figure 5.5)**.

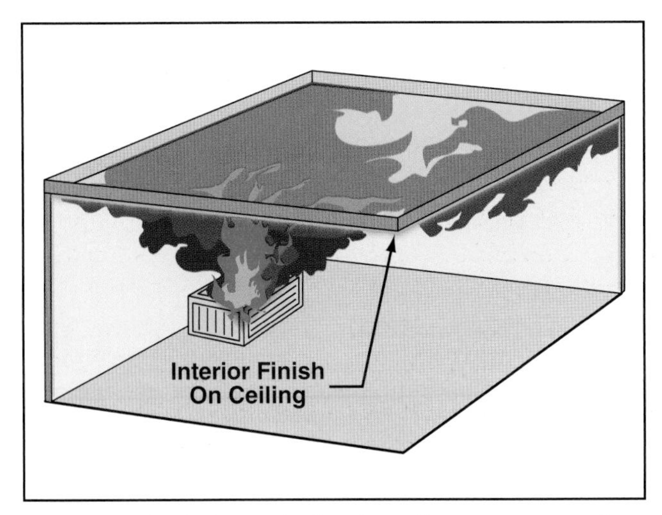

Figure 5.4 As the heat of a fire rises, flames can spread across a ceiling.

Figure 5.5 Apparatus used for the Steiner Tunnel test. *Courtesy of Underwriters Laboratories, Inc.*

Deriving the numerical rating. To derive the numerical flame spread rating, the flame travel along the test material is compared to two standard materials: asbestos cement board and red oak flooring. Asbestos cement board is assigned a flame spread rating of 0 and red oak is assigned a flame spread rating of 100. According to the test protocol, the flame will travel along the oak flooring 24 feet (7.3 m) in 5½ minutes. The flame spread of other materials during the test is compared to that of red oak. Obviously, the higher the flame spread rating, the more rapidly flame will spread. **Table 5.2** shows the flame spread ratings for a few materials.

Table 5.2

Material Rating	Flame Spread
Gypsum Wallboard	10-15
Treated Douglas Fir Plywood	15-60
Mineral Acoustical Tile	15-25
Walnut-faced Plywood	171-260
Veneered Woods	515 (Approximate)

Flame spread ratings over 200 are not permitted by Code.

Significance of the Flame Spread Rating. The flame spread rating developed in the tunnel test is a means of comparing the surface flammability of a material to standard materials under controlled test conditions. It is NOT an absolute measure of the spread of fire travel. The flame spread rating may not produce an accurate correlation with the actual behavior of a material in a fire. This is due to the effect upon surface burning of such factors as room shape and dimensions and the fuel load in a room. Furthermore, differences between field applications and test conditions may result in different behavior in the field.

The thickness of the test specimen has an effect on the flame spread rating because a thicker material has different thermal insulating properties than a thin material. When interior finish materials are intended to be used in varying thicknesses, they must be tested in those thicknesses.

The means used to attach the test material to the roof of the test furnace also affects the flame spread rating. For example, if a thin layer of a combustible adhesive is applied to a noncombustible backing material such as concrete, it will create little additional hazard. If the same adhesive is used on a combustible backing, however, it will likely increase the hazard. For test results to be accurate, test specimens must be attached with the same materials and methods used in the actual installations.

Limitations of test findings. The methodology used in the tunnel test can be difficult to use for accurate evaluation of some materials. Some plastic materials, for example, can melt and drip onto the floor of the test furnace. This behavior produces one result under test conditions and very different results in actual field applications.

Floor Coverings

The flame spread rating developed in the tunnel test does not apply to floor coverings. If a floor covering such as carpeting is used for wall or ceiling finish, however, it must meet the same flame spread criteria as other wall and ceiling finishes.

The actual determination of the flame spread rating requires the test procedures of the ASTM E-84 tunnel test. Unfortunately for fire inspectors, there is no way the flame spread rating can be positively determined in the field. Relative surface burning characteristics can be assumed for some interior surfaces such as concrete block or plaster. Other materials, such as corrugated paper, can usually be assumed to have an objectionably high flame spread rating. On the other hand, the flame spread rating of many materials, especially composite materials, simply cannot be determined when encountered in the field unless the manufacturer can be identified and contacted. This can pose difficult enforcement problems for inspectors in the field.

Smoke Developed Rating

In addition to the flame spread rating, the tunnel test provides an additional measure of flammability: the *smoke developed* rating. The smoke developed rating is a measure of the relative visual obscurity created by the smoke from a tested material. It is measured by a photoelectric cell and a light source located at the end of the tunnel furnace. As with the flame spread rating, red oak is used as a standard testing material and has been assigned a smoke developed rating of 100. Therefore, under test conditions, a material with a smoke developed rating of 200 produces smoke that is twice as visually obscuring as red oak. Codes limit the maximum smoke developed to 450.

It is very important to remember that the smoke developed rating is *not* an indication of the toxicity or volatility of the products of combustion of the interior finish materials. The tunnel test does not detect or measure a completely transparent product of combustion such as carbon monoxide. Furthermore, the tunnel test does not measure the combined effects of heat, irritation, and toxicity.

Toxicity — Ability of a substance to do harm within the body.

Volatility — Ability of a substance to vaporize easily at a relatively low temperature.

The smoke developed rating is a measure of the visual obscurity caused by the smoke from a tested material. The rating does not indicate the toxicity or volatility of the material.

In the case history cited earlier, the products of combustion of the polyurethane to which the occupants were exposed involved more than the density of the smoke. The occupants were subjected to a synergistic effect involving several combustion products at high temperature.

Fire-Retardant Coatings

The flame spread rating of some interior finishes, most notably wood materials, can be reduced through the use of *fire retardant* coatings. Several types of fire-retardant coatings that may be available include the following:

Fire Retardant — Any substance, except plain water, that is applied to another material or substance to reduce the flammability of fuels or slow their rate of combustion by chemical or physical action.

- Intumescent paints
- Mastics
- Gas-forming paints
- Cementitious and mineral fiber coatings

Different types of coatings behave in different ways. For example, the intumescent paints expand upon exposure to heat to create a thick, puffy coating that insulates the wood surface from heat and excludes oxygen from the wood. The mastic coating forms a thick, noncombustible coating over the surface of the wood. Several fire-retardant coatings have been tested and listed by the testing laboratories.

Fire-retardant coatings are valid treatments for the reduction of surface burning when applied as directed but they are susceptible to misuse. They must be applied at a specified rate of square feet per gallon (square meter per liter) and may require more than one coat. They also may not have a permanent effect if used in exterior applications or environments with high humidity. In any case, products that have not been tested by a reputable laboratory should not be trusted.

The difference between surface flammability and fire resistance is sometimes misunderstood. For example, fire-retardant coatings only affect the coated surface. They do not affect the untreated back side of a panel. In addition, a material that is listed as a fire-retardant coating does not increase the fire resistance of structural components or assemblies unless it has also been tested and listed for use in a fire-resistive assembly. However, because fire retardant coatings can be field applied with a paint brush, an attempt may be made to use them outside their listings as an alternative to other required fire protection. For example, attempts have been made to substitute a fire-retardant coating for structural fireproofing.

Fire-retardant coatings only affect the coated surface and not the untreated portion of a material. They cannot be substituted for structural fireproofing.

Large-Scale Testing

The ASTM E-84 test procedure to measure the surface burning characteristics of materials (the tunnel test) is useful because it provides reproducible results and is a widely recognized standard. The limitations of the traditional tunnel test have been recognized and efforts have been made to improve upon this method.

Some materials will produce a fire hazard greater than indicated by the tunnel test when they are installed in an environment that approximates a real room. This increased hazard is especially true of foam plastic materials. There are two reasons for this increased hazard:

- Flame spread is generally different over a vertical surface than across a horizontal surface **(Figure 5.6)**.
- The walls and ceiling of a room provide for re-radiation of heat between the intersecting surfaces.

Considerable effort has been made over the years to develop test procedures that incorporate the size and shape of real rooms. These methods are collectively known as *corner tests.* Early corner tests consisted of a ceiling and two intersecting sidewalls. The walls and ceiling of the assembly were lined with the material to be tested. Different configurations have been used with different size walls and ceiling heights.

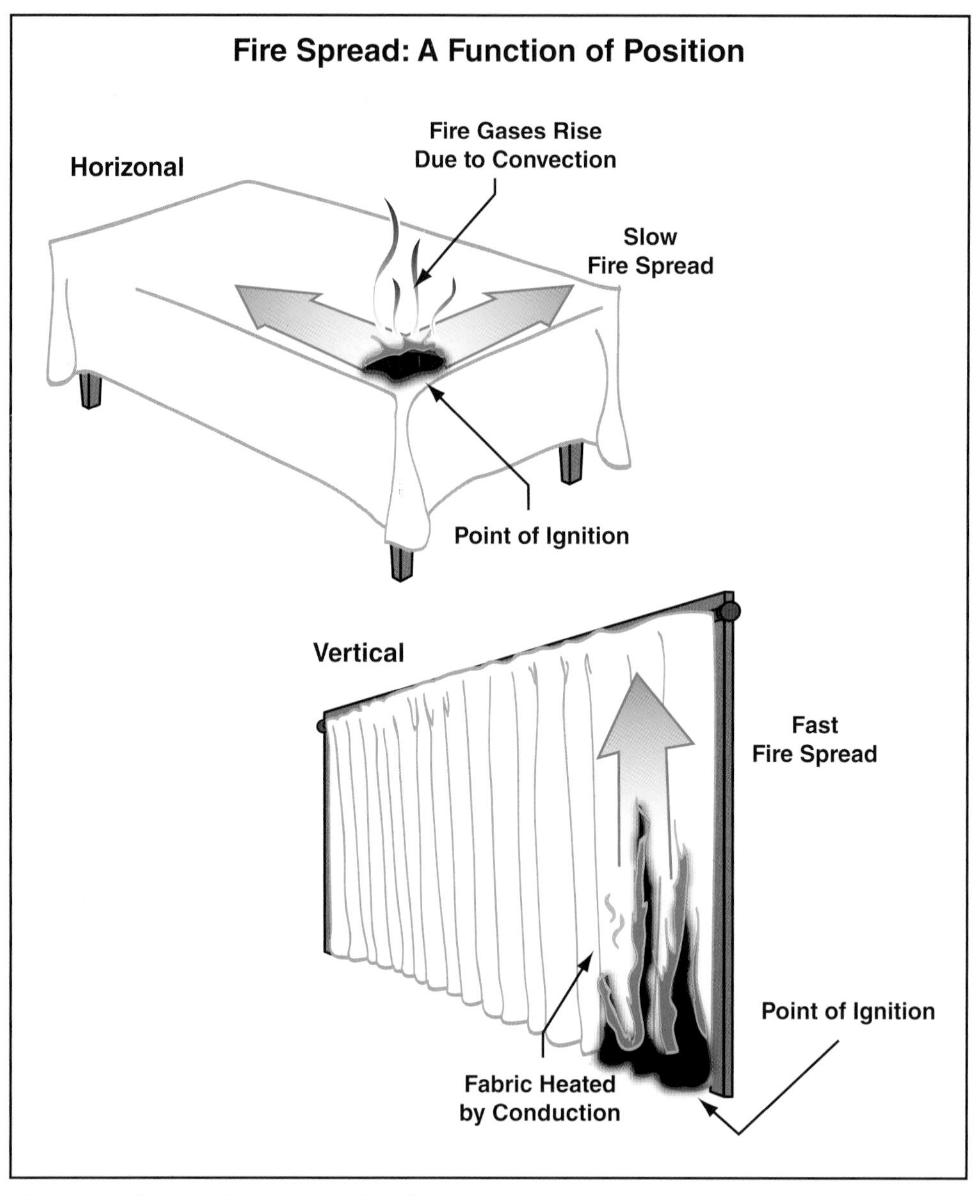

Figure 5.6 The position of a material affects the way it burns.

One large-scale test was developed for evaluating the fire performance of textile wall coverings. This test, known as NFPA® 265, *Standard Methods of Fire Tests for Evaluating Room Fire Growth Contribution of Textile Coverings on Full Height Panels and Walls,* was originally developed when carpet-like textiles began to be used as wall coverings. A more recently developed large-scale test is NFPA® 286, *Fire Tests for Evaluating Contribution of Wall and Ceiling Interior Finish to Room Fire Growth.* This test has been developed to handle materials that may not remain in place during the tunnel test, such as plastic materials that may melt and drip.

Both the 265 and 286 tests use a room enclosure as shown in **Figure 5.7**. The material to be tested is placed on three of the walls. (The surface of the wall containing the door opening is not covered). In the 286 test, the material is also placed on the ceiling if it is intended to be used in that manner. The materials are subjected to two different-size gas flames.

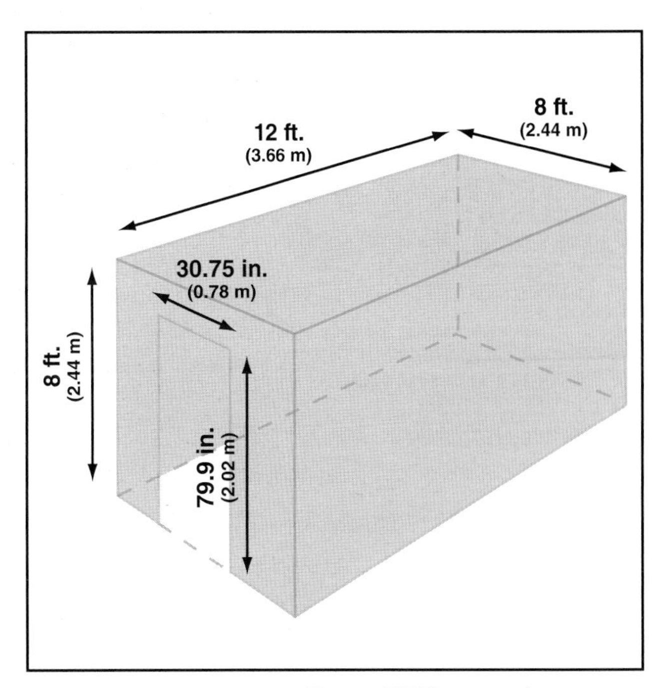

Figure 5.7 Room enclosure dimensions for the interior finish test.

These tests do not provide a numerical test result such as the flame spread rating derived from the tunnel test. Instead, the test material is judged either satisfactory or unsatisfactory depending on the extent of fire growth that occurs within the test room upon exposure to the two gas flames. For example, the *International Building Code* allows interior finish materials to be tested in accordance with NFPA® 286 instead of the tunnel test. However, the code then establishes specific acceptance criteria including the following:

- Noting whether or not the flames spread to the ceiling of the test chamber
- The flame cannot spread to the outer extremity of the sample
- Flashover cannot occur
- The peak rate of heat release cannot exceed 800 kW
- A maximum amount of smoke released

Fire and Smoke Containment

In the field of building fire protection there are two basic concepts that complement each other: active fire protection and passive fire protection. *Active fire protection* consists of equipment such as an automatic sprinkler system or fire alarm system that requires a power source for operation. This power source may be manual, such as is the case with portable fire extinguishers. *Passive fire protection* relies on building construction and materials to contain fire or products of combustion. Fire walls or stair enclosures are examples of passive fire protection. In another sense, active fire protection extinguishes or controls a fire while passive fire protection limits the spread of fire.

During construction, there are several main methods used to achieve passive fire protection. Chief among these are compartmentation by the use of fire walls, fire partitions, enclosure and shaft walls, and fire doors.

Compartmentation Systems — Series of barriers designed to keep flames, smoke, and heat from spreading from one room or floor to another; barriers may be doors, extra walls or partitions, fire-stopping materials inside walls or other concealed spaces, or floors.

Rated Assembly — Assemblies of building components such as doors, walls, roofs, and other structural features that may be, because of the occupancy, required by code to have a minimum fire-resistance rating from an independent testing agency. Also called Labeled Assembly.

Shelter in Place — Having occupants remaining in a structure or vehicle in order to provide protection from a rapidly approaching hazard (fire, hazardous gas cloud, etc...). Opposite of evacuation.

Compartmentation

The subdivision of a building or the floor levels of a building by fire-rated walls or partitions is generally referred to as *compartmentation*. In some occupancies, such as hotels with many rooms, a degree of compartmentation is inherent in the architecture of the building. In other occupancies, such as a large warehouse, there is little or no compartmentation. A lack of compartmentation can result in a rapid spread of fire horizontally and vertically through a building. Examples of avenues for fire spread are open vertical shafts and undivided attics.

Building codes contain explicit requirements for fire-rated walls and partitions in various occupancies. Typical requirements include corridor enclosures and vertical shaft enclosures. **Figure 5.8** shows a simple floor plan with typical required fire-rated partitions. The fire-rated floor and ceiling assemblies required by building codes in multistory buildings provide not only structural fire resistance but also act to prevent vertical spread of fire.

Fire-rated partitions can provide areas of refuge for occupants when immediate or rapid evacuation is not possible, such as in a hospital. Codes typically require fire-rated partitions to subdivide patient floors so patients can be moved from the area of fire origin to a protected part of the floor. This concept of providing an area of refuge is referred to as *defending-in-place* or *sheltering in place*. Occupants of high-rise residential buildings have survived fires simply by remaining in another apartment separated from the fire by fire-rated partitions. This protection, of course, requires that intervening doors be closed **(Figures 5.9 a and b)**.

Almost any floor and ceiling assembly or wall construction will act as a barrier to fire to some degree, but not every wall or partition in a building is fire rated. Partitions separating individual rooms within an apartment, for example, are not fire rated. However, assemblies that have a fire-resistive rating as described in Chapter 2 have the advantage of providing a known level of protection. The degree of fire resistance required of a wall or partition will depend on its purpose. Building code requirements are changing and requirements for compartmentation are being seen less and less.

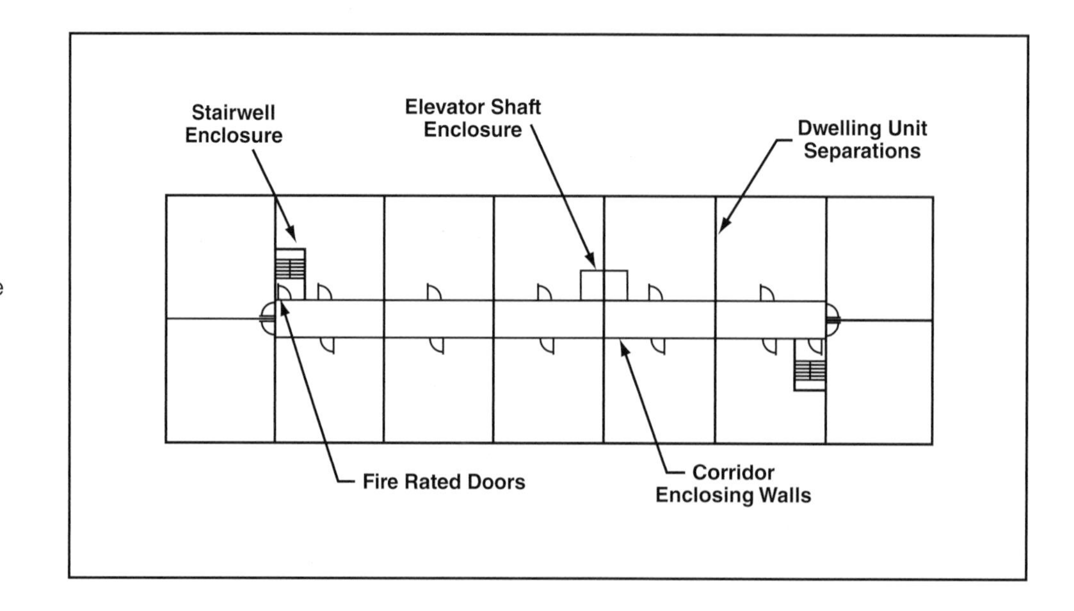

Figure 5.8 Typical location of fire-rated partitions in the upper floor of a residential occupancy.

Figures 5.9 a and b A — Apartment side of a corridor door in a nonsprinklered building closed at the time of a fire. B — When the door is opened, note the extent of fire exposure from the corridor side. *Courtesy of Ed Prendergast.*

Figure 5.10 A fire wall blocked the spread of fire in a warehouse building. Note destruction beyond the opening in the fire wall. *Courtesy of Ed Prendergast.*

Fire Walls

Fire walls, or area separation walls, are erected to limit the maximum spread of fire. They are constructed with sufficient fire resistance and structural stability to act as an absolute barrier to a fire under conditions of total burnout on either side **(Figure 5.10)**. Fire walls subdivide a building into smaller areas so that a fire in one portion of a building is limited to that area and does not destroy the entire building. For example, fire walls could divide a 100,000 square foot (9 290 sq m) warehouse into 25,000 square foot (2 323 sq m) areas.

Fire Wall — Fire-rated wall with a specified degree of fire resistance, built of fire-resistive materials and usually extending from the foundation up to and through the roof of a building, that is designed to limit the spread of a fire within a structure or between adjacent structures.

The containment of a fire greatly reduces potential economic loss and can enable a stricken industry to recover more quickly. Fire walls can also separate various functions within a plant so that the loss of one area will not result in total loss of the facility. For example, a hazardous chemical mixing operation can be separated from shipping, warehousing, and other departments of a chemical factory.

Fire walls are not popular with designers because they increase structural costs and may interfere with the free movement of material through a factory or warehouse. Fire walls can also be architecturally unattractive in occupancies such as shopping malls or airline passenger terminals where an expansive interior is desired. Developers may circumvent requirements for fire walls by resorting to free-standing structures, like single-family residences, that are separated but built very close to one another. Some single-family homes can be as close as 18 inches (450 mm) apart.

Building codes typically allow elimination of fire walls when a building is equipped with an automatic sprinkler system if it meets criteria for occupancy, height, and other code requirements. It is not unusual, therefore, to find very large industrial, warehouse, and mercantile facilities constructed without fire walls.

A properly constructed and maintained fire wall is an important ally to tactical fire fighting forces. Fire walls act to contain a major fire in situations where available fire fighting resources would otherwise be unable to stop a fire. When a section of a building on one side of a fire wall becomes heavily involved, the fire wall is a natural line along which to establish a defense. With fire doors closed, one or two handlines can be positioned to check for any spread of fire at cracks or around door edges. This can be accomplished with a minimum of personnel, freeing other firefighters to protect exposures or to attack the main body of fire.

Firefighters can use an opening through a fire wall as a protected vantage point from which to attack the main body of fire **(Figure 5.11)**. Great care must be exercised when opening fire doors under these circumstances. If the situation becomes untenable and firefighters are forced to withdraw, fire doors must be closed.

Under severe fire conditions, structural collapse must be anticipated on either side of a fire wall; therefore, fire walls must meet special structural requirements. Fire walls can be constructed either as "freestanding walls" or as "tied walls." Freestanding walls are self-supporting and are independent of the building frame **(Figure 5.12)**. Freestanding fire walls are usually found in buildings of wood-frame or wood-joisted masonry (Type III or IV) construction, although they may also be used in noncombustible buildings. Freestanding walls must be designed to resist a lateral load of at least 5 pounds per square foot (.24 kPa) and are self-supporting with respect to vertical loads. They are independent of the basic building frame although the building frame may provide some horizontal support.

Figure 5.11 Firefighters can attack a fire from this vantage point and close the door if they need to withdraw.

Figure 5.12 A commonly seen fire wall. *Courtesy of Dave Coombs.*

Tied fire walls are erected at a column line in a building of steel-frame or concrete frame construction. In a steel-frame building, any steel members, such as columns, that may be incorporated into the fire wall must be provided with the same degree of fire resistance required for the fire wall itself. The structural framework must have sufficient strength to resist the lateral pull of the collapse of framework on either side.

Originally fire walls were required to have a fire-resistive rating of four hours, but recent building codes permit fire walls with fire-resistive ratings of 2, 3, or 4 hours, depending on the occupancy. A 4-hour rated firewall usually must be constructed of masonry or concrete and have 3-hour-rated openings. In most buildings fire walls must be constructed of noncombustible materials. Building codes may contain exceptions in this regard for wood-frame (Type V) buildings. Fire walls with lesser fire ratings can be constructed of other fire-resistive materials.

The *International Building Code* (IBC) also permits combustible structural members to be framed into a masonry or concrete fire wall from opposite sides provided there is a 4-inch (100 mm) separation between the ends of the structural members **(Figure 5.13)**.

Fire walls must extend beyond walls and roofs to prevent the radiant heat of flames on one side of a fire wall from igniting adjacent surfaces. This is accomplished by continuing the fire wall through the roof with a parapet. The parapet height above a combustible roof is determined by the building code and varies from 18 to 36 inches (450 mm to 900 mm). Some building codes contain exceptions that permit the elimination of parapets under certain conditions. An example of this is where a fire wall can terminate at the underside of a noncombustible roof that has a covering of low combustibility (Class B).

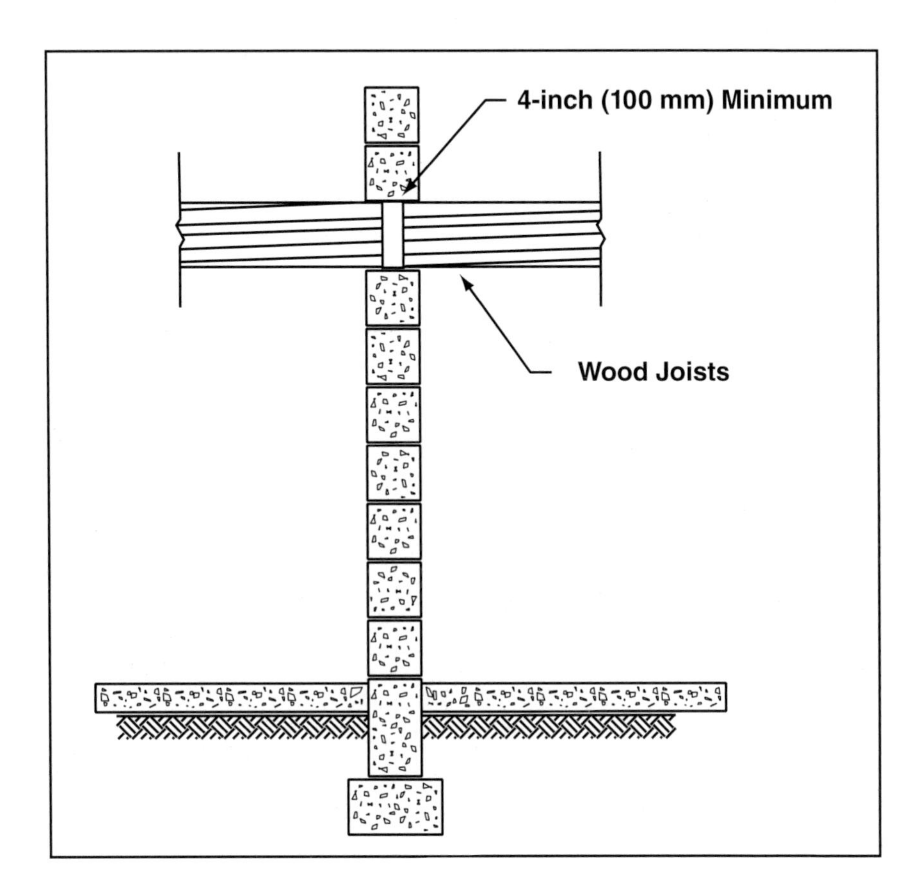

Figure 5.13 Allowable framing of combustible structural members into a fire wall.

Fire walls of the type shown in Figure 5.12 are easily identifiable. Their presence can be noted and incorporated into tactical fire fighting decisions. However, non-free-standing firewalls that do not have parapets are not readily identifiable from the outside of a building. This is particularly true for fire walls with a 2-hour rating that can be constructed of materials other than masonry.

Fire Partitions

Fire Partition — Fire barrier that extends from one floor to the bottom of the floor above or to the underside of a fire-rated ceiling assembly. A fire partition provides a lower level of protection than a fire wall. An example is a one-hour rated corridor wall.

Fire partitions are interior walls used to subdivide a floor or area of a building that do not qualify as fire walls. Fire partitions may not extend continuously through a building. A fire partition is usually erected from a floor to the underside of the floor above. Fire partitions are typically not required to have as much fire resistance as a fire wall. For example, partitions enclosing an exit corridor may have a 1-hour fire rating.

Fire partitions can be constructed from a wide variety of materials including lath and plaster, gypsum wallboard, concrete block, and combinations of materials. The material chosen depends on the required fire resistance and the construction type of the building. For example, the partition wall separating adjacent units in an apartment building may be required to have a 1-hour fire resistance. A common method used to accomplish this in a fire-resistive structure is to use ⅝ inch (16 mm) fire-rated gypsum board applied to both sides of 2½-inch (64 mm) steel studs. Such an assembly may or may not be used for a load-bearing application.

Figure 5.14 Common use of fire-rated glazing in a hospital.

If the ⅝ inch (16 mm) gypsum wallboard were applied to both sides of 2 x 4-inch (50 mm x 100 mm) wood studs, the partitions would have a 1-hour fire resistance and could be used a for load-bearing partition in a Type III or Type V building. The wood stud partition could not be used for a load-bearing partition in a fire-resistive building (Type I or Type II) because the structural components in fire-resistive building are themselves required to be fire-resistive.

Fire-rated glazing can be used for fire partitions where visibility is desired and a fire rating is required. A typical application is a corridor separation in a healthcare facility that permits the staff to observe patients and still have a protected exit passage **(Figure 5.14)**.

Enclosure and Shaft Walls

Enclosure walls are used to enclose such vertical openings as stairwells, elevator shafts, and pipe chases that extend from floor to floor in a building. The purpose of enclosure walls is to block the vertical spread of fire through a building and, in the case of stairwells, to protect a means of egress. The construction of enclosure walls is similar to that of partition walls. The main difference between the two designations is in their function.

Enclosure walls are required to have a fire-resistance rating of one or two hours depending on the height of the building. Stairwells in buildings three stories or less, for example, are required to have a 1-hour enclosure. In buildings taller than three stories, they are required to have 2-hour enclosures. Enclosure walls are usually non-load bearing, although load-bearing masonry stair enclosures are found in older mill buildings. Most common construction materials are used for enclosure walls. These walls can be constructed of gypsum board with steel or wood studs, lath and plaster, or concrete block **(Figure 5.15)**. Hollow clay tile enclosure walls can be found in older fire-resistive buildings.

Figure 5.15 Enclosure walls must have the appropriate degree of fire resistance to serve as a fire barrier. *Courtesy of McKinney (TX) Fire Department.*

Windows with fire-rated glazing can be used with stair enclosures. As in the case of partition walls, the use of fire-rated glazing provides a fire barrier while permitting observation of the stair enclosure, which can enhance security.

Adequate protection of vertical openings is especially important because of the natural tendency of products of combustion to rise vertically through a building by means of stairwells and shafts.

Glazing — Glass or thermoplastic panel in a window that allows light to pass.

Buildings were once commonly provided with light shafts or interior courts for light and ventilation **(Figure 5.16)**. Windows from interior rooms would open into a light shaft to facilitate ventilation. Light shafts could be provided in an individual building but were frequently provided between adjacent buildings. These light shafts can prove troublesome for firefighters because they can provide a means of vertical communication of fire from window to window.

Figure 5.16a and b Light shafts are designed to let in natural light. Like other vertical shafts, they can serve as an avenue for the spread of fire. *Courtesy of Gregory Havel, Burlington, WI.*

Curtain Walls

When a building is constructed using a structural frame for its main structural support, the exterior wall functions only to enclose the building and is known as a curtain wall. The design function of the curtain wall is to separate the interior environment from the exterior environment. To that extent, the curtain wall must resist wind, rain, and snow. Curtain walls must also be designed to control heat loss, noise transmission, and solar radiation.

Curtain Wall — Nonbearing exterior wall attached to the outside of a building with a rigid steel frame. Usually the front exterior wall of a building intended to provide a certain appearance

The concept of the curtain wall came into existence with the development of the steel-framed high-rise building. Because the main structural support was provided by the frame, the exterior wall did not need to be load bearing. In fact, the exterior wall itself could be supported by the frame at each floor of the building. Curtain walls are not limited to buildings with steel frames. They are also frequently used in buildings with concrete frames.

NOTE: Refer to Chapter 10, Concrete Construction, for more information on concrete buildings.

Curtain walls are often constructed using a combination of glass and steel, stainless steel, or aluminum **(Figure 5.17)**. They may also be constructed with lightweight concrete, plastic, fiberglass, and a variety of metal panels with core materials such as expanded paper honeycombs and compressed glass fiber.

Figure 5.17 A curtain wall in a commercial building. *Courtesy of Ed Prendergast.*

Two aspects of curtain walls are significant to the firefighter: their degree of fire resistance and the extent to which they permit vertical communication of fire. Because a curtain wall is non-load-bearing, it lacks the inherent fire resistance provided by a more massive load-bearing wall. Some curtain wall assemblies, such as those made of aluminum and glass, have no fire resistance. Nonetheless, building codes may require that exterior walls – including curtain walls – have some degree of fire resistance to reduce the communication of fire between buildings. The required fire resistance depends on the separation distance between buildings and the building occupancy.

It is not uncommon for such structures as office buildings and high-rise apartments to be constructed with curtain walls that are noncombustible but have no fire resistance. This is possible because they are either spaced far enough from other buildings or face an open area such as a street so that the building code does not require any fire resistance rating.

Fire Stop — Solid materials, such as wood blocks, used to prevent or limit the vertical and horizontal spread of fire and the products of combustion in hollow walls or floors, above false ceilings, in penetrations for plumbing or electrical installations, in penetrations of a fire-rated assembly, or in cocklofts and crawl spaces.

Curtain walls are usually supported by the building frame at the edge of the floor assembly **(Figure 5.18)**. It is not unusual for a gap to be created between the edge of the floor and the curtain wall. This opening may be several inches wide and will provide a path for communication of fire up the inside of the curtain wall. To prevent this occurrence, suitable fire stopping must be provided to maintain the continuity of the floor as a fire-resistive barrier.

Curtain walls that are not fire-resistive frequently extend from the floor of one level to the ceiling. If a room becomes heavily involved in fire, it is possible for the flames to overlap the edge of a fire-resistive floor slab and expose the story above. It is also possible for a fire to communicate vertically up the outside of a building by this means **(Figure 5.19)**.

Fire Doors

For a fire wall or partition to be effective it must provide a continuous barrier to fire. In the real world, however, it is necessary for people and equipment to be able to pass through walls and partitions. A stairwell enclosure

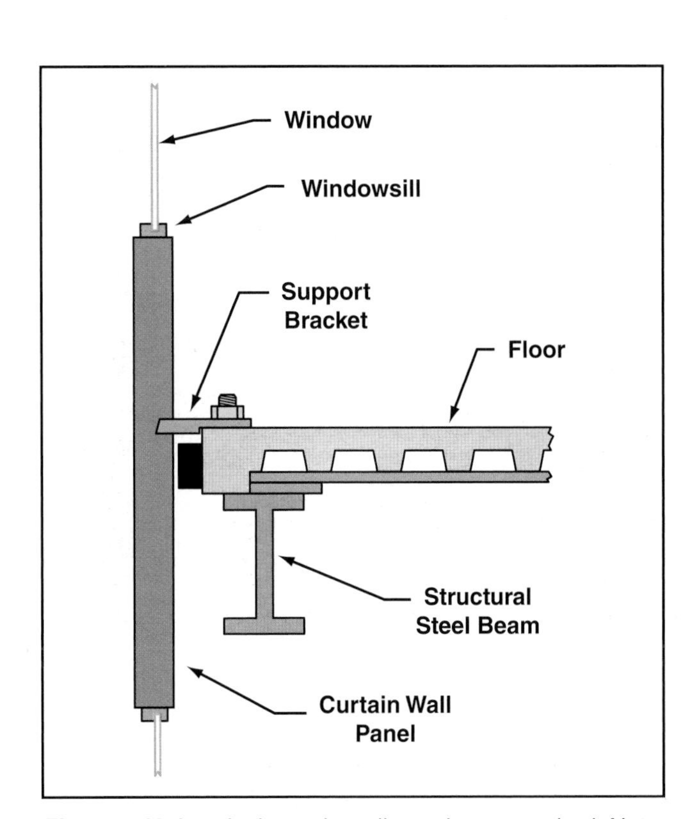

Figure 5.18 A typical curtain wall attachment method. Note the gap created between the curtain wall and the building. The support brackets are not usually continuous and therefore do not create any effective firestop.

Figure 5.19 Without adequate firestopping, fire can communicate up the side of a building *Courtesy of Ron Jeffers.*

Figure 5.20 A fire door in an older mill-style building. *Courtesy of McKinney (TX) Fire Department.*

Figure 5.21 Some fire codes require an automatic fire door that covers the elevator door when activated.

Fire Door — A specially constructed, tested, and approved fire-rated door assembly designed and installed to prevent fire spread by automatically closing and covering a doorway in a fire wall during a fire to block the spread of fire through the door opening.

would defeat the purpose of the stairwell if occupants could not get *into* the stairwell; therefore, openings must be provided through fire walls and fire partitions. Protection must be provided for these openings.

The most common means for protecting openings through fire-rated walls is by the use of *fire doors*. The use of fire doors to block the spread of fire is an established fire protection technique; they can be found in industrial buildings dating back to the end of the nineteenth century **(Figure 5.20)**. As with other aspects of passive fire protection, fire doors do not prevent or extinguish fires. When they are maintained and operated properly, however, fire doors are very effective in limiting total fire damage. Fire doors differ from ordinary or non-fire doors in their construction, their hardware, and the extent to which they may be required to close automatically **(Figure 5.21)**.

Fire Door Classifications

To effectively block the spread of fire, a fire door must have some degree of fire resistance similar to fire-rated walls. Lightly constructed panel doors or glass doors cannot act as a barrier to the high temperatures developed in a fire. Furthermore, doors constructed of different materials will have differing degrees of fire resistance.

A fire door is rated for its fire resistance in a manner somewhat similar to that used for fire-resistive structural assemblies (See Chapter 2). Fire doors are rated as 4 hours, 3 hours, 1½ hours, 1 hour, ¾ hour, ½ hour, and 20 minutes. Existing fire doors, however, have been classified by using a letter designation, an hourly rating, or both. Thus, a fire door may be classified either A, B, C, D, or E. It may also be classified with an hourly rating as a "3-hour" door.

The classification of a fire door is known as its "fire protection rating" and is *not* to be confused with the fire-resistance rating used to describe structural components.

The letter designations that have been used to classify fire doors actually were meant to describe the type of opening for which a door was intended, but historically the letters have been used to describe the door itself. Although the letter designations are no longer used, they are included here because they may be encountered in existing buildings. The letter designations are as follows:

- Class A - Openings in fire walls
- Class B - Openings in vertical shafts and openings in 2-hour-rated partitions
- Class C - Openings between rooms and corridors having a fire resistance of 1 hour or less
- Class D - Openings in exterior walls subject to severe fire exposure from the outside of a building
- Class E - Openings in exterior walls subject to moderate or light exposure from the outside

Thus, a fire door classified as a Class B door is intended to protect the openings in stairwells. A fire door may also be found with a combination classification, such as a Class B 1½-hour rating. This designation means that the door is intended to protect an opening in a vertical shaft and has a 1½-hour rating. The ½-hour and ⅓-hour doors are primarily used in smoke barriers and openings to corridors.

Building codes determine the fire protection rating of fire doors that are required for various openings. Codes typically require 4- or 3-hour rated doors in fire walls of greater than a 2-hour rating. Doors rated at 1½ hours are normally required for 2-hour rated vertical enclosures. One-hour doors are used for 1-hour vertical shaft enclosures and exit enclosures.

A few apparent inconsistencies may be encountered in this area of fire protection and the fire official should be aware of them. For example, a code may permit an opening in a 2-hour stairwell enclosure to be protected with a 1½-hour fire door rather than a 2-hour door. A code may also require two 3-hour fire doors to protect an opening in a 4-hour wall and not permit a 3-hour door to be used in combination with a 1½-hour door to satisfy the requirement.

Some of these inconsistencies rest on the assumption that the fire exposure at a door opening is decreased by the clear space necessary to maintain access to the opening. In other words, because combustible materials will not be piled against a door, the fire rating of a door can be less than that of the wall. The requirement for two 3-hour doors on a 4-hour fire wall may at first seem like overkill. However, some authorities view it as a means for increasing the reliability of the protection for the wall opening in that the requirement increases the probability that at least one of the doors will close.

Testing of Fire Doors

Fire doors are tested in accordance with the procedures contained in NFPA® 252, *Standard Methods of Fire Tests of Door Assemblies,* which is also designated ASTM E-152. The test procedure uses a furnace to expose the fire doors to the same time-and-temperature curve used to establish the fire resistance rating of structural assemblies. It should be noted that the conditions for passing the test for door assemblies are not as rigid as those required for fire-rated walls.

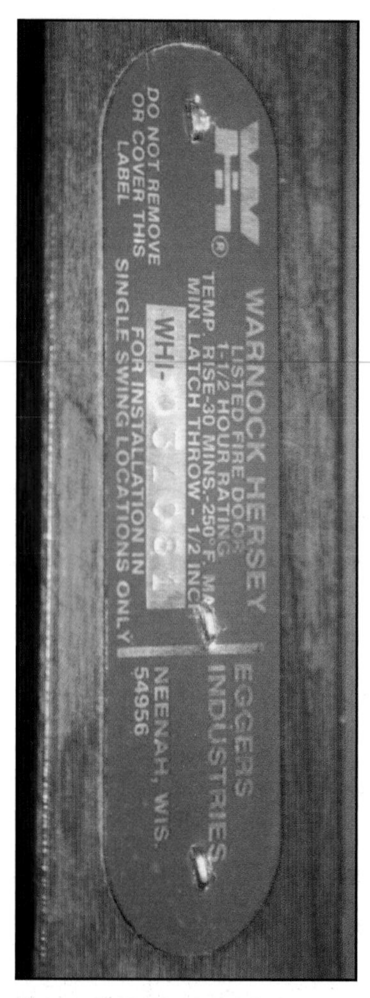

Figure 5.22 The label for a swinging fire door is placed on the edge of the door. *Courtesy of Ed Prendergast.*

Criteria for acceptability. For fire doors, the primary criterion for acceptability is that the fire door must remain in place during the test. Some warping of the door is permitted as well as intermittent passage of flames after the first 30 minutes of the test. There is also no maximum surface temperature rise on the unexposed side of the door for most of the doors tested. In fact, metal doors may actually glow red from the heat of the test fire. The fact that fire doors are allowed to get very hot reinforces the importance of keeping combustibles away from the immediate vicinity of fire doors in fire walls if they are kept permanently closed. If an opening through a fire wall is no longer needed for passage, the opening should be filled in with construction materials equivalent to the wall.

The second criterion for acceptability is that the fire door assembly must remain in place when subjected to a hose stream immediately following the fire test. The use of a hose stream subjects the door assembly to cooling and impact effects that might accompany fire fighting. Doors with a ⅓-hour rating may not be subjected to the hose test depending on their intended application.

Fire doors are tested by many of the same laboratories that are involved in the testing of building materials. Underwriters Laboratories, Inc. publishes a list of fire doors that have passed the test in the *Fire Resistance Directory.*

Identifying rated fire doors. Rated fire doors are identified with a label indicating the door type, hourly rating, and the identifying label of the testing laboratory **(Figure 5.22)**. The fire door labels facilitate the identification of fire doors by building and fire inspectors in the field. It is not uncommon for labels to be painted over during building maintenance and there has also been at least one case of counterfeit laboratory labels appearing in the field.

Fire Door Frame and Hardware

For a fire door to effectively block the spread of fire, it must remain closed and attached to the fire wall under fire conditions. Therefore, a fire door must be equipped with hardware that holds the door closed under the stresses of fire exposure. In addition, when fire doors are installed in a frame, the frame must also withstand exposure to a fire. The testing of fire doors includes the frames as well as the hardware, which are also listed by the testing laboratories for use with fire doors.

The hardware used on fire doors is referred to as either "builder's hardware" or "fire door hardware." Builder's hardware is applied to swinging fire doors and includes such items as hinges, locks and latches, bolts, and closers **(Figure 5.23)**. Builder's hardware can be shipped to a job site separate from the fire doors.

Fire door hardware is used on both sliding and swinging fire doors. Typical fire door hardware components are illustrated in **Figure 5.24**. Fire door hardware is normally shipped with the fire doors.

Glazing and Louvers in Fire Doors

A fire door is often provided with vision panels. The vision panels enhance safety and security by permitting observation through a closed door. Any glazing material used in a fire door is fire rated. This glazing is often not provided by the door manufacturer and may be installed at the jobsite by a glazing con-

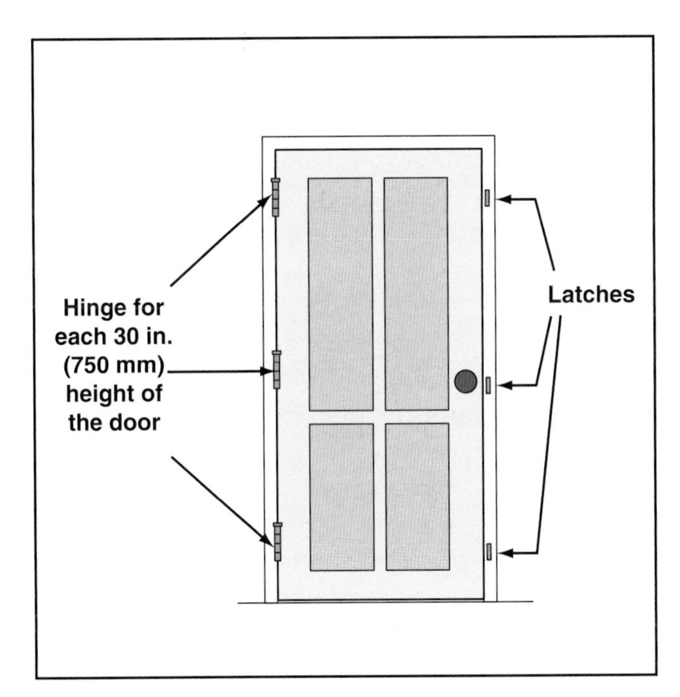

Figure 5.23 Builder's hardware examples.

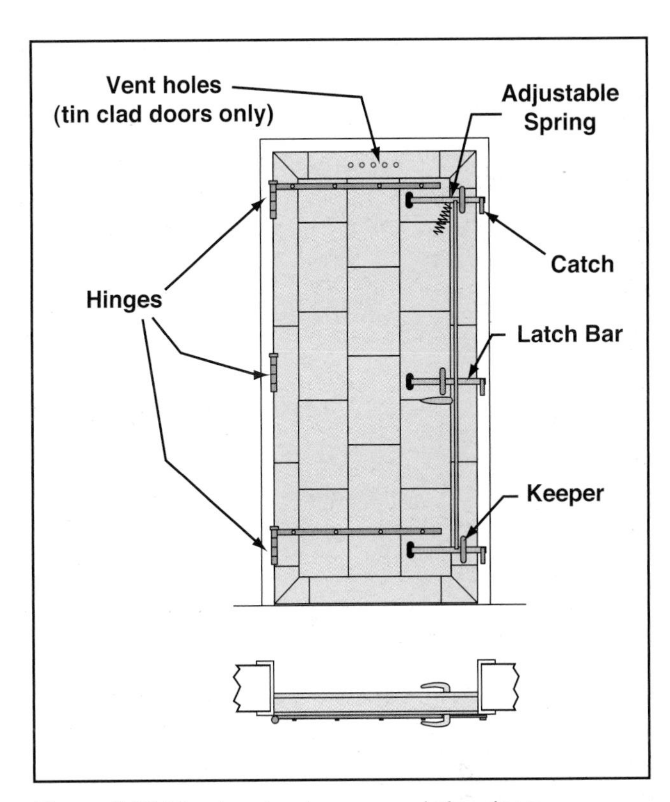

Figure 5.24 Fire door hardware on existing door.

tractor. NFPA® 80, *Standard for Fire Doors and Fire Windows,* requires that each piece of glass installed in a fire door have a listing mark that is visible after installation.

At one time the only fire-rated glazing available was wired glass **(Figure 5.25)**. Wired glass is a sheet of glass in which a steel wire net has been imbedded. The steel net distributes the heat throughout the glass and helps hold the glass in place. Wired glass will crack when exposed to a fire but will remain in place until it begins to soften and fall out.

In recent years fire-rated glazing has been developed that does not rely on imbedded steel wire. These products can provide higher hourly fire ratings than wired glass. They are somewhat more visually appealing and can be used as windows or sidelights in fire-rated walls as well as in fire doors. Some of the recently developed fire-rated glazing is impact-resistant, so can also be used for security purposes.

Figure 5.25 At one time, wired glass was the only product that provided the required fire resistance in a fire door.

Until recently, only fire doors with ratings up to 1½ hours were permitted to have glass panels but now doors with ratings up to 3 hours can be equipped with them. There are restrictions on the allowable area of glass in fire doors. Fire doors with ratings of 1, 1½, and 3 hours can have glass panels up to 100 square inches (64,500 sq mm) in area per door. Fire doors with ratings of ¾ hour can have a total glass area consistent with their listing, but an individual piece cannot exceed 1,296 square inches (836,100 sq mm). Fire doors with ratings of ½ or ⅓ hour can have fire-rated glass up to the maximum area to which they were tested.

It is also sometimes desirable to install louvers in a fire door to permit ventilation while the door is closed, such as in the case of a furnace room enclosure. The louvers in a fire door must close in case of fire to protect the opening. Usually, louvers are closed by means of a fusible link. Swinging fire doors with ratings up to 1½ hours can be equipped with louvers. The louvers themselves may not be produced by the door manufacturer and are listed separately by the testing laboratories. Louvers cannot be arbitrarily installed in fire doors. Only those fire doors that are listed for the installation of louvers can have louvers installed.

Fire Door Closing Devices

To perform its function, a fire door must be closed when a fire occurs. However, it is frequently desirable for fire doors to remain open or to be readily openable to allow for ordinary movement of building occupants. Fire doors can be either automatic or self-closing. An automatic door is normally held open and closes automatically under fire conditions when an operating device is activated. A self-closing door is normally closed and will return to the closed position if it is opened and released. The devices that operate fire doors include door closers, door holders, and door operators.

- ***Fire Door Closers*** — A fire door closer is used for either sliding or swinging fire doors. It can incorporate a hold-open device or can be self-closing. **Figure 5.26** illustrates one type of swinging fire door closer; it incorporates a fusible link device that holds the door open and releases the door under fire conditions. Self-closing fire door closers are commonly used for such applications as stairwell doors and doors that separate hotel rooms from corridors. One commonly used self-closer uses a spring hinge to close the door when released.
- ***Door Holders*** — A door holder can be used with swinging, sliding, or rolling fire doors. It is intended to be used with a suitable door closer **(Figure 5.27)**. The electromagnetic door holder can be used in conjunction with a smoke detector that releases the holder. This arrangement is very useful in areas with a large volume of traffic such as school stair enclosures. Having fire doors held open prevents the practice of blocking the doors in an open position. The smoke detectors are sufficiently sensitive that the doors quickly close under fire conditions. Electromagnetic holders are often used in health care occupancies where they can be released by operation of a fire alarm system.
- ***Door Operators*** — A door operator is intended for use with sliding fire doors that are mounted on either a level or inclined track. This device consists of an electric operator that opens and closes the door for normal use. Under fire conditions, a fusible link disconnects the door from the operator and allows it to close by means of a spring-powered door closer or a system of suspended weights.

For a fire door that is normally in the open position to close, some type of operating device must first sense a fire or smoke from a fire. The oldest and simplest device is a fusible link that melts from the heat of a fire. A fusible link has the advantages of being inexpensive, relatively rugged, and easy to maintain. A disadvantage is that it is slower to operate than devices that react to smoke or rise in the rate of temperature. A significant amount of smoke may flow through a door opening before a fusible link can release a fire door.

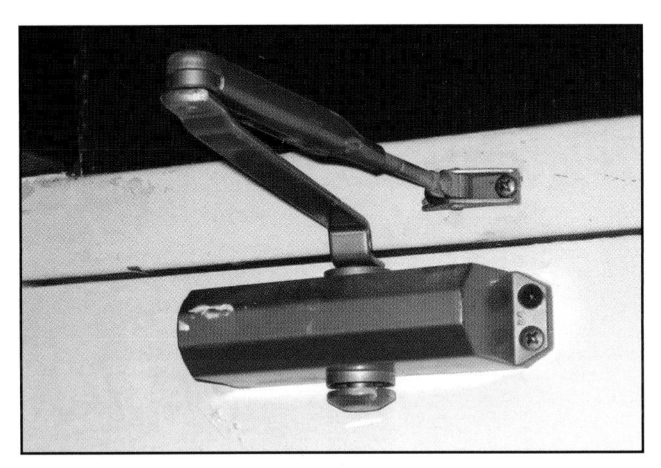

Figure 5.26 A typical self-closing swinging fire door mechanism found on stairway doors.

When a smoke detector is used to operate a fire door, the door closes more quickly. It also permits easy testing of the fire door. A smoke detector costs more and requires periodic cleaning. As with all smoke detectors, they must be properly positioned with respect to dead air spaces or ventilation ducts.

Types of Fire Doors

There are a number of types of fire doors designed for use in a particular setting. Among them are rolling doors, overhead doors, horizontal doors, as well as other types.

Figure 5.27 A typical arrangement for electromagnetic door holders.

Rolling steel fire doors. An overhead rolling steel fire door is often used to protect an opening in a fire wall in an industrial occupancy. It is also frequently used at the opening in a wall separating buildings. An overhead rolling steel fire door may be used on one or both sides of a wall opening. One architectural advantage of an overhead rolling fire door is that it is relatively inconspicuous and out of the way.

Rolling steel fire doors are constructed of interlocking steel slats with other operating components including a releasing device, governor, counterbalance mechanism, and wall guides **(Figure 5.28, p. 176)**. An overhead rolling door ordinarily closes under the force of gravity when a fusible link melts but motor-driven doors are also available **(Figure 5.29, p. 176)**.

Overhead rolling doors may be found installed across corridors. Without a conventional swinging door at the same location, the overhead door may create a dangerous dead-end corridor when it closes.

Overhead doors can also create a dangerous condition for firefighters who may not be able to see the door installation through heavy smoke. Firefighters advancing on a fire must use care when moving through an opening protected with overhead rolling fire doors to avoid being struck by a closing door. In addition, a door that closes after firefighters have passed through it can trap firefighters in the fire area, cutting off their escape path, restricting water through pinched hoselines, and disorienting members who do not realize that the door came down across the corridor behind them.

CAUTION

Firefighters must take caution that overhead doors do not close behind them and restrict movement of personnel or equipment.

Figure 5.28 A typical overhead fire door in a warehouse. *Courtesy of McKinney (TX) Fire Department.*

Figure 5.29 A motor-driven overhead door. *Courtesy of Ed Prendergast.*

Horizontal sliding doors. Horizontal sliding fire doors are often found in older industrial buildings **(Figure 5.30)**. These doors are usually held open by a fusible link and slide into position along a track either by gravity or by the force of a counterweight.

A horizontal sliding door can be constructed of several different materials. A common type of sliding door is a metal-covered, wood-core door. The wood core provides thermal insulation while the sheet metal covering protects the wood from the fire. Because wood undergoes thermal decomposition when exposed to heat, a vent hole is usually provided in the sheet metal to vent the gases of decomposition.

The metals used to cover the wood core include steel, galvanized sheet metal, and *terneplate,* which is a metal composed of tin and lead. Smooth galvanized sheet metal is used on wood-core doors known as "kalamein doors." Fire doors made with galvanized steel or terneplate are commonly referred to as "tin clad" doors although strictly speaking the metal used is not pure tin.

Swinging fire doors. It is a very common design practice to use swinging fire doors for such applications as stairwell enclosures and corridors. Although a swinging fire door has the disadvantage of requiring a clear space around the door to ensure closure, it is a good choice where the door is frequently closed and provision must be made for pedestrian traffic.

An awkward architectural situation can arise where it is necessary to provide a fire door on either side of a wall and the swinging doors are in an exit path so that they are required to swing in the direction of exit travel **(Figure 5.31)**. This can be accomplished by making use of a vestibule between the doors that is of fire-resistive construction. A pair of fire doors may be installed to close off a corridor at a fire wall, with one door swinging in each direction to accommodate exit travel in both directions. These are used in some health care facilities, apartment buildings, and schools.

Swinging fire doors are available with ratings of 3 hours to 20 minutes. Just as with sliding doors, swinging doors can be constructed from a variety of materials, including the metal-clad wood style shown in **Figures 5.32 a and b**. Many companies in the US produce listed swinging fire doors, so a large variety of them are available.

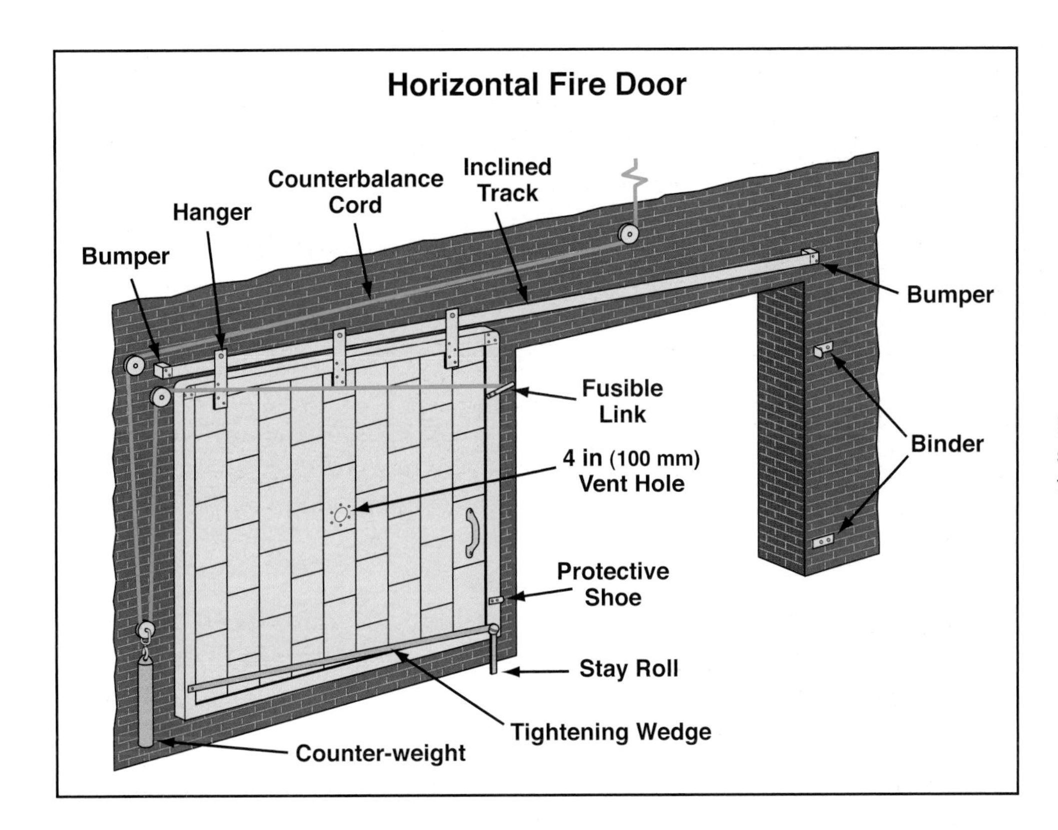

Figure 5.30 A horizontal sliding fire door is likely to be found in an older building.

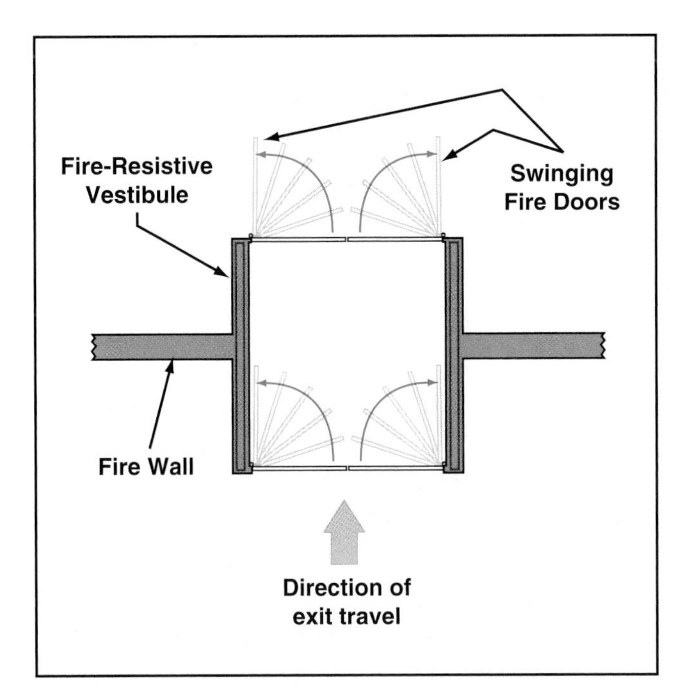

Figure 5.31 When fire doors are located on either side of wall, providing a vestibule can help ensure the necessary degree of fire protection.

Figures 5.32 a and b Old-style metal clad fire door. Note that the manufacturer's embossed label above the UL label has been painted over and does not indicate an hourly rating due to its age. *Courtesy of Gregory Havel, Burlington, WI.*

Figure 5.33 A horizontal folding fire door may be used where a fixed wall is not desired. *Courtesy of Ed Prendergast.*

Special fire doors. Special types of fire-rated doors are available for applications such as freight and passenger elevators, service counter openings, security (bullet-resisting doors), dumbwaiters, and chute openings.

A horizontal folding fire door has also been developed. This door is motor driven and requires electrical power for operation. A signal from a smoke detector or fire alarm system initiates the door closing. A battery powers the motor if the regular power supply is interrupted. This type of door is frequently used where a fire-rated partition is required and the designer does not wish to provide a fixed wall to create an unobstructed floor plan. An example of this situation would be a corridor separation in the lobby of a health care facility **(Figure 5.33)**.

Maintenance of Fire Doors

For fire walls to be effective, the doors protecting the openings must operate correctly and close under fire conditions. Failure of fire doors to close properly is a very common occurrence in actual fires. Reasons for failure include damage to the door closer, the door itself, or door guides. Overhead fire doors, which are used infrequently, are especially subject to damage **(Figure 5.34)**. Fire doors of all types are frequently obstructed. Stairwell doors are often blocked or held in the open position. **Figure 5.35** shows an unusual but not untypical situation.

Proper fire door operation requires that the doors be properly maintained. The closing mechanisms on overhead doors are more complicated and typically more inconspicuous than those on swinging doors. Inspection, testing and proper maintenance, therefore, are always important. Chapter 15 of NFPA® 80, *Standard for Fire Doors and Other Opening Protectives*, contains extensive information on the maintenance of fire doors.

Figure 5.34 This damaged fire door failed to close under fire conditions. *Courtesy of Ed Prendergast.*

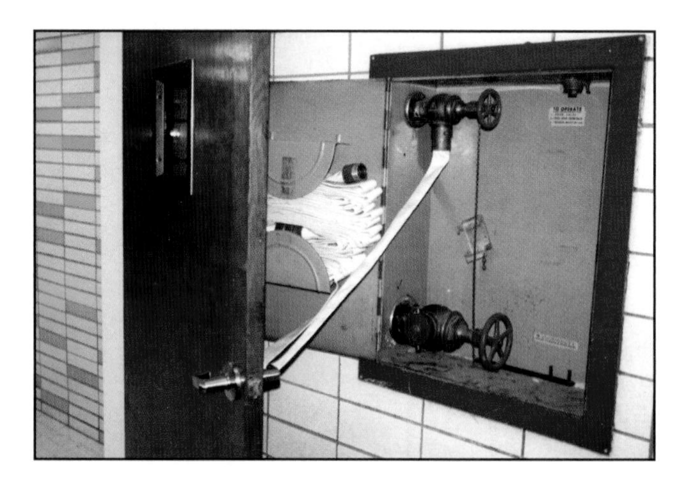

Figure 5.35 A novel use of standpipe hose – holding a stairwell fire door open. Not only does this practice defeat the purpose of a fire door, note that the hose nozzle is also missing. *Courtesy of Ed Prendergast.*

Firefighters should be aware that any fire doors can fail and become non-operational during a fire. Fire fighting tactics should take this possibility into account so firefighters do not become trapped.

Summary

The fire safety of a building is determined by features that go beyond the basic construction of the building. These include interior finishes, compartmentation, HVAC systems, and fire protection systems. The surface flammability of materials used for interior finish can be evaluated in a laboratory and is known as the flame spread rating. Based on their flame spread ratings, materials can be classified and their application controlled by building codes.

Compartmentation is provided in buildings as a means of retarding or blocking the spread of fire and smoke. Building codes contain specific requirements for fire walls, fire partitions, and enclosure of vertical openings. Openings through fire-rated walls must themselves be fire rated to block the spread of fire. The doors used to protect openings must be provided with closers and be properly maintained.

Fire protection systems are typically required by code provisions. The existence of a fire protection system, especially an automatic sprinkler system, will affect building design. Codes may reduce other aspects of building design, such as required fire resistance or allowable area, when a sprinkler system is provided.

Review Questions

1. What is a tunnel test?
2. What is the purpose of a curtain wall?
3. What are the requirements for a Class D fire door?
4. List three fire door closing devices.
5. How is manual smoke control accomplished with an HVAC system?

References

1. NFPA® Handbook 19th ed. Chap 3. Section 12.
2. NFPA® 265, *Fire Tests for Evaluating Room Fire Growth Contribution of Textile Coverings on Full Height Panels and Walls*, 2002.
3. NFPA® 286, *Standard Methods of Fire Tests for Evaluating Contribution of Wall and Ceiling Interior Finish to Room Fire Growth*, 2006.
4. International Building Code®, 2006.
5. NFPA® 92A, *Smoke Control Systems*, 2006.

NFPA is a registered trademark of the National Fire Protection Association, Quincy, Mass.

Chapter Contents

Divider page photo courtesy of McKinney (TX) Fire Department.

Key Terms

FESHE Objectives

Fire and Emergency Services Higher Education (FESHE) Objectives: *Building Construction for Fire Protection*

5. Identify the principle structural components of buildings and demonstrate an understanding of the functions of each.

Foundations

Learning Objectives

After reading this chapter, students will be able to:

1. Explain how different types of surface material affect the types of foundations and the types of buildings that can be built on them.
2. Describe the types of foundations and the conditions that determine which type is used.
3. Describe the construction of foundation walls and the concerns related to cracking.
4. Explain the differences between uniform and differential settlement.
5. Discuss shoring and underpinning and their potential impact on fire department operations.

Chapter 6
Foundations

Case History

Event Description: A fire department was dispatched to a working fire in a church. Upon arrival on scene, firefighters noticed heavy fire on the front face of the church and started fire attack in that location. When the fire was out, firefighters entered the structure through the front and began to overhaul into the ceiling looking for fire extension. The fire had made way through the ceiling so firefighters had to go deeper into the church to find the last of the extension.

About 10 minutes into the fire, one firefighter's low air on his SCBA began to sound, so he and his partner began to make their way out of the building. About 2-3 steps outside the front door, the floor in the church collapsed. The fire had began in the basement, burned through the foundation, and burned up the front of the church. The progression of the fire had made it seem as though the fire had started on the outside of the building, not in the basement. The firefighters could easily have fallen through the floor into the flames below and been hurt severely or killed.

Lesson Learned: While making an interior attack, don't just sound the floor on the way in, but on the way out as well.

The function of a foundation is to transfer the structural load of a building to the ground. Ordinarily, the structural problems of foundations are of little interest to the firefighter. However, the failure of a foundation can create or aggravate structural problems within the building supported by the foundation. Supports that shift or settle will alter the forces on the structural members above the foundation of the building. In severe cases, the frame of a building may be distorted, floors may slope, walls and glass may crack, and doors and windows may not work properly **(Figure 6.1, p. 184)**. In some cases, automatic fire protection system piping can be damaged. These altered load patterns may hasten structural collapse under fire conditions. Additionally, when a foundation collapse occurs firefighters may become involved in a rescue operation **(Figure 6.2, p. 184)**.

Surface Materials

Various types of surface materials are found throughout different regions of the world. They range from loose sand at one extreme to solid granite at the other. The properties of these materials affect the type of foundation and the

Figure 6.1 A settling or uneven foundation can cause structural instability even at the roof.

Figure 6.2 Shifting of a foundation during street repairs resulted in collapse of this building wall. *Courtesy of Ed. Prendergast.*

type of building that can be built at a given location. For example, the allowable foundation pressure for sandy clay is 1,500 pounds per square foot (71.8 kPa). For bedrock it is 12,000 pounds per square foot (574 kPa).

Different types of soil may be found at different depths in a given location. Before the planning of a building can proceed, civil engineers must determine the properties of the soil that lies beneath the site. This is done either by digging a test pit or by using test borings. Test pits are usually used up to a depth of about 8 feet (2.5 m). The use of a test pit permits the *strata* of the soil to be observed and measured. Depending on the soil type, the location of the water table may also be observed.

Strata – Identifiable layers of different soils.

Test borings are used at depths greater than those of test pits. A test boring can provide information on the bearing capacity of soil through the use of a driving hammer. The number of impacts of a standard driving hammer required to drive the boring tube a given distance provides information about the soil strata. The boring tube can also return information about the water table and soil samples to be analyzed by a laboratory. Usually a number of test borings will be taken across a given site.

Types of Foundations

The type of foundation required for a project depends on the type of building and the soil conditions at the site. For example, a small garage or shed requires only a simple foundation whereas a high-rise building requires a foundation that extends 100 feet (30 m) or more into the ground.

Naturally, a foundation must support the dead load of a building and the live load of its contents. In addition to these loads, a foundation may need to be designed to resist such other forces as the following:

- Wind loads that may apply lateral or uplifting forces to a building
- Soil pressure
- Uplifting forces from underground water
- Thrusts resulting from the support of arches, domes, and vaults
- Seismic forces

NOTE: Loads were covered in Chapter 3.

The main factors that determine the type of foundation to be used for a building are the soil conditions and the structural configuration. Other factors include the following:

- The working space available
- Environmental factors
- Impact on adjacent property
- Building codes and regulations

Shallow Foundations

Foundations are divided into two types: shallow and deep. A shallow foundation transfers the weight of the building to the soil at the base of the building. A shallow foundation can be used where the load-bearing ability of the soil directly under the building is adequate to support the building. Deep foundations penetrate the layers of soil directly under a building to reach soil at a greater depth that can support the weight of the building **(Figure 6.3)**.

Shallow foundations usually make use of footings to transmit the load to the soil. A *footing* is a widened base at the bottom of a column or foundation wall. The increased area of the footing reduces the compressive stress on the soil. The footings are placed on undisturbed soil or on soil that has been carefully compacted to prevent excessive settling.

Footing — That part of the building that rests on the bearing soil and is wider than the foundation wall.

A *wall footing* is a continuous strip of concrete that supports a wall **(Figure 6.4)**. Wall footings can take several forms, such as:

- Increased thickness of a floor slab at its edges (known as a monolithic floor)
- A widened strip of concrete under a wall that supports a raised floor with a crawl space
- A widened strip of concrete under full story-high walls that create a full or partial basement **(Figure 6.5, p. 186)**

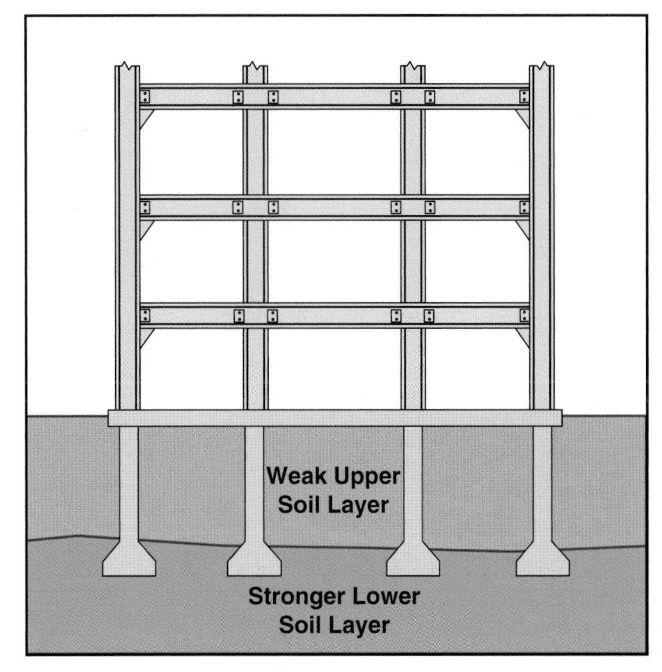

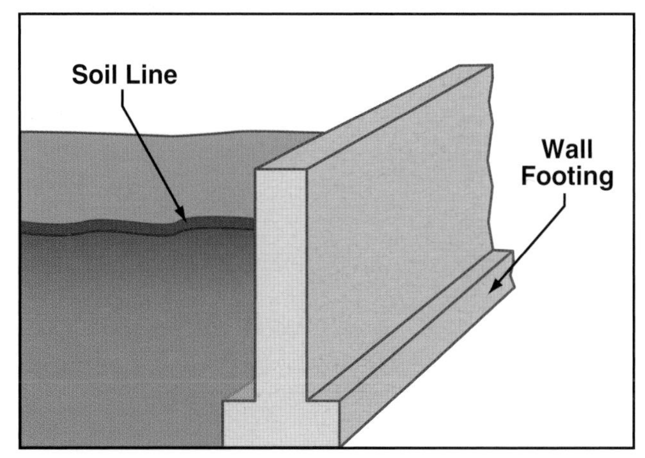

Figure 6.4 A wall footing is continuous beneath the length of a wall.

Figure 6.3 A deep foundation can be used to carry the weight of a building to the soil layer that can support the building.

Figure 6.5 Wall footings can take several forms.

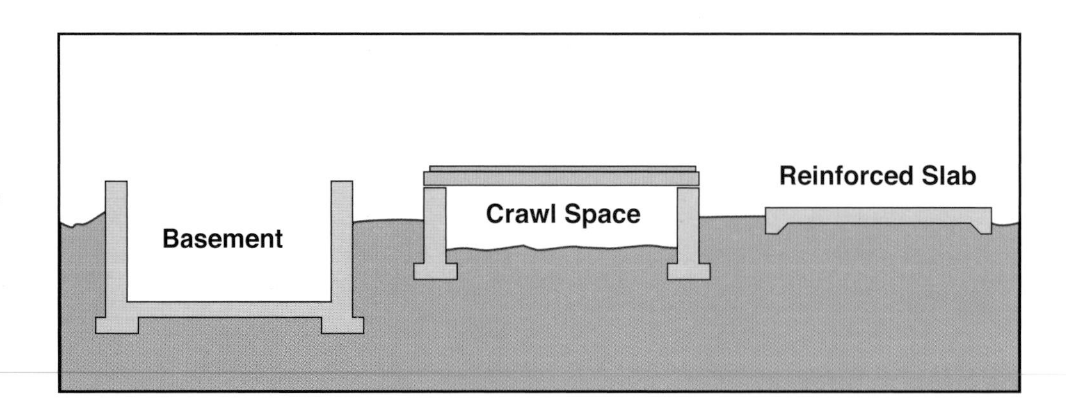

Column Footing — Square pad of concrete that supports a column.

Grillage Footing — Footing consisting of layers of beams placed at right angles to each other and usually encased in concrete.

Mat Foundation — Thick slab beneath the entire area of a building. A mat foundation differs from a simple floor slab in its thickness and amount of reinforcement.

A *column footing* is a square pad of concrete that supports a column. Although most column footings are reinforced concrete pads, when a column is supporting a large load, the footing may be designed as a grillage footing. A *grillage footing* consists of layers of beams placed at right angles to each other that are usually encased in concrete **(Figure 6.6)**. The beams that make up the grillage footing distribute the load of the column over the area of the footing.

When the load-bearing capacity of the soil beneath a building is low, the footing must be large in area and a *mat foundation* may be used. A mat foundation is a thick slab beneath the entire area of a building. A mat foundation differs from a simple floor slab in its thickness and amount of reinforcement. A mat may be several feet thick and heavily reinforced, whereas a slab may be only a foot thick.

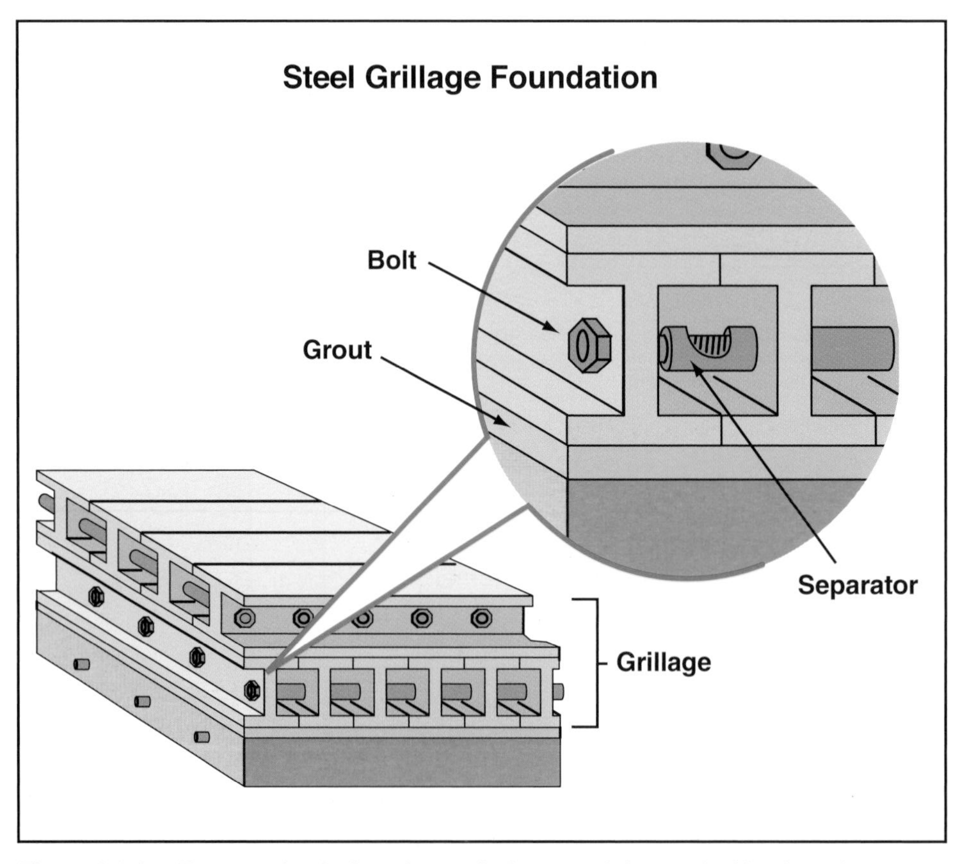

Figure 6.6 A grillage can be designed to assist in transmitting vertical loads to the footing.

In some cases where soil strength is low, a type of foundation known as a *floating foundation* may be used. A floating foundation is the same as a mat foundation except that it is located beneath a building at a depth such that the weight of the soil removed is equal to the weight of the building. By this means the total weight supported by the soil beneath the building remains the same before and after excavation and settlement is minimized **(Figure 6.7)**. One story of soil can be equal to five to eight stories of building. This, of course, will vary with the density of the soil and the construction of the building.

Floating Foundation — Foundation for which the volume of earth excavated will approximately equal the weight of the building supported. Thus, the total weight supported by the soil beneath the foundation remains about the same, and settlement is minimized because of the weight of the building.

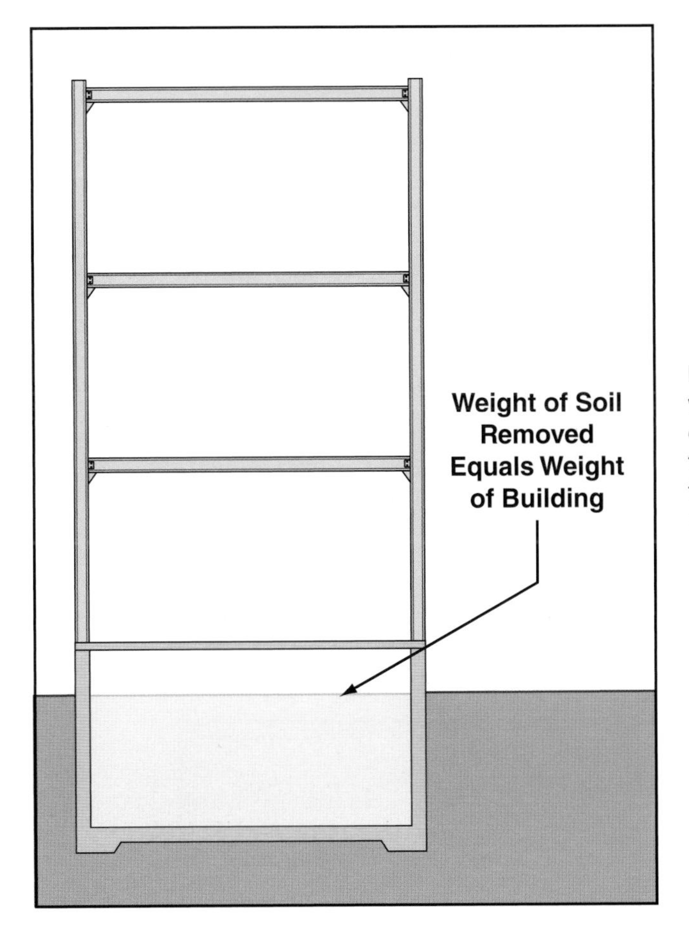

Figure 6.7 Because the weight of the soil removed equals the weight of the building, a floating foundation is designed to minimize settlement.

Deep Foundations

Deep foundations are more costly than shallow foundations, so are used only where shallow foundation cannot be used. Deep foundations take the form of either *piles* or *piers* **(Figure 6.8, p. 188)**.

Piles are driven into the ground and develop their load-carrying ability either through friction with the surrounding soil or by being driven into contact with rock or a load-bearing soil layer. Piles can be timber, steel, or precast concrete. Timber piles have been used for centuries and still may be used to support light loads. The use of timber for piles is limited by several factors, including the possibility of decay and the length of available trees from which the timbers are cut.

Piers are constructed by first drilling or digging a shaft and then filling it with concrete. When a pier is designed with a footing, it is known as a "belled" pier. Piers are sometimes referred to as caissons. More accurately, the *caisson* is the protective sleeve used to keep water out of the excavation for the pier.

Piles — Used to support loads, piles are driven into the ground and develop their load-carrying ability either through friction with the surrounding soil or by being driven into contact with rock or a load-bearing soil layer.

Pier — Load-supporting member constructed by drilling or digging a shaft, then filling the shaft with concrete.

An unusual problem arises when a building is to be built on sloping ground. This situation often occurs with beachfront property or property located on the side of a canyon or steep slope. As the slope of the ground increases, the foundation design becomes more critical. When the ground slope angle is severe, piles or piers are placed in the more dense lower soils. These deep elements act as vertical cantilevers to resist the lateral force exerted by the building **(Figure 6.9)**. The vertical elements are connected with a tie beam.

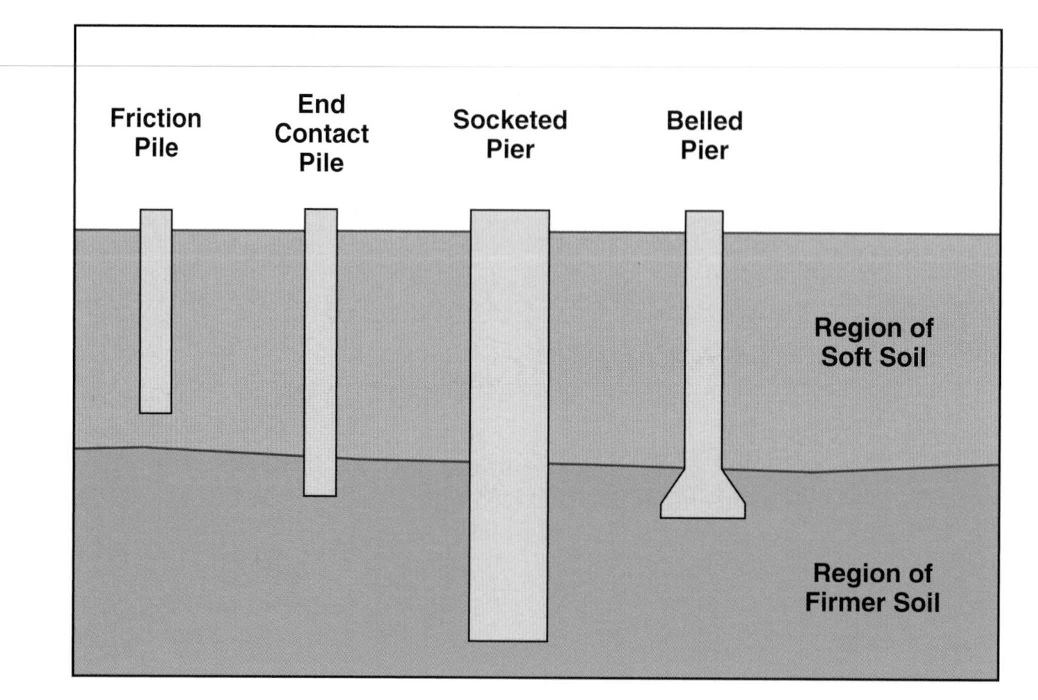

Figure 6.8 Deep foundations can be either piles or piers.

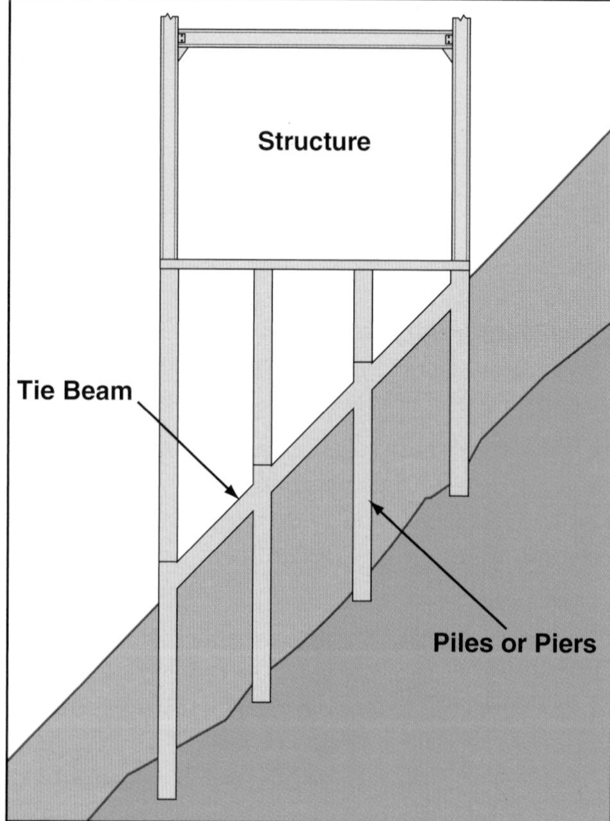

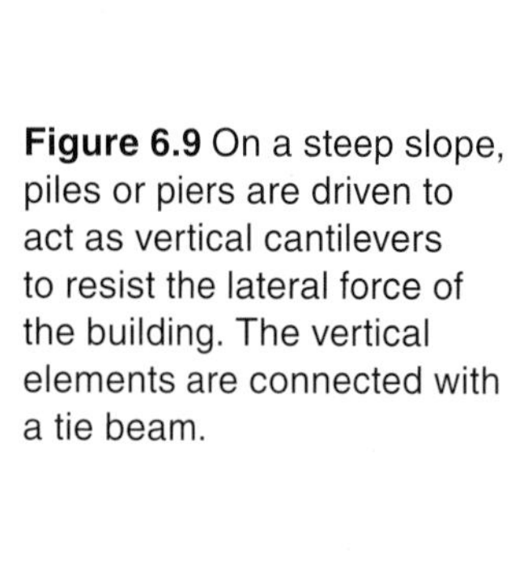

Figure 6.9 On a steep slope, piles or piers are driven to act as vertical cantilevers to resist the lateral force of the building. The vertical elements are connected with a tie beam.

Foundation Walls

When a building is to have a basement, the foundation must include walls to enclose it. Concrete is the material most commonly used for foundation walls. Concrete is durable and is resistant to moisture and insects.

Many concrete foundations walls develop visible cracks for varying reasons. These cracks usually do not significantly affect the ability of the wall to support or distribute the load it is carrying. When a structure is inspected for stability, however, any change in size or extension of cracks or fissures should be given close attention. Any vertical or horizontal misalignment along the length of a crack in a foundation wall indicates a movement or shift in the structure, which may mean a change in the imposition of loads on structural members.

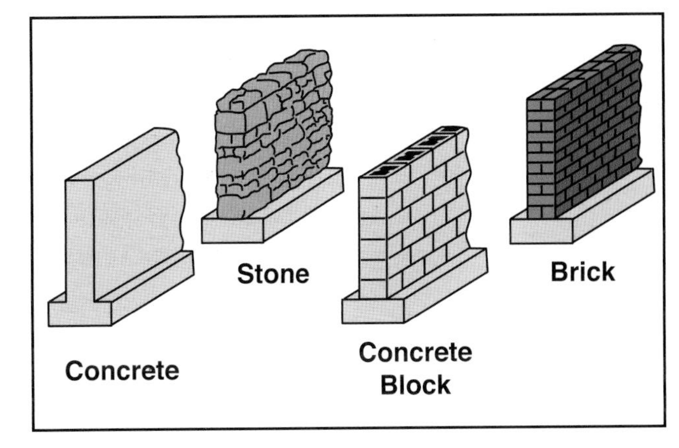

Figure 6.10 Types of foundation walls.

In addition to concrete, it is possible to construct foundation walls of either concrete block, stone, or brick **(Figure 6.10)**. Concrete block is a contemporary material for basement walls. Stone basement walls are found only in older buildings. A distinguishing aspect of stone foundations is that they were often constructed without using any bonding mortar or cement. The stones were carefully quarried and meticulously assembled to form an amazingly tight-fitting and strong foundation.

Wood is used occasionally for foundation walls **(Figure 6.11)**. The wood must be treated with preservatives to resist decay. Wood foundations are used with wood-frame buildings (See Chapter 7). Thus a labor savings can be ac-

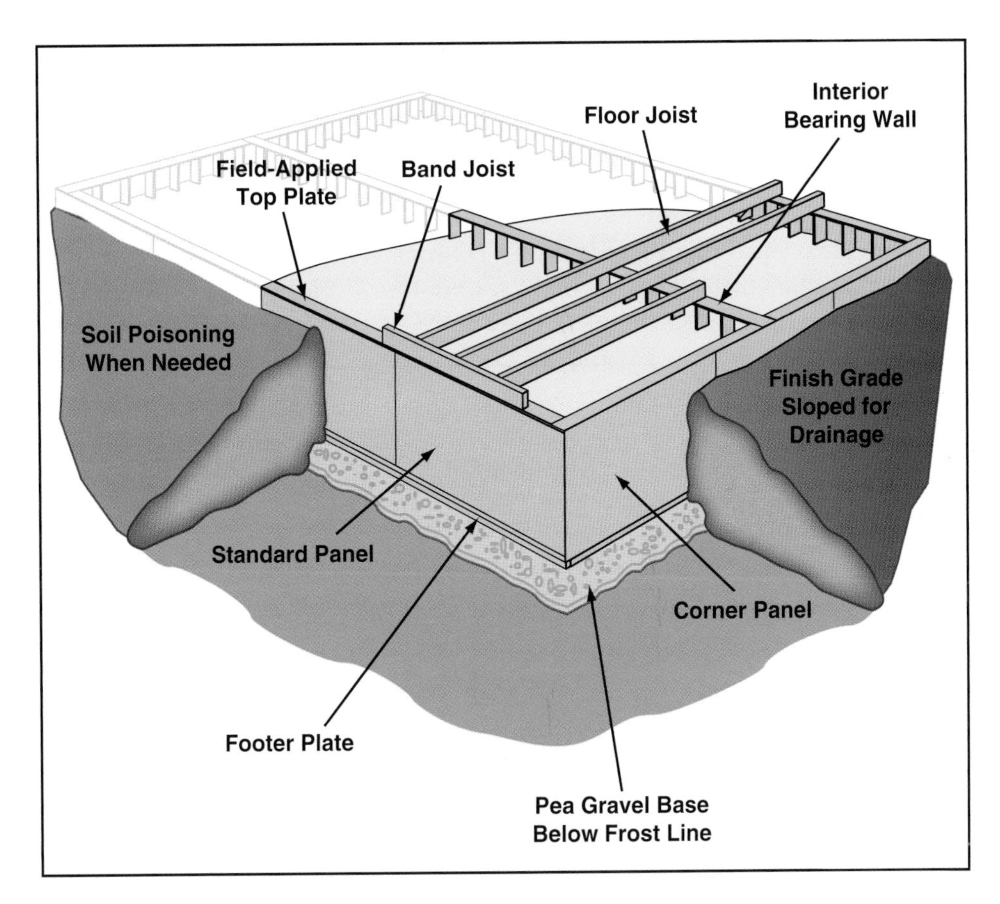

Figure 6.11 Wood foundation walls.

complished with a wood foundation because the same carpenters can be used to erect the rest of the building. Wood foundations are also easier to insulate where that is a consideration.

A layer of expanded polystyrene (EPS) is sometimes placed on the outside of foundation walls for insulation. The EPS conserves heat in cold climates and reduces the energy needed for air conditioning in warm climates. The use of EPS also permits foundations to be dug to shallower depths resulting in a cost saving. This is due to the fact that the reduced heat transfer through the EPS to the surrounding soil results in a frost line that is closer to the surface.

Building Settlement

After buildings are constructed, they will move either because of various unforeseen factors or as a result of construction defects. These movements may range from minor (and may be only cosmetically identifiable) to structurally significant. Movement may be downward as in the case of settlement, upward as in the case of heaving, or outward as in the case of lateral displacement. Settlement of foundations is the most frequent building movement.

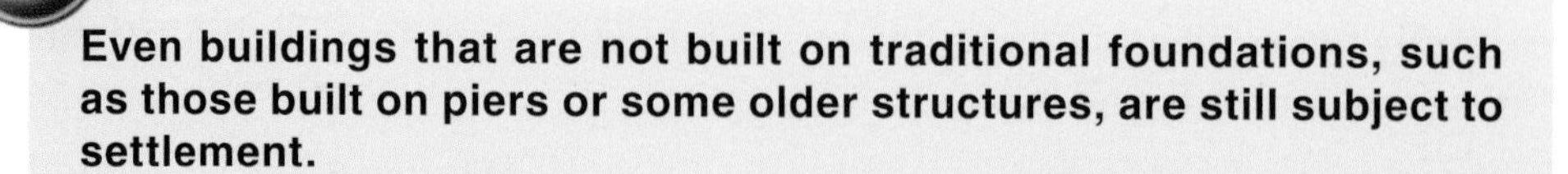

Even buildings that are not built on traditional foundations, such as those built on piers or some older structures, are still subject to settlement.

Despite soil testing and engineering analysis, all buildings settle to some extent due to the compaction of the soil beneath them. Settlement may be only a few millimeters or it may be several inches depending on soil conditions. If the foundation rests on bedrock, the settling will be minimal. If the foundation rests on clay, an alarming amount of settlement can occur.

Settlement of a foundation can be either uniform or differential. In *uniform settlement,* parts of a foundation settle at the same rate and misalignment between structural members is minor **(Figure 6.12)**. *Differential settlement* takes place when different parts of the foundation settle by different amounts.

Figure 6.12 Differential settlement produces distortion between members.

Differential settlement of a foundation is more troublesome than uniform settlement because it can result in significant misalignment of structural members.

Differential settlement of a foundation can result from several conditions, including the following:

- Nonuniform soil conditions under the foundation
- Footings of different size
- Footings placed at different elevations
- Unequal loads on footings

Shoring and Underpinning

Situations may arise where it is necessary to strengthen and stabilize an existing foundation. This process is known as *underpinning.* The need for underpinning can result from several causes such as:

- Excessive settlement because of an inadequate foundation
- An increase in the load on a foundation because of the construction of additional stories
- Erosion of soil from under or around the foundation

In addition, when excavation occurs on adjacent property for basements, sewers, or subways, it is frequently necessary to provide underpinning for the existing structure.

The terms shoring and underpinning are sometimes used interchangeably. However, *shoring* refers to temporary supports. *Underpinning* refers to permanent supports. Shoring an existing structure is frequently necessary to support the structure until underpinning can be put into place **(Figures 6.13 a and b)**.

Underpinning — Process of strengthening an existing foundation.

Shoring — General term used for lengths of timber, screw jacks, hydraulic and pneumatic jacks, and other devices that can be used as temporary support for formwork or structural components or used to hold sheeting against trench walls. Individual supports are called shores, cross braces, and struts.

Access Pit
Shoring To Support Load During Work
Access Pit
Existing Foundation
Existing Foundation
Underpinning
Underpinning

Figure 6.13 a Two techniques for supporting an existing wall with shoring so that permanent underpinning can be put into place.

Figure 6.13 b An example of shoring used to support a wall.

The placement of shoring and underpinning is difficult and often dangerous work. It frequently involves excavation by hand because it may not be possible to position power equipment when the work is done in limited spaces such as basements. If a collapse occurs in connection with this work, the fire department typically is called upon to conduct a rescue operation. Such rescue operations always require extreme caution and coordination.

NOTE: Refer to the IFSTA **Rescue** and **Technical Rescue for Structural Collapse** manuals for more information about shoring techniques.

Summary

Although foundations are normally not a significant factor in fire fighting, they do affect overall structural stability. Because soil conditions differ in different locations, engineers will use different types of foundations. Foundations can be shallow or deep depending on the soil properties and the building design. Foundations may also occasionally need to be reinforced through underpinning and shoring. Accidents involving underpinning and shoring may involve the fire department in rescue operations.

Review Questions

1. What is the purpose of a test pit?
2. What are some of the forces a foundation may be designed to resist?
3. List several materials that may make up foundation walls.
4. What is differential settlement?
5. Why is the placement of shoring and underpinning difficult and dangerous?

References

1. Edward Allen and Joseph Iano, *Fundamentals of Building Construction, Materials and Methods, 4th Ed.* John Wiley and Sons, 2004, Chapter 2.
2. James Ambrose, *Building Structures, 2nd Ed.* John Wiley and Sons, Part Seven.
3. *Design Guide For Frost-Protected Shallow Foundations,* US Department of Housing and Urban Development, 1994.

Chapter Contents

Divider page photo courtesy of McKinney (TX) Fire Department.

Key Terms

FESHE Objectives

Fire and Emergency Services Higher Education (FESHE) Objectives: *Building Construction for Fire Protection*

1. Demonstrate an understanding of building construction as it relates to firefighter safety, buildings codes, fire prevention, code inspection and firefighting strategy and tactics.
8. Identify the indicators of potential structural failure as they relate to firefighter safety.

Wood Construction

Learning Objectives

After reading this chapter, students will be able to:

1. Discuss the material properties of the wood products used in construction.
2. Explain the variables that affect the combustibility of wood used as a construction material.
3. Describe the methods of treating wood with a fire retardant.
4. Describe the framing systems constructed of wood and the purpose of fire stops in those framing systems.
5. Describe the materials used to construct the exterior and interior walls of a wood-frame building.
6. Discuss the considerations related to collapse, ignition-resistance, and deterioration as they relate to wood-frame construction.

Chapter 7
Wood Construction

Case History

Event Description: We responded to a structure fire and the incoming engine reported heavy smoke from several blocks away. On arrival, the first-in engine reported heavy smoke from the second floor of a 2-story wood-frame residence. The first-arriving engine and truck companies made entrance through the front door and encountered heavy smoke and fire conditions on the first floor. The fire was extinguished in the front room of the first floor with one 1¾-inch preconnect line and a quick search was made. PPV was in place at the front door.

During the search, the truck company noticed that the upstairs floor was sagging. Engine and truck company crews proceeded upstairs, still experiencing heavy smoke and fire conditions. We noticed that the floor was burnt out and notified Command, advising that the integrity of structure was compromised. We continued primary search, avoiding the burnt-out section. Fire had traveled into the attic because of the balloon frame construction. The truck crew was pulling down some ceiling to suppress some of the fire when the front room floor collapsed, sending a Captain, a Lieutenant, and a firefighter to the floor below. Another engine company was on the first floor and assisted the truck company and engine company personnel out of the structure. At this time, a personnel accountability report was called. No injuries occurred during this incident.

Lessons Learned: We had a post-incident debriefing and found that several companies working in the structure heard popping noises minutes before the collapse. This should have been communicated over the radio system to the companies above. The companies that were on the second floor had no chance to exit the room at the time of the collapse. We had ground ladders set at two sections of the second floor for egress.

Source: National Fire Fighter Near-Miss Reporting System.

Wood has been used as a basic construction material for centuries and continues to be a fundamental structural material. Wood is used in a wide variety of building applications in all localities. This chapter deals with the properties of wood as a building material and its fire-related behavior. Various structural systems that use wood and methods of construction are also described.

Material Properties of Wood

Wood has a unique position among building materials. Because it is a plant product, most of its "manufacture" actually occurs in nature. Consequently wood is cheap to produce and renewable; however, little control can be exercised over wood as a product. As a building material wood has some disadvantages. For one thing, wood is never dimensionally true. In addition, weather conditions can change its size and shape and wood does not shrink or swell uniformly. Wood can also have such defects as knots, knotholes, decay, insect damage, splits, and warping.

The varied properties of wood affect its use in construction. The strength of wood varies significantly with species, grade, and direction of load with respect to grain. Wood is stronger in a direction parallel to the grain than against the grain. For example, the allowable compressive strength parallel to the grain varies from 325 to 1,700 psi (2 275 kPa to 11 900 kPa) for commercially available grades and species of framing lumber. This represents a variance of a factor of five.

Wood has a useful tensile strength. In fact, on the basis of strength to unit weight, wood has a tensile strength comparable to steel. Typical defects in wood, however, greatly reduce this comparison, so the allowable tensile strength of wood is about 700 psi (4 900 kPa).

The strength of wood is also affected by its moisture content. Wood in a living tree contains a large amount of water. When the tree is cut, the water begins to evaporate. As the water leaves the wood, either naturally or through drying, the wood begins to shrink in size and increase in strength. It is possible to dry lumber to any moisture content, but most structural lumber has a moisture content of 19 percent or less.

Lumber — Lengths of wood cut and prepared for use in construction.

Lumber can be defined as lengths of squared wood (also known as dimensional wood) used for construction. Lumber is graded for both structural strength and appearance. Lumber that has a higher grade is more costly. In a given structure, though, only a few critical columns or beams may require a high structural grade. The grading of lumber permits it to be used more economically by allowing the designer to specify a higher grade where it is needed and a less expensive, lower grade in less-critical members **(Figure 7.1)**.

A large variety of wood products are available for use in the construction industry. Some of the specific forms include the following:

- Solid lumber
- Laminated wood members
- Structural Composite Lumber
- Panels
- Manufactured components

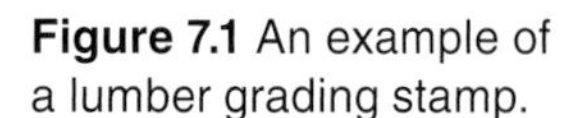
Figure 7.1 An example of a lumber grading stamp.

Solid Lumber

Solid lumber includes boards, dimension lumber, and timbers. Boards have a nominal thickness of 2 inches (50 mm) or less. Dimension lumber has a nominal thickness of 2 to 4 inches (50 mm to 100 mm) or more. Timbers have a nominal thickness of 5 inches (125 mm) or more. Dimension lumber is available in lengths from 8 to 18 feet (2.44 m to 6 m) in 2-foot (0.6 m) increments. In addition, members for use as rafters can be supplied in lengths up to 24 feet (8 m).

Table 7.1
Dimensions of Lumber

Nominal Size inches (mm)	Actual Dimension inches (mm)
2 x 4 in (50 mm x 100 mm)	1½- x 3½ in (38 mm x 89 mm)
4 x 4 in (100 mm x 100 mm)	3½ x 3 9/16 in (89 mm x 90.5 mm)
6 x 6 in (150 mm 105 mm)	5½ x 5 5/8 in (139.7 mm x 143 mm)
2 x 6 in (50 mm x 150 mm)	1½ x 5 5/8 in (38 mm x 143 mm)

Lumber sizes are typically given in nominal dimensions. The actual dimensions of a finished piece of wood will be smaller than the nominal dimension. Thus a nominal 2 x 4- inch (50 mm x 100 mm) board will have finished dimensions of 1½ x 3½ inches (38 mm x 89 mm). The difference arises because of variations in the machinery used to cut a log into rough lumber. A piece of rough lumber will have its dimensions reduced to a standard size in the finishing process. **Table 7.1** illustrates the nominal and actual sizes of typical softwood lumber products.

Laminated Wood Members

Laminated wood members are produced by joining flat strips of wood with glue **(Figures 7.2 a and b)**. The beams produced by this method are known as *glulam beams*. The thickness of the individual laminations varies from ¾ inch to 2 inches (19 mm to 50 mm). The advantage to producing laminated structural members is that sizes and shapes can be produced that are not available from solid pieces cut from logs. Glulam beams can be formed into curves or given varying cross sections **(Figure 7.3, p. 200)**.

Glulam — Short for glue-laminated structural lumber.

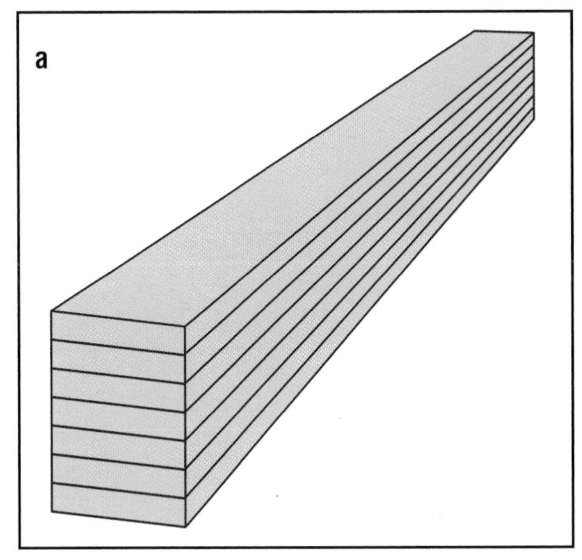

Figures 7.2 a and b Glulam beams are produced from individual boards that are glued together. *Photo courtesy of McKinney (TX) Fire Department.*

Figure 7.3 Curved glulam beams used to support a church roof.

Laminated members can be produced in depths ranging from 3 to 75 inches (75 mm to 1 875 mm) and lengths up to 100 feet (33 m). When necessary, short pieces of lumber can be joined to obtain the required length during the manufacturing process. Three types of joints can be used to join laminated members end-to-end. These are the butt joint, the scarf joint, and the finger joint **(Figure 7.4)**. The butt joint is easy to produce but cannot be used where tensile forces are to be transmitted along the length of the beam. The scarf and finger-type joints can be used to transmit tensile forces.

NOTE: These joints are also used for standard lumber.

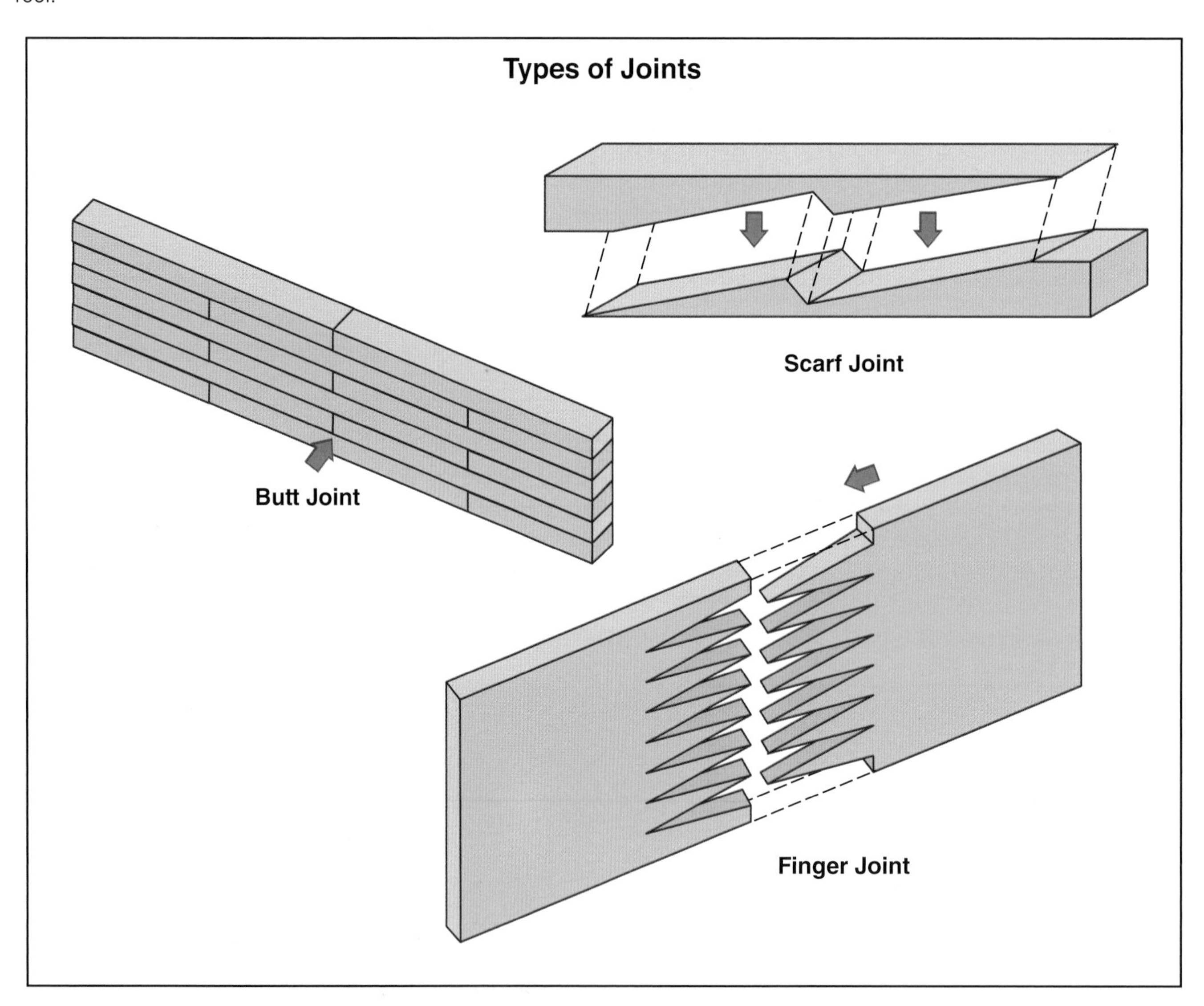

Figure 7.4 These standard methods of joining wood are also used to secure laminated beams.

Structural Composite Lumber

Over the years the lumber industry has developed new products that increase the efficiency of the tree-harvesting process. One group of products is known as structural composite lumber (SCL). These products allow the use of the outer fibers of a log as well as the inner portions traditionally used. The products include laminated veneer lumber (LVL), parallel strand lumber (PSL), and laminated strand lumber (LSL).

Veneer — Surface layer of attractive material laid over a base of common material; for example, a veneered wall (faced with brick) or a veneered door (faced with a thin layer of hardwood).

Laminated veneer lumber is produced by peeling sheets of *veneer* from the outer portion of a log. These sheets of veneer are laminated in parallel alignment. LVL finds application in I-joists and beam sections 1¾ to 3½ inches thick (44.5 to 89 mm).

Parallel strand lumber is made from the outermost veneers peeled from a log that are not as uniform as those used in LVL. These veneers range from 2 to 8 feet in length (0.61 to 2.44 m) and produce odd-shaped strands that are coated with an adhesive and cured under pressure. PSL can be produced in a variety of standard sizes. PSL is the strongest of the three SCL products and can be used for heavily loaded columns and long spans.

Laminated strand lumber is produced by taking long strands of wood up to 12 inches (300 mm) in length and bonding them with a resin in a steam pressing process. LSL is typically used for short-span beams and columns.

Panels

Wood panel products include plywood, nonveneered panels, and sandwich or composite panels.

Plywood

Plywood panels are made up of several thin layers or veneers that are rotary sliced from rotating logs and glued together. The direction of the grain of adjacent veneers is at right angles to each other. The exterior veneers have their grain oriented in the long dimension of the sheet. The thickness of individual layers varies from 1/16 inch to 5/16 inch (1.58 mm to 7.9 mm). The nominal thickness of panels varies from ¼ inch to 1⅛ inches (6.35 mm to 28.5 mm). Plywood is produced in standard sheets measuring 4 feet by 8 feet (1.22 m x 2.44 m). Plywood is used for several applications including sheathing, concrete formwork, the webs of composite beams, and even the hulls of ships **(Figure 7.5)**.

Plywood — Wood sheet product made from several thin veneer layers that are sliced from logs and glued together.

Figure 7.5 Plywood used for the sheathing in a wood-frame building. *Courtesy of Ed Prendergast.*

Nonveneered Panels

Nonveneered products include oriented strand board (OSB), particleboard, and waferboard.

Oriented Strand Board (OSB) — Construction material made of many small wooden pieces (strands) bonded together to form sheets, similar to plywood.

Oriented strand board. Oriented strand board uses long, strand-like wood particles that are compressed and glued into three to five layers. The strands are oriented with each layer in the same direction, similar to the grain in the veneer layers in plywood. Controlling the direction of the strands results in a panel that is stronger and stiffer than waferboard or particleboard. OSB is widely used for sheathing and subflooring in wood-frame buildings **(Figure 7.6)**.

Particleboard. Particleboard is made from wood particles bonded with synthetic resins under heat and pressure. The individual wood particles can range in size from 1 inch (25 mm) to very fine. Particleboard is sometimes referred to as flakeboard, chipboard, or shavings board. Particle board panels may be single layer or multilayer. Particleboard can be manufactured in sizes up to 8 by 40 feet (2.4 m x 12.2 m). Particleboard is not generally used for structural applications although it may be used in lateral-force-resisting diaphragms. Because it can be manufactured in large sheets, particleboard is used for flooring in manufactured and mobile homes.

Waferboard. Waferboard is similar to particleboard but uses wafer-like pieces of wood that are larger than those used in particleboard. For structural purposes, waferboard has been largely replaced by OSB.

Composite or Sandwich Panels

Composite Panels — Produced with parallel external face veneers bonded to a core of reconstituted fibers.

Composite panels (or sandwich panels) consist of a face and back panel such as plywood or OSB bonded to a central core material. The core can be a variety of materials such as paper honeycomb or plastic foam. One type of composite panel is the structural insulated panel (SIP). A structural insulated panel consists of outer wood panels (usually OSB) with a plastic foam core between the panels **(Figure 7.7)**. The foam plastic is usually expanded polystyrene. SIP's are factory made with holes precut for electrical and plumbing risers.

Structural insulated panels are used where energy efficiency is desired. The inner core can be up to 1 foot (.3 m) thick. Once in place, the inner surface of the SIP can be covered with wallboard and the outer surface covered with a siding. Therefore, a building constructed with SIP would be difficult to identify until the wall is opened **(Figures 7.8 a and b).**

Span Rating

Wood panel products are graded for their structural use and their exposure durability. A grade stamp appears on the back of a structural panel that indicates its intended structural application and its suitability for exposure to water. **Figure 7.9, p. 204,** illustrates a grade stamp for

Figure 7.6 Oriented strand board being used for the sheathing in a wood-frame dwelling. *Courtesy of Ed Prendergast.*

Figures 7.7 a and b Section of composite panel consisting of a plastic foam core and OSB outer layers. *Photo a courtesy of Ed Prendergast; photo b Courtesy of Greg Havel.*

Figure 7.8 a SIP roof panels installed on laminated wood arches. One of the connecting splines is visible where the last panel is missing. *Courtesy of Gregory Havel, Burlington, WI.*

Figure 7.8 b: SIP roof panels with firring strips installed so that electrical conduits and cables can be concealed. *Courtesy of Gregory Havel, Burlington, WI.*

sheathing. The span rating of 32/16 indicates that the panel may be used as roof sheathing on rafters 32 inches (800 mm) apart or as subflooring on joists 16 inches (400 mm) apart.

The difference in span rating for panels used for either roof sheathing or subflooring is noteworthy to a firefighter. It indicates the difference between the amount of support and load-carrying capability that may be found in roof construction versus that used in floor construction. Greater distances typically exist between the supporting members in roofs than in floors.

Manufactured Components

Manufactured members are prefabricated from components such as dimension lumber, panels, adhesives, and metal fasteners and shipped to the construction site for erection. Manufactured members include trusses, box beams, I-beams, and the panel components previously discussed. Manufacturing members away from the job site permits greater quality control and more efficient use of materials than assembling them at the construction site.

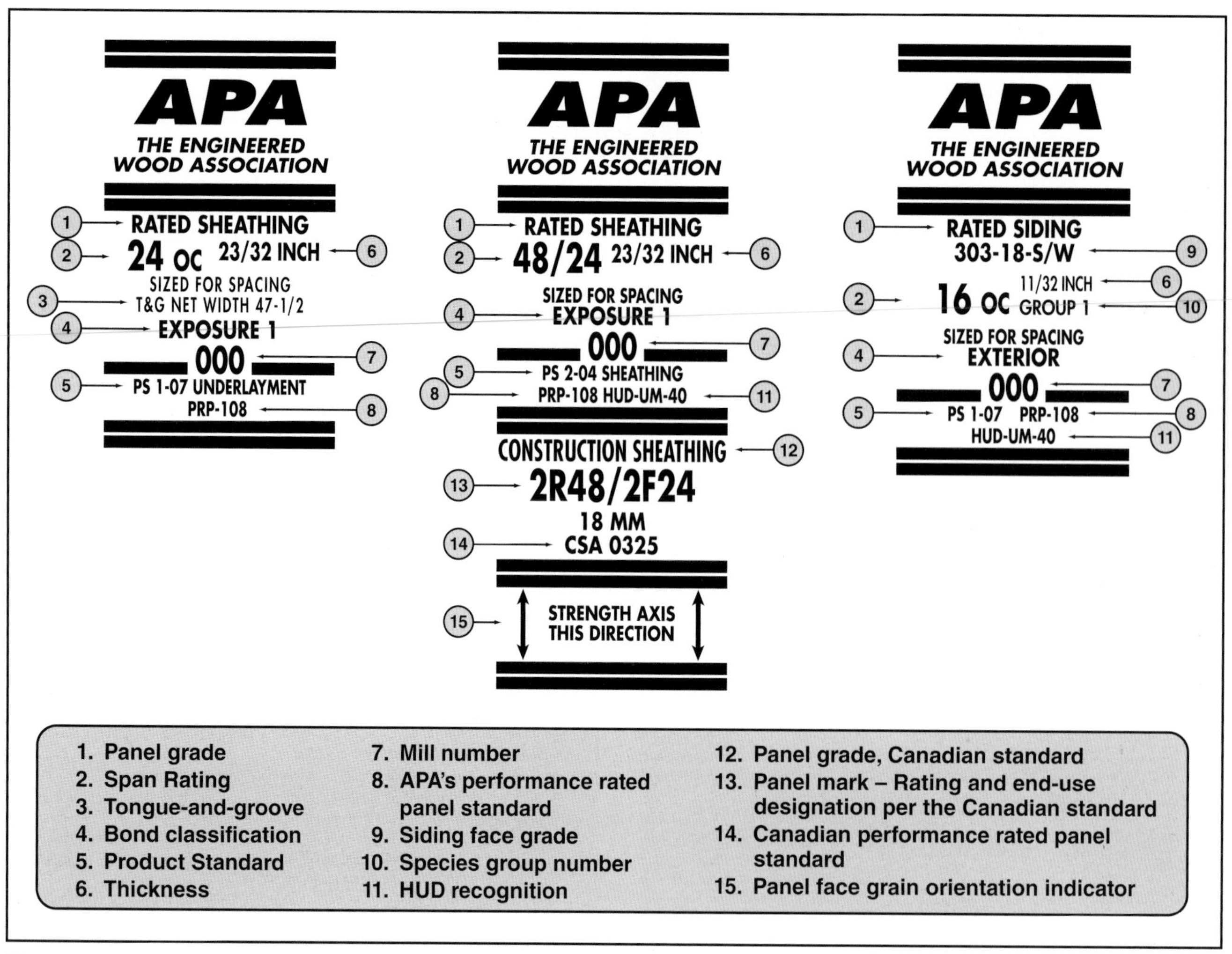

Figure 7.9 A grade stamp for sheathing. *Source: American Plywood Association.*

Gusset Plates — Metal or wooden plates used to connect and strengthen the intersections of metal or wooden truss components roof or floor components into a load-bearing unit.

Trusses

Wood trusses are sometimes categorized as light-frame trusses, heavy timber, or split-ring trusses. Light-frame trusses are made up of 2-inch (50 mm) nominal members that are all in the same plane **(Figures 7.10 a and b)**. The truss members are usually connected by metal-toothed plates driven into the wood members although nailed plywood gusset plates can also be used. Light-frame trusses have become very popular for roof framing where the spans are small or moderate.

Light-frame wood trusses pose a special problem for firefighters because the relatively slender wood members used to construct the truss will fail earlier in a fire than the heavier members used in the past. (See following discussion under Combustion Properties of Wood). Concern has also been expressed about the possible early failure of the metal connector plates. Wood and steel behave very differently when exposed to the heat of a fire. Some authorities have theorized that the thermal expansion of the steel teeth of the metal connectors can cause the teeth to work loose from the wood and result in early failure. However, this tendency may be offset by the ability of the steel connector to reflect some of the thermal radiation of the surrounding fire.

Heavy-timber trusses are made up of members up to 8 or 10 inches (200 mm to 250 mm). Heavy-timber trusses were once very common before the advent of steel and many remain in use. In current practice, however, heavy-timber trusses are used mainly for their appearance. **Figure 7.11** illustrates a heavy-timber truss used to support a roof.

Timber trusses with large members pose special problems for designers. A problem arises at the connection points between the members where the forces must be transferred from one member to another. (Note the connection points of the truss in Figure 7.11). A number of different styles of connection can be found. In modern heavy-timber trusses steel gusset plates with through-bolts are frequently used at the connection points.

A split-ring truss makes use of a short circular piece of steel within and between two adjacent wood members to transfer the load between the members. Simply using screws or bolts to connect wood members has a serious limitation: steel screws and bolts are stronger than wood. They tend to concentrate the load application to the wood member over the surface area of the screw or bolt. The result is deformation of the wood at the point of application and an inefficient use of the wood.

Figure 7.10 a and b Lightweight trusses under construction and the metal gusset plates used to attach them. *Courtesy of McKinney (TX) Fire Department.*

Figure 7.11 Heavy-timber truss used for roof support. *Courtesy of Ed Prendergast.*

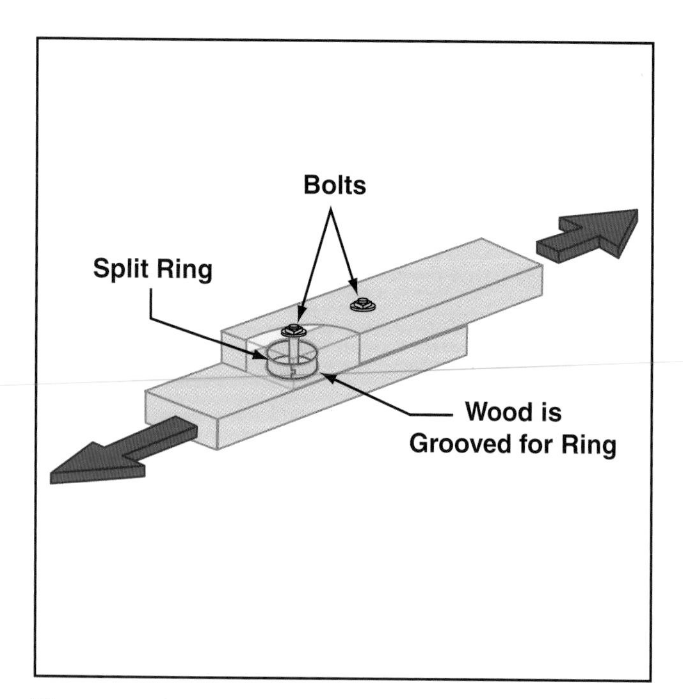

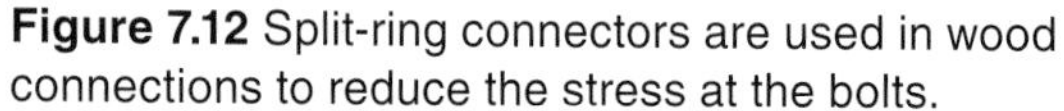

Figure 7.12 Split-ring connectors are used in wood connections to reduce the stress at the bolts.

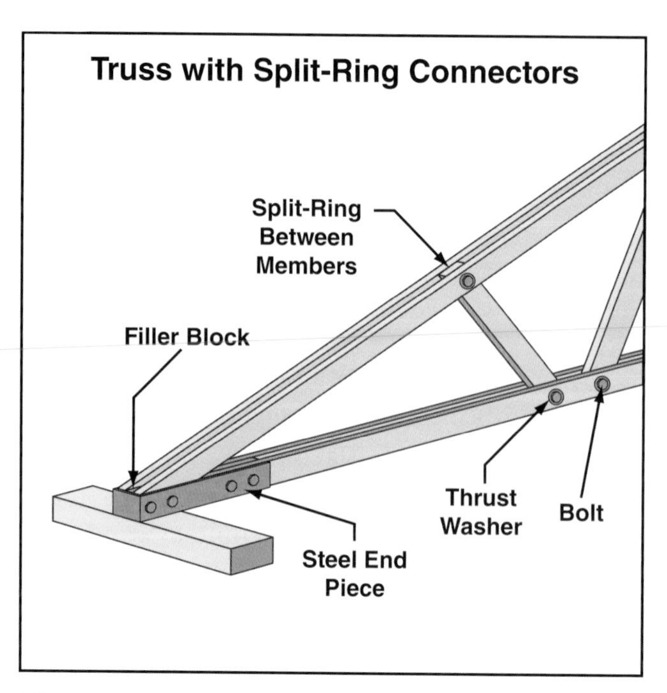

Figure 7.13 Wood truss that would make use of split rings.

With a split-ring connector, a bolt with a thrust washer is run through the split ring to hold the members together **(Figure 7.12)**. The split-ring connector provides a larger bearing surface for the transfer of load between the two members. **Figure 7.13** shows a wood truss that would make use of split-rings at the connection points. Notice the steel piece used to join the members at the end of the truss.

Box Beams and I-Beams

Box beams and I-beams can be manufactured in several ways **(Figures 7.14 a and b)**. The vertical webs use various thicknesses of plywood, laminated veneer lumber, or oriented strand board. The flanges of the beam can be made of laminated veneer lumber or solid wood lumber. The wood I-beams are frequently used for floor joists. Wood I-beams are also used for rafters in the framing of roofs.

Combustion Properties of Wood

One serious and fundamental drawback to wood as a building material is its combustibility. Despite its numerous advantages as a construction material, a building constructed of wood can be completely destroyed when a fire occurs **(Figure 7.15)**. The structural members provide a large amount of fuel for combustion. Furthermore, the voids created in floor, roof, and wall cavities result in many square feet (square meters) of combustible surface area surrounded by large volumes of air for combustion. The relative hazard posed by a combustible material such as wood is a function of several variables including ignition temperature, heat of combustion, and the ratio of the surface area to mass.

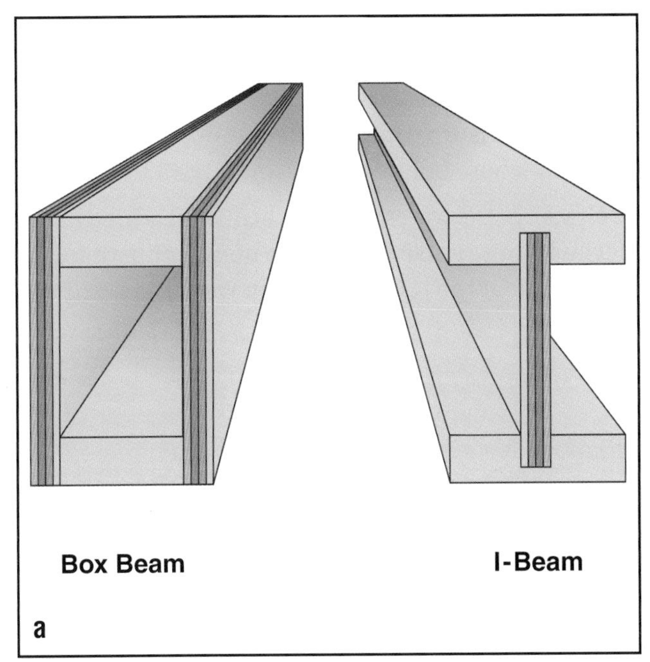

Figures 7.14 a and b Box beams and I-beams are made of a variety of materials. *Photo courtesy of Dave Coombs.*

Figure 7.15 Wood structures are susceptible to damage from fire. *Courtesy of Donny Howard.*

Ignition Temperature

Materials with relatively low ignition temperatures are easier to ignite than materials with high ignition temperatures. However, the burning of wood is not a simple process.

Ignition temperatures for wood are difficult to measure precisely because the ignition temperature of wood is affected by such variables as the following:

- Density of the wood
- Size and form of the wood
- Moisture content
- Rate of heating
- Nature of the heating source
- Air supply

Pyrolysis — Thermal or chemical decomposition of fuel (matter) because of heat that generally results in the lowered ignition temperature of the material. The pre-ignition combustion phase of burning during which heat energy is absorbed by the fuel, which in turn gives off flammable tars, pitches, and gases. Pyrolysis of wood releases combustible gases and leaves a charred surface.

Ambient Temperature — Temperature of the surrounding environment.

Heat of Combustion — Total amount of thermal energy (heat) that could be generated by the combustion (oxidation) reaction if a fuel were completely burned. The heat of combustion is measured in British Thermal Units (Btu) per pound or kilojoules per gram.

Ignition of wood occurs when an ignition source of sufficient intensity is applied to the wood. The ignition source itself (such as a flame) must produce enough heat to raise the temperature of the wood to a point where a process known as *pyrolysis* begins. Pyrolysis is the thermal decomposition of wood and begins at a temperature somewhere below approximately 392°F (200°C).

The initial products of pyrolysis are the water that may still be retained in the wood and carbon dioxide. Because these two products are noncombustible, the ignition process would be reversible if the ignition source were removed.

There is some evidence that pyrolysis and ignition can occur at temperatures lower than normally required if the wood is subjected to a temperature higher than *ambient* but lower than the ignition temperature. For example, framing in a wall close to a wood stove chimney may not ignite for years, but may eventually ignite after exposure to heat at a temperature that is below the original ignition temperature.

As the temperature of wood is increased above the beginning of pyrolysis, the products involved are combustible and increase in quantity. Eventually a point is reached where the chemical process liberates enough heat that it is self-sustaining and flame is produced.

Heat of Combustion

The heat of combustion of a fuel is the total amount of thermal energy that could be released if the fuel were completely burned. The heat of combustion is measured in British Thermal Units (Btu) per pound or kilojoules per gram (kJ/g).

The significance of the heat of combustion of wood can be illustrated with a simple example. A ½-inch (13 mm) sheet of plywood measuring 4 x 8 feet (1.22 m x 2.44 m) contains 1.33 cubic feet (0.4 m^3) of wood. If the wood has a density of 30 pounds per cubic foot (480 kg/$m^{3)}$, the sheet would weigh about 40 pounds (18 kg). If a minimum heat of combustion of 8,000 Btu per pound is assumed, the burning of this one sheet of plywood would release about 320,000 Btu of thermal energy (33,760 kJ).

Example of Heat of Combustion

Using a 12-pound natural Christmas tree for an example at 8,000 Btu per pound would equal 96,000 Btus of heat energy released if the tree is fully consumed.

A structure built of wood contains tons of wood in its structural system. The thermal energy potentially available in a fire becomes very large. By contrast, the potential thermal energy available from the structural components in fire-resistive or noncombustible buildings (such as Type I or Type II construction) is very limited.

The severity of a fire is not determined only by the total amount of fuel available. The rate at which a fuel is consumed and, therefore, the rate at which energy is released determines fire growth rate and severity.

Surface Area and Mass

The combustion of wood is a surface phenomenon. Flaming combustion takes place at the surface of the wood where the gaseous products of the pyrolysis of wood can combine with the surrounding air **(Figure 7.16)**. A greater surface area for a given mass of wood permits a more rapid combining of fuel vapors and air for combustion and, therefore, an overall greater rate of burning.

The relationship between surface area and mass is the *surface-to-mass ratio* **(Figure 7.17)**. The surface area-to-mass ratio of wood has great significance in fire fighting. The relatively slender pieces of structural lumber used in light framing are consumed and fail much more quickly than heavy timbers. Decking made of plywood fails more quickly than decking made from heavier planks or boards. This basic relationship accounts for the fact that the more massive timbers used in heavy-timber construction burn slowly and can retain their structural integrity longer than light-frame members.

Surface-To-Mass Ratio — The ratio of the surface area of the fuel to the mass of the fuel.

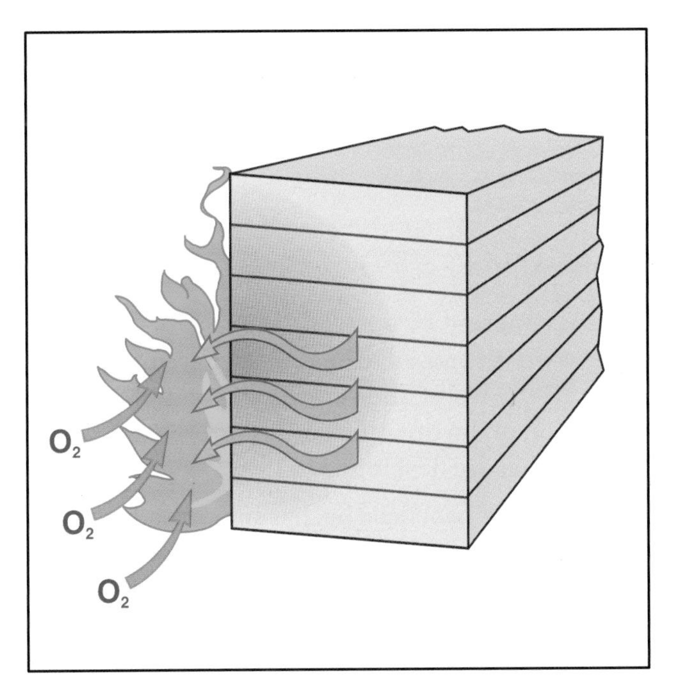

Figure 7.16 Flaming combustion occurs at the surface of wood products. The fuel products of pyrolysis evolve from the wood and combine with oxygen in the air.

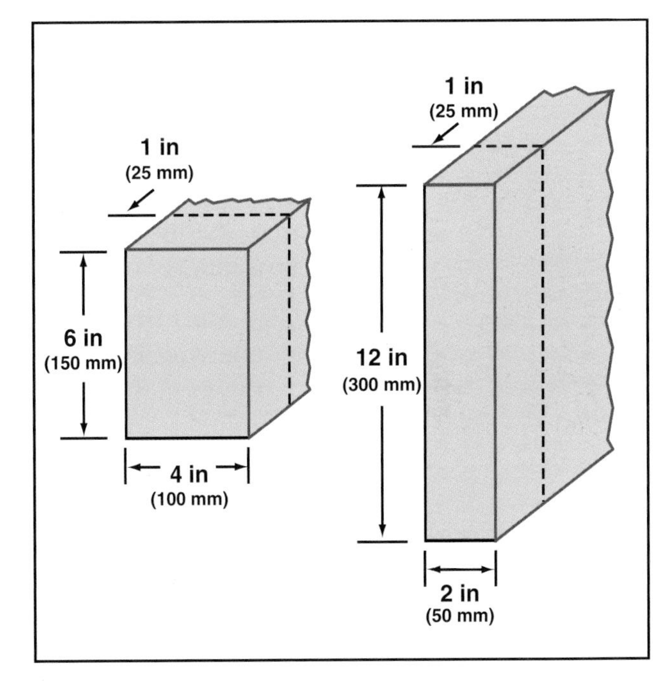

Figure 7.17 These two pieces of wood have the same mass, but the shape of the piece on the left gives it a lower surface area to mass ratio. The piece on the right, with a higher surface-to-mass ratio, would fail in a fire before the one on the left.

Fire-Retardant Treatment of Wood

Wood can be treated to greatly reduce its combustibility. Building codes permit the use of fire-retardant-treated wood for certain applications in fire-resistive and noncombsutible construction (Type I and II). For example, the *International Building Code* allows fire-retardant treated wood in non-load bearing partitions where the required fire resistance is two hours or less.

Fire Retardant — Any substance, except plain water, that is applied to another material or substance and that is designed to reduce the flammability of fuels or slow their rate of combustion by chemical or physical action.

Fire retardant-treated-wood resists ignition and has increased fire endurance when compared with nontreated wood. However, wood that has received a fire-retardant treatment is not completely noncombustible and should not be confused with materials that are fire-resistive as described in Chapter 2.

The two main methods of fire-retardant treatment are pressure impregnation and surface coating. Surface coating is used primarily to reduce the surface burning of wood i.e., the flame spread rating (see Chapter 4 for more information on surface coating).

Pressure impregnation of wood is performed by placing the wood to be treated in a large cylinder in which a vacuum is created. The vacuum draws air out of the cells of the wood. A solution containing the fire-retardant chemical is introduced into the cylinder and the cylinder is pressurized. The pressure forces the fire-retardant chemicals into the cells of the wood. Pressure impregnation has the advantage of producing a treatment that is permanent when used under proper conditions.

A number of fire-retardant chemicals are available. They are proprietary products, however, and their exact formulations are not available. The fire-retardant treatments most commonly used are combinations of inorganic or organic salts. Any of the following chemicals may be used in the treatment:

- Ammonium phosphate
- Ammonium sulfate
- Boric acid
- Zinc chloride
- Sodium dichromate
- Borax

Most fire-retardant chemicals operate by accelerating the formation of charring in the wood when the wood is exposed to heat. The charring reduces the formation of volatile gases in the wood and retards the actual flaming combustion. The chemical reaction of the fire retardant is intended to occur at temperatures lower than those that are developed in fires.

The fire-retardant treatment of wood is beneficial but has some disadvantages and limitations. Some treatments may use chemicals that are water soluble and cannot be used for exterior applications. Others can be used in interior applications where high humidity does not exist. Fire-retardant treated wood is required by the codes to be labeled indicating its suitability for exposure to moisture. Fire-retardant wood is also stamped, but after it is placed in the structure cannot be identified as such **(Figures 7.18 a and b)**.

The fire-retardant treatment of wood will reduce its strength. Designers using fire-retardant treated wood must reduce the allowable stresses within treated members to values specified by the company that provides the treatment.

Fire-retardant treatments used in the 1980's were somewhat hygroscopic, meaning that fire-retardant treated plywood could react with the moisture in the air under conditions of elevated temperature and become brittle and crumbly. These conditions sometimes occurred in the attic spaces beneath roofs. Where fire-retardant treated plywood had been used as roof sheathing, the deterioration of the plywood would result in structural failure. Unevenness and sagging of the roofs would occur three to five years after installation.

Deterioration of plywood used in roofs poses an additional danger for firefighters. Firefighters should not work on any roof where an indication of roof deterioration (such as the sagging or unevenness) is observed.

NOTE: Because fire-retardant treatments and wood products are continually changing, it is advisable that firefighters continue to stay current on the materials used in their response areas.

Thermoplastic Composite Lumber

Thermoplastic composite lumber is a wood-like product produced from wood fiber and polyethylene or polyvinylchloride (PVC). It is a product developed as an alternative to preservative-treated lumber. Thermoplastic composite lumber is not intended to be used in the structural framing of a building such as beams, trusses, or studs. Its main application is in the construction of outside decks and railings **(Figure 7.19, p. 214)**. It is manufactured as boards in sizes comparable to sawn lumber and in various shapes as architectural trim. Its main advantage is its resistance to weathering.

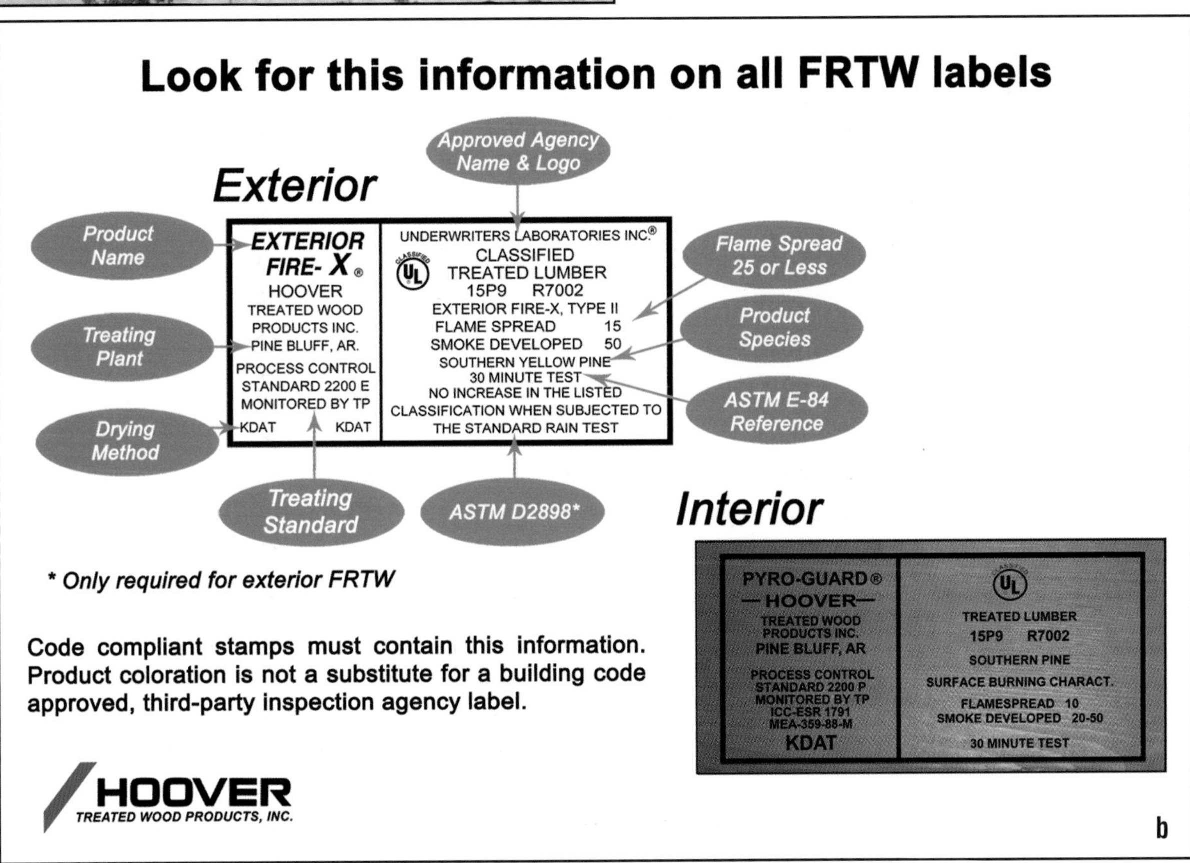

Figures 7.18 a and b (a) Stamp used on fire-retardant wood; (b) explanation of elements in fire-retardant stamp. *Photo a courtesy of Greg Havel, photo b courtesy of Hoover Wood Products.*

Figure 7.19 Thermoplastic composite lumber looks much like ordinary lumber but will behave like plastic under fire conditions. *Courtesy of Donny Howard.*

Thermoplastic composite lumber is a combustible product with a flame spread rating of 80. Firefighters need to be aware that decks constructed of thermoplastic composite lumber may look at first glance like wood but under fire conditions will melt like plastic. These types of decks also burn very black and sooty, unlike wood.

Wood Structural Systems

Wood is very commonly used in frame structural systems. Perhaps the only use of wood for solid wall construction is the use of solid logs in a log cabin. It is probably for this reason that buildings using a wood structural system are often referred to by firefighters as "frame buildings" with the modifier "wood" omitted. In fact, though, buildings can have a structural frame of wood, steel, or concrete.

The wood framing systems most frequently encountered can be broadly classified into two basic types: timber framing and light-wood framing. Other wood construction types that may be encountered include pole construction, log construction, and prefabricated panel construction. Smaller wood-frame structures, such as private garages and single-family dwellings, can be constructed using only the tried-and-true techniques of carpentry without engineering analysis.

Figure 7.20 Engineered wood-frame structures can be several stories high. *Courtesy of Ed Prendergast.*

NOTE: Wood timber framing should not be confused with masonry heavy-timber buildings that make use of exterior masonry walls (Type IV construction).

Large or custom-designed wood structures require engineering analysis. Engineered wood structures can be built several stories high **(Figure 7.20)**. Because of labor costs and limitations in the basic strength of wood, most wood-frame buildings do not exceed three stories.

Heavy-Timber Framing

Heavy timber framing evolved from hand-hewn wooden timbers that were painstakingly cut from logs. Until the development of water-powered sawmills approximately 200 years ago, the production of individual boards was a slow and laborious procedure.

In heavy-timber design, the basic structural support is provided by a framework of beams and columns that are made of wooden timbers **(Figure 7.21)**. Either trusses or beams can be used to support the roof. The exterior walls are non-load-bearing panels with an exterior siding that may be any of several materials. Ordinary corrugated sheet metal is sometimes used for the exterior walls of small storage or industrial buildings if energy conservation is not important. In heavy-timber framing, the columns are not less than 8 x 8 inches (200 mm x 200 mm) and the beams (except roof beams) are not less than 6 x 10 inches (150 mm x 250 mm).

Figure 7.21 Heavy-timber framing uses columns not less than 8 x 8 inches (200 mm x 200 mm). *Courtesy of Vermont Timber Works.*

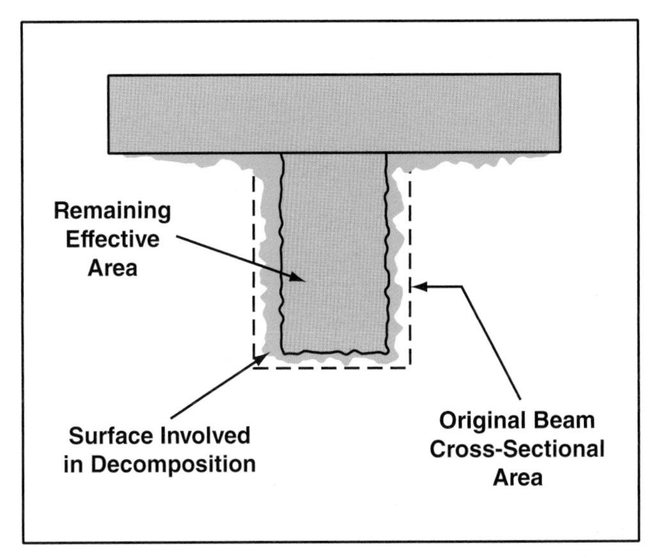

Figure 7.22 The cross-sectional area of a wood member gradually decreases as it burns.

Because they are more massive and have a lower surface-area-to-mass ratio, heavy timbers display greater structural endurance under fire conditions than members used in light-frame construction. The integrity of a wood structural member, of course, deteriorates as the wood is consumed. As a wood member burns, the cross-sectional area gradually diminishes until the member can no longer support the applied loads **(Figure 7.22)**. Equations have been developed that can be used to estimate the fire resistance of heavy-timber columns and beams when exposed to the standard ASTM E-119 fire described in Chapter 2. One such equation that is contained in the model building codes is as follows:

$$t = 2.54ZB[4 - (B/D)]$$

This equation is used for beams exposed to fire on three sides. For members exposed to fire on four sides or for columns a different equation would be used. In the equation

t = fire resistance in minutes

B = width of beam in inches before fire exposure

D = depth of beam in inches before fire exposure

Z = load factor which accounts for the loading on the beam. Values for Z are provided in the building codes.

The above equation is applicable to timber members with nominal dimensions 6 inches (150 mm) or greater.

As an example of the use of the above equation, consider a beam exposed to fire on three sides with an 8-inch (200 mm) width and a 14-inch (350 mm) depth. If the beam were loaded to 75 percent of its allowable load Z, the load factor, would be 1.1. The equation becomes:

$$t = (2.54)(1.1)(8)(4 - 8/14) = 76.6 \text{ minutes}$$

These equations are mainly of value for analytical purposes. They do not account, for example, for the connectors and fasteners that may be used in a timber assembly. If a certain degree of fire resistance were required in a timber assembly, any exposed connectors would have to be suitably protected. Furthermore, they do not address the fire resistance of any supported assembly such as a floor deck.

From the standpoint of the Incident Commander, the values needed for an analytical solution are simply not available in the course of an event. The conditions of fire loading, actual fire growth, and even timber dimensions are not precisely known in the middle of a fire fighting operation. Therefore, the structural integrity of heavy-timber framing in an actual fire situation *must* always be viewed conservatively.

NOTE: Because heavy-timber "mill" construction is combined with masonry walls, its fire-behavior characteristics are covered in Chapter 8, Masonry.

The overall integrity of wood-frame systems is affected by the methods used to join the joists, beams, and columns. In the design of connections for heavy-timber construction, the engineer must take into account factors that are unique to wood. These include the following:

- Specific gravity of the wood
- Shrinkage of the wood
- Position of fasteners, such as bolts, relative to the grain of the wood
- Relative size of the wood members and the fasteners

The firefighter, of course, is concerned with the integrity of timber connections under fire conditions and the likelihood of the collapse of roof or floor decks.

A connection between members must be capable of transferring a load from member to member. In light-frame wood construction, nails, staples, metal plates, or screws may be adequate. In the case of heavy-timber framing, however, the loads carried are greater and the connection usually incorporates through bolts, special brackets, or the bearing of one member directly on another. **Figure 7.23** illustrates several techniques used to connect timber members.

Mortise — Notch, hole, or space cut into a piece of timber to receive the projecting part (tenon) of another piece of timber.

Tenon — Projecting member in a piece of wood or other material for insertion into a mortise to make a joint.

Older timber construction made use of a type of joint known as a *mortise and tenon* joint. In this method, one timber member is cut to fit into a recess in a mating member **(Figure 7.24)**. This method of joining members is highly labor intensive and therefore costly. In modern construction, mortise and tenon joints are used only in rare cases where the designer desires a particularly artistic or quaint appearance.

The glulam beams discussed earlier are used frequently in heavy-timber construction where greater lengths are required. Glulam beams behave in the same manner under fire conditions as solid timbers. Information available from the United States Department of Agriculture indicates that heat of a fire has essentially no effect on the adhesives that are used in contemporary glulam beams.

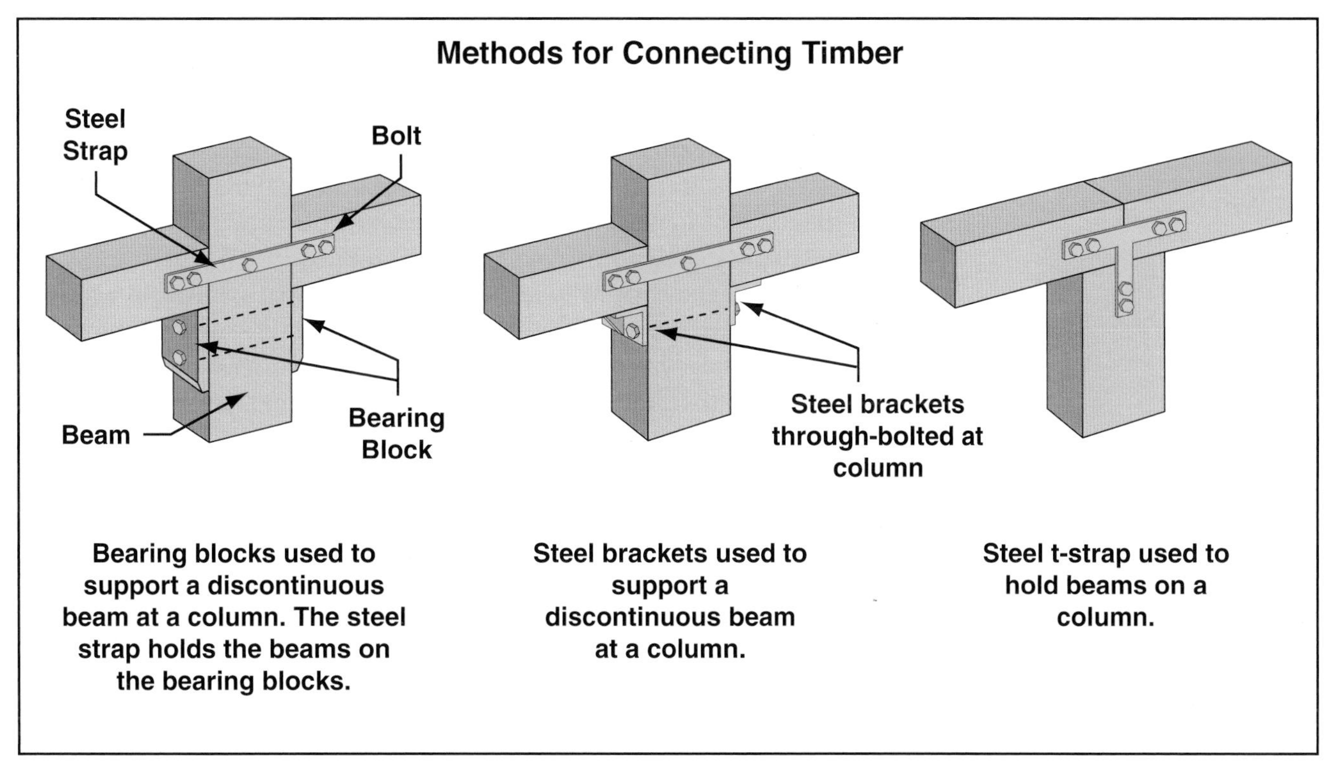

Figure 7.23 Methods for connecting timber.

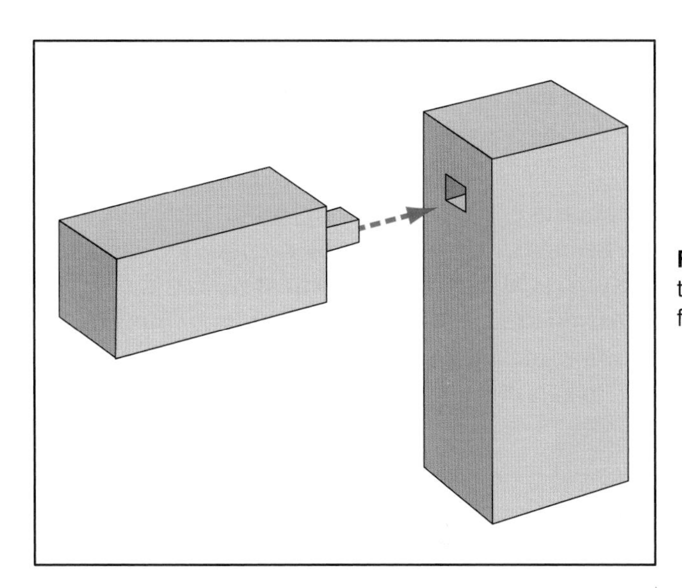

Figure 7.24 A mortise and tenon joint is likely to be found in older construction

Figure 7.25 Post and beam framing is heavier than light-frame construction but smaller than heavier-timber construction. *Courtesy of Ed Prendergast.*

Post and Beam Framing

Post and beam framing is a form of wood-frame construction in which the columns (termed the posts) and the beams are of dimensions less than those used in heavy-timber framing but greater than those used in light-frame construction. The posts are usually 4 x 4 inches (100 x 100 mm) or 6 x 6 inches (150 x 150 mm). The posts are usually spaced 4 to 12 feet (1.3 m to 4 m) **(Figure 7.25)**. Post and beam construction was once fairly common but is usually more labor intensive than light-frame construction. This method of construction has enjoyed some resurgence and is frequently used in rustic-style dwelling and small storage buildings.

The posts and beams used in the framing create square or rectangular shapes that must be braced to provide diagonal stability. This is accomplished by using diagonal bracing at the intersection of the post and beam or through the use of wall panels. In both heavy framing and post and beam framing the interior wood surface is usually left exposed. Architecturally, the exposed warm and rustic surface of the wood creates an attractive finish. From a fire fighting viewpoint, leaving the wood framing exposed offers some advantage because it eliminates combustible voids that can provide avenues for communication of fire.

Light Wood Framing

The most popular form of wood framing is known as light wood-frame construction. Light wood-frame construction evolved in the 19th century when the development of the sawmill made it possible to expeditiously cut boards from logs. Light wood framing makes use of 2-inch (50 mm) nominal lumber such as 2 x 4s or 2 x 8s (50 mm x 100 mm or 50 mm x 200 mm). The walls are formed from vertical members known as "studs" that are 2 x 4's or 2 x 6's (50 mm x 100 mm or 50 mm x 150 mm) spaced 12, 16, or 24 inches (300 mm, 400 mm, or 600 mm) on center. The floors are supported by solid joists, truss joists, or wood I-joists. Inclined roofs are supported by "rafters" or light trusses **(Figures 7.26 a and b)**. The two basic types of light wood framing are balloon framing and platform framing **(Figures 7.27 a and b)**.

Balloon-Frame Construction — Type of structural framing used in some single-story and multistory wood frame buildings wherein the studs are continuous from the foundation to the roof. There may be no fire stops between the studs.

Balloon Framing

In balloon-frame construction, the exterior wall studs are continuous from the foundation to the roof. The joists that support the second floor are supported by ribbon boards that are recessed into the vertical stud. Historically, the term *balloon frame* came from the fragile appearance of the thin, closely spaced studs compared to the more massive members used in the earlier timber construction. They were said to be as fragile as a balloon.

Figure 7.26 a and b Lightweight wood trusses and framing for a residence. *Photo a courtesy of Ed Prendergast, Photo b courtesy of McKinney (TX) Fire Department.*

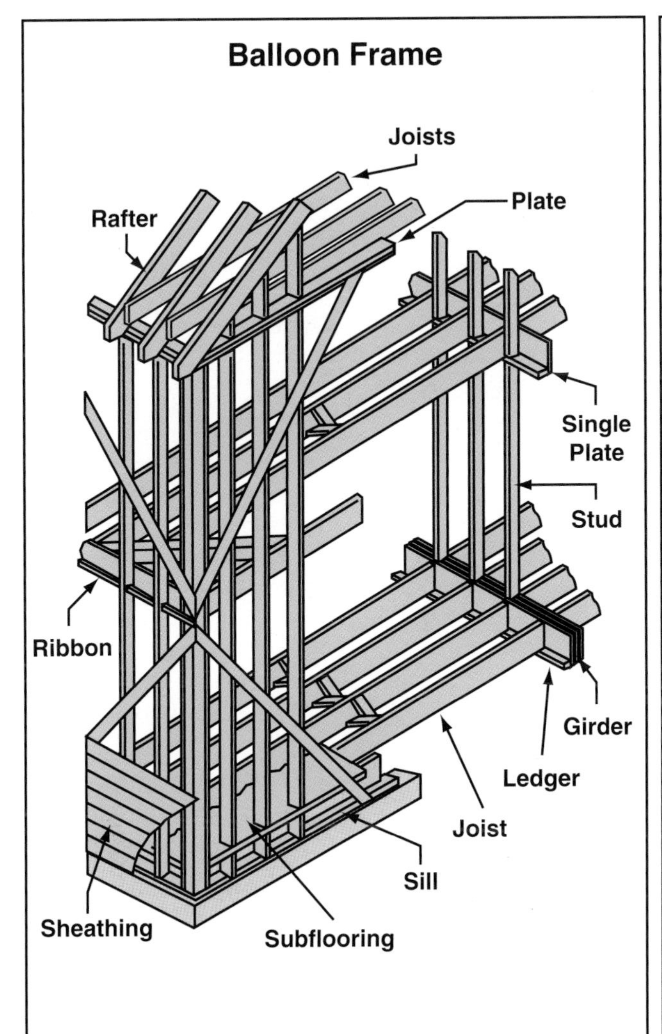

Figure 7.27 a Balloon-frame construction can have open channels from the foundation to the attic permitting rapid fire spread.

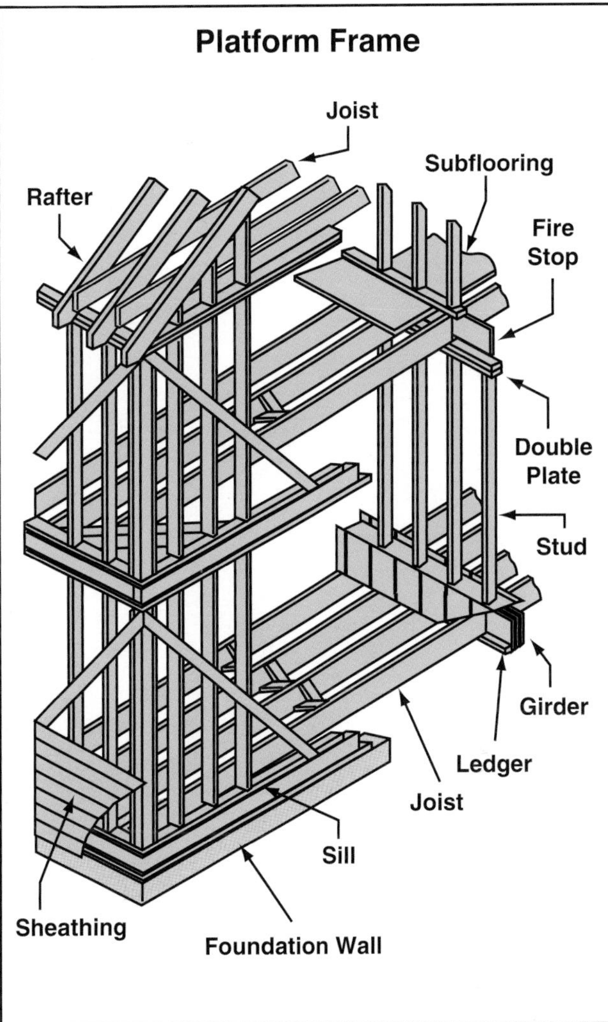

Figure 7.27 b Platform-frame construction has each floor constructed on its own platform, reducing open channels in the walls.

The vertical combustible spaces between the studs in balloon-frame construction provide a channel for the rapid communication of fire from floor to floor. Unlike timber framing, light-wood framing is usually not left exposed. The framing is usually covered with an interior finish of plaster or drywall. The interior finish will act to retard the spread of fire into the stud spaces; however, a fire may penetrate the interior finish through penetrations for electrical fixtures, plumbing, or heating.

Once the fire spreads originates in or spreads into the stud space, it can readily spread from the vertical cavity into the floor joists and into the attic space **(Figures 7.28 a and b, p. 220)**. Therefore, a fire in a balloon-frame building can be difficult to control. For example, fire issuing from an attic may give arriving firefighters the impression that the fire originated in the attic when it may have originated in the basement and communicated through the stud wall.

Balloon framing has the advantage of minimizing the effects of lumber shrinkage that can occur over time as the lumber dries and loses its moisture content. Shrinkage in lumber occurs to a greater degree in the cross-sectional dimensions than in its length. Therefore, less pronounced shrinkage effects

Figures 7-28 a and b (a) Note the open channels in a ballon-frame house that can contribute to fire spread. (b) Another balloon-frame structure after a fire. *Photo a courtesy of Wil Dane; photo b courtesy of Ed Prendergast.*

occur with the use of the continuous studs of balloon framing. With the increased use of dried lumber, however, the advantages of balloon framing have diminished. Balloon framing has not been widely used since the 1920s, although many balloon-frame buildings remain.

Platform Framing

Platform Frame Construction — Type of framing in which each floor is built as a separate platform and the studs are not continuous beyond each floor. Also called Western Frame Construction.

In platform framing (also sometimes known as Western framing), the exterior wall vertical studs are not continuous to the second floor. The first floor is constructed as a platform upon which the exterior vertical studs are erected. The second floor is also constructed as a platform and the second floor studs are erected on the second floor.

The construction of a platform-frame building begins with a wood sill being attached to the foundation, usually with bolts **(Figure 7.29)**. A header and floor joists or trusses are attached to the sill. Subflooring is then attached to the floor joists to form a floor deck. The first floor wall framing with a top and bottom plate is usually laid out horizontally on the floor deck and then raised into its vertical position. When the first floor walls are in position and braced, the second floor joists are erected on the top plates of the first floor walls.

From a construction standpoint, platform-frame buildings are easier to erect than balloon-frame buildings. This type of construction uses shorter, more easily handled lengths of lumber **(Figure 7.30)**. The flooring of each story can be used as a platform on which to work while erecting additional walls and partitions. Platform framing is more prone to shrinkage than balloon framing. This is because platform framing makes use of more horizontal members in its frame than a balloon-frame building. The shrinkage can produce greater vertical movement at different points. This vertical movement can cause undesirable effects such as cracking of plaster and misalignment of door and window openings.

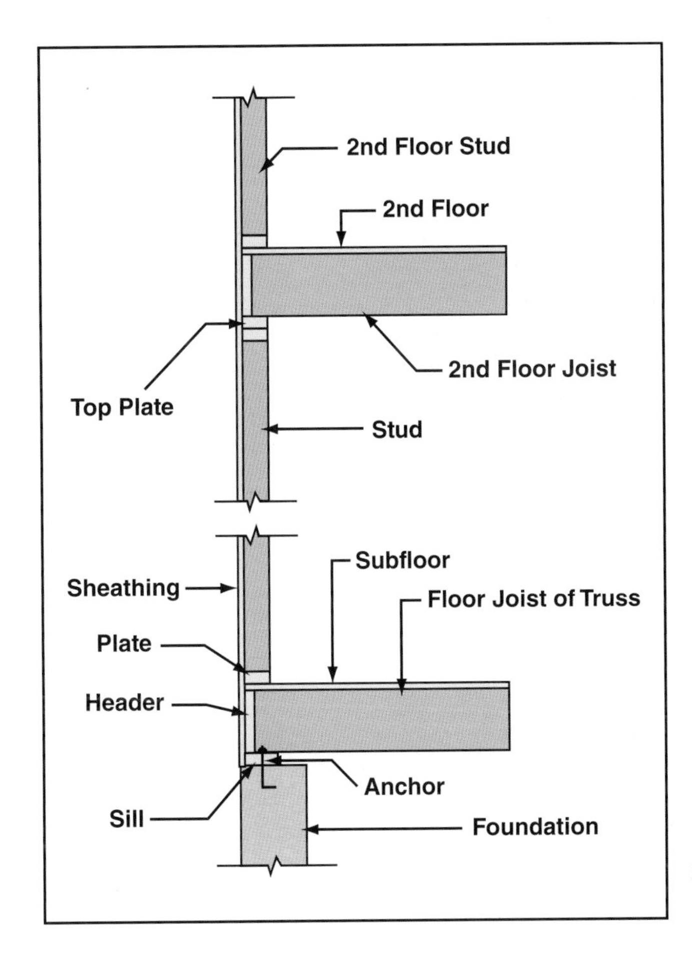

Figure 7.30 Platform frame building under construction. Notice that the studs are not continuous from floor to floor. *Courtesy of Ed Prendergast.*

Figure 7.29 Details of a wall for platform frame construction.

Firestopping

To prevent the rapid spread of fire through the concealed spaces within combustible construction, building codes require firestopping (or fireblocking). Fire stopping can consist of various materials including 2-inch (50 mm) nominal lumber, gypsum board, cement fiber board, and batts or blankets of mineral wool, glass fiber, or other approved materials **(Figures 7.31 a-c, p. 222)**.

Fire Stop — Solid materials, such as wood blocks, used to prevent or limit the vertical and horizontal spread of fire and the products of combustion in hollow walls or floors, above false ceilings, in penetrations for plumbing or electrical installations, in penetrations of a fire-rated assembly, or in cocklofts and crawl spaces.

The building codes contain specific requirements for the placement of firestopping. A few locations where firestopping is required include the following:

- Within stud walls at the ceiling and floor levels
- At the interconnections between vertical stud wall spaces and horizontal spaces created by floor joists or trusses
- Between stair stringers
- Behind soffits

In a platform-frame building, the plate installed on the top of the studs provides an inherent fire stop that will act to block the spread of fire from floor to floor within the walls. In balloon-frame buildings, however, firestopping must be provided in addition to the structural members.

Of particular concern to firefighters is the spread of fire through combustible attics and concealed spaces, especially those with light-frame trusses. Building codes require *draft stopping* in the attic spaces of combustible construction. The usual requirement is for the attic space to be subdivided into

Figures 7.31 a-c. Firestopping can be accomplished by using gypsum board or by placing lumber horizontally to help stop vertical fire spread. *Photos a and b courtesy of McKinney (TX) Fire Department; photo c courtesy of Dave Coombs.*

Draft Stops — Dividers hung from the ceiling in large open areas that are designed to minimize the mushrooming effect of heat and smoke. Also called Curtain Boards and Draft Curtains.

Sheathing — (1) Covering applied to the framing of a building to which siding is applied. (2) First layer of roof covering laid directly over the rafters or other roof supports. Sheathing may be plywood, chipboard sheets, or planks that are butted together or spaced about 1 inch (25 mm) apart. Also called Decking or Roof Decking.

areas of 3,000 square feet (278.7 sq. m). The draft stops can be constructed of various materials including gypsum board, wood structural panels, 1-inch (25 mm) nominal lumber, or cement fiberboard.

Exterior Wall Materials

In addition to the structural framing, the exterior walls of a wood-frame building include sheathing, building paper, insulation, and siding.

Sheathing

Sheathing is installed on the outside of the studs to provide structural stability, insulation, and an underlayer for the siding. The most common sheathings are plywood, OSB, particle board, or exterior gypsum sheathing **(Figures 7.32 a and b)**.

Building Paper

A layer of building paper is provided between the sheathing and the siding to act as a vapor barrier. The paper acts to reduce the infiltration of moisture and air.

Figures 7.32 a and b Commonly used sheathing materials: Oriented Strand Board for a residence and exterior gypsum sheathing for a commercial structure. *Photo a courtesy of Dave Coombs; photo b courtesy of McKinney (TX) Fire Department.*

Insulation

The use of foam plastics as insulation has attracted considerable attention in recent years. Because foam insulation is combustible and because flame spreads rapidly over its surface, building codes impose stringent regulations on its use. Typically, a code will require that foam insulation be faced with a thermal barrier such as gypsum wallboard to prevent or retard surface ignition of the foam.

The extent to which the presence of a foam insulation in a wood-frame wall will increase fire spread within the wall depends on the existence of an air space. If an air space exists between the foam and the wall surface, fire development within the wall space will be rapid because the fire will spread over the plastic surface and have air to feed it available from within the space. If the space is completely filled with the foam, however, the fire would have to burn upward through the material and would progress much more slowly. The use of a combustible insulation does somewhat increase the possibility of a fire starting within the wall. This is due to the possibility of an electrical malfunction within the wall igniting the insulation.

Noncombustible materials can be used for insulation. These include glass wool and rock wool in the form of batts or blankets, or fiber glass **(Figure 7.33)**. The paper or foil coverings used on fiber glass insulation are combustible, however.

Insulation can also take the form of loose-fill material such as granulated rock wool, granulated cork, mineral wool, and glass wool. The loose wool insulation can be either blown into stud spaces or packed by hand. Cellulose fiber and shredded wood can also be used as loose insulation material. These can be treated with water-soluble salts to reduce their combustibility. A fire in such material will progress in a slow smoldering manner. Thus, good fire fighting tactics require that the wall be opened and the insulating material thoroughly checked.

Figure 7.33 Fiber glass, covered with either paper or foil, is a very common way to provide insulation. *Courtesy of McKinney (TX) Fire Department.*

Siding Materials

Siding provides the exterior cladding of a wood-frame building. It contributes to the appearance of a building and provides weather protection. Various materials can be used for siding, including wood boards, plywood, wood shingles, aluminum, stucco, asphalt, stone, cement board, and vinyl **(Figures 7.34 a and b)**. Some siding materials such as stucco and stone are noncombustible; other materials are combustible.

The combustibility of a siding material can affect the fire behavior of a building in two ways. One way is through the exterior communication of fire from window and door openings **(Figure 7.35)**. The other way is by being ignited by an exposing fire.

When a building is remodeled, a new siding material is frequently applied over the existing siding. It is not unusual, therefore, for older buildings to have multiple layers of siding. One common method is to place a vinyl siding over an existing siding.

Asbestos sidings were commonly used from the 1930's until the 1970's; they may be covered with another siding or still exposed. The hazardous nature of asbestos can cause problems during overhaul. Decontamination of gear and notification of environmental officials may be required in these cases.

Figures 7.34 a and b (a) Wood shingles used as exterior siding. (b) vinyl and stone siding. *Photo a courtesy of Ed Prendergast.*

Figure 7.35 Fire can spread up the exterior of a building from window and door openings. Notice that the aluminum siding has melted, exposing older siding underneath. *Courtesy of Ed Prendergast.*

Brick Veneer

A wood-frame building, like other buildings, can be provided with an exterior facing of brick. Such construction is termed *brick veneer.* Brick veneer construction provides the architectural styling of brick at less cost than a full masonry wall. The brick veneer adds little to structural support and must be tied to the wood frame wall at intervals of 16 inches (400 mm). Over time, the metal ties may corrode. The brick veneer does add to the thermal insulating value of the wall **(Figure 7.36)**.

Under fire conditions, the external brick layer of a brick veneer building protects the wood frame from external exposure. Because the main structural support is still provided by an internal wood frame, there is little difference between a brick veneer building and an ordinary wood-frame building in terms of fire behavior. Fire spread can also occur in the void between the brick and the building paper.

From the outside it can be difficult for a firefighter to visually determine if a building has brick bearing walls or brick veneer walls. One frequently used rule is that in a brick bearing wall every sixth course of brick is a header course with the ends of the brick facing out (See Chapter 8). However, this is only a general rule because some masonry bearing walls may make use of horizontal ties instead of a header course. Occasionally brick veneer walls are constructed with half-bricks that resemble a header course.

Figure 7.36 Brick veneer being applied to a wood frame. *Courtesy of Ed Prendergast*

Interior Finish Materials

The interior walls of wood-frame buildings can be left exposed with no interior finish. This is often done in small buildings such as garages or sheds. Such buildings would be classified as Type V-B under the building codes. For most occupancies, however, an interior finish is provided. An interior finish enhances appearance and provides a measure of thermal insulation.

The interior of a wood-frame building can be finished using a number of materials including ordinary plywood or OSB. Where a degree of fire resistance is required, the interior finish materials most commonly used are gypsum board or plaster **(Figure 7.37)**. Plaster is not often used in modern construction because it is relatively labor intensive. When the structural framing of a wood-frame building, including the floor and roof construction, is provided with protection to achieve a 1-hour fire resistance, the building can be classified as

Figure 7.37 Gypsum board is a very common interior finish material that provides a measure of fire resistance. *Courtesy of Greg Havel, Burlington, WI.*

Type V-A (See Chapter 2). A typical means of achieving this is through the use of 5/8 inch (15.8 mm) gypsum board attached to the studs and ceiling joists. A number of specific methods for utilizing gypsum board are listed in the Underwriters Laboratories *Fire Resistance Directory.*

In most cases the gypsum board used for interior finish can be easily breached by firefighters during overhaul. Opening the wall facilitates the location of fire travel in light wood-frame construction. In some occupancies such as schools and correctional facilities, an impact-resistant gypsum board may be used. This product is produced with a fiberglass mesh to resist impacts. Impact-resistant gypsum board is indistinguishable from ordinary gypsum board, but will prove more difficult to penetrate with hand tools.

If a wood-frame building (Type IV-A) is equipped an automatic sprinkler system, the codes may permit the elimination of the 1-hour structural fire resistance. Additionally, a code may permit the elimination of the structural fire resistance for roof members located more than 20 feet (6.1 m) above the floor level.

Figure 7.38 Exposed wood floor construction in a residential basement. *Courtesy of Ed Prendergast.*

Collapse Considerations of Wood-Frame Construction

The combustible voids inherent in light wood-frame construction provide paths for the rapid spread of fire through a structure. A fire-resistive interior finish such as gypsum board will delay the entrance of fire into the void spaces if it is not penetrated. Once the framing members become exposed to a fire, however, fairly rapid failure of the structural system must be anticipated.

It can be difficult for firefighters to determine the extent of fire spread above or below the area where they may be working. Where truss joists are used in floor construction it is possible for fire to spread in four directions; parallel to and perpendicular to the truss joists. Frequently, especially in residential construction, the flooring will consist of plywood or OSB subflooring with a carpet or tile floor surface. The thin subflooring will fail within minutes when exposed to significant fire.

A further area of concern is the possible existence of an unfinished floor over a basement space. In **Figure 7.38** wood I-beams are being used to support a floor of OSB. The I-beams are not protected. The wood structural members would be immediately exposed to any fire in the basement. Notice further that the floor joists are being supported by an unprotected steel beam.

Case History

On August 13, 2006 a 55-year-old Wisconsin firefighter was killed and another was injured when they both fell through the floor into the basement of a single-family dwelling.

The fire occurred in a two-story, 7-year-old single family residence of ordinary construction. It encompassed approximately 3,500 square feet of living space above grade and 2,100 square feet below grade. The floor system in the structure consisted of wood trusses and engineered wood I-beams.

The fire originated in an unfinished section of the basement that was approximately 750 square feet in area **(Figure 7.39, p. 228)**. The ceiling in the room of origin consisted of exposed floor trusses supported by wood beams and metal posts. The basement was separated into two sections by a hallway. One side of the hallway (the south side) consisted of a concrete block wall. The north side of the hallway consisted of a framed partition possibly with drywall. The portion of the basement south of the hallway was finished.

The fire occurred at approximately 1223 hours; an engine, a truck, and an ambulance were dispatched and arrived at 1227 hours. Arriving firefighters encountered light smoke showing from the structure. The arriving engine company began to lay a line. As the companies began operations they detected smoke coming from a vent next to the front door that appeared to be coming from the basement. Subsequently black smoke and burning particles were seen emitting from another vent near the ground at the rear of the structure.

The engine company advanced a 1¾ inch line through the front door to a basement stairway at the rear of the building. (See diagram). Two firefighters were directed to conduct a primary search of the first floor at approximately 1234 hours. They entered through the front door and took a couple of steps to the left. The two firefighters were forced to go to their knees due to heavy smoke conditions. As they crawled forward they heard a load crack and then both firefighters fell through the floor into the basement.

Both firefighters fell into the room of origin. This portion of the basement immediately became engulfed in fire. The firefighter who survived was able to shield herself from the fire with some debris. As other firefighters directed a hose stream into the basement from above, the surviving fire firefighter was able to make her way into the next basement room. She was able to make her way to the rear of the basement where other firefighters assisted her through a window.

The Incident Commander activated a rapid intervention team (RIT). The RIT entered the basement by means of a set of stairs in the garage portion of the building. However, they were not able to enter the room of origin. The body of the deceased firefighter was recovered the next day.

Lesson Learned: Of significance in this incident was the fact that the floor trusses over the room of origin supporting the floor were unprotected. Failure of the floor over the finished portion of the basement was not indicated. Unfinished basement ceilings are commonly encountered in residential construction. Early failure of unprotected wood floor trusses must be anticipated whenever there is an indication of significant fire. These types of structural members must be identified during preincident planning.

Source: NIOSH

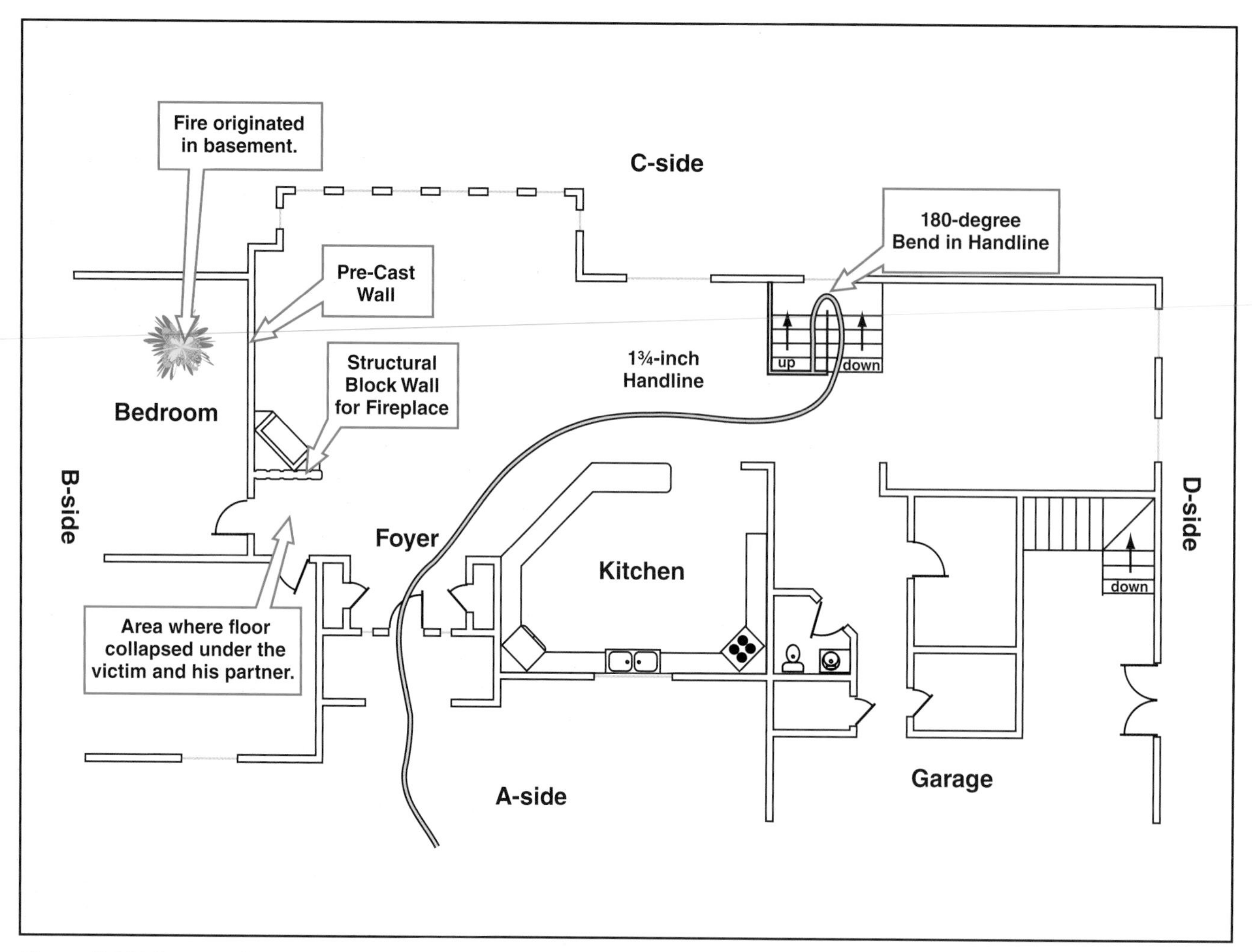

Figure 7.39 Diagram of structure where firefighters died. *Source: NIOSH*

Figure 7.40 Structures in the wildland-urban interface are in special danger of destruction due to fire. *Courtesy of Dave Coombs.*

Ignition-Resistant Construction

As urban development moves into rustic wildland areas, a serious problem develops. Fires in the often highly flammable vegetation can rapidly engulf and destroy structures. Structures of light wood-frame construction are especially vulnerable **(Figure 7.40)**. The problem has been particularly serious in the western part of the United States. To protect property in these areas, jurisdictions have adopted code requirement to make buildings ignition-resistant. The International Code Council publishes the *International Wildland-Urban Interface Code* which can be adopted by local jurisdictions.

The specific requirements for ignition-resistant construction will vary with the severity of the hazard in a given area. Factors such as ground slope, clear space around property, water supply, and climate are used to evaluate the hazard severity in a given location.

Ignition-resistant construction should not be confused with the *fire-resistive* construction described in Chapter 2. Ignition-resistant construction is intended to decrease the vulnerability of structures to exposure from wildland fires. Requirements for ignition-resistant construction include features such as fire-resistant roof coverings to protect against flaming embers (See Chapter 11). Depending on the hazard severity, an ignition-resistant building may also be required to have an exterior wall that is either noncombustible or that has a 1-hour fire-resistive rating when exposed from the outside. Other requirements include limitation on the size of attic and under-floor vents and their protection with corrosion-resistant screens.

Deterioration of Wood Buildings

Over time, wood is vulnerable to deterioration from several causes unique to wood. These include insects, decay, and shrinkage. In addition, wood structures are subject to the same forces that affect buildings constructed from other materials, such as settling, erosion, and weathering. All of these forces can undermine the structural stability of a wood-frame building, thus compounding problems for firefighters. Indications of the deterioration of a wood-frame building are often readily apparent from the outside **(Figures 7.41 a and b)**. Any building showing exterior signs of deterioration should be approached cautiously.

Figures 7.41 a and b Wood buildings frequently show indications of deterioration from age and weather.

Summary

The distinguishing characteristic of a wood-frame building is that the basic structural system is combustible. The fundamental combustibility of wood contributes fuel to a fire that occurs in a wood-frame building. In addition, the structural integrity of the wood framing members is lost as the wood is consumed, and structural failure will occur. The trend toward lighter weight, more precisely engineered wood assemblies increases the speed at which they can fail during a fire.

Wood-frame buildings, especially light wood-frame buildings, have numerous concealed spaces within the walls, attics, and floor spaces. These concealed spaces provide an avenue for the spread of fire. The concealed spaces must be opened in the course of fire fighting to check for extension of fire. In addition, the concealed spaces contain heating ducts, electrical wiring, plumbing, cooking exhausts, and chimneys. These building components give rise to the possibility of fires originating within the concealed spaces.

Review Questions

1. What are glulam beams?
2. How is pressure impregnation of wood performed?
3. What is the fire fighting advantage of post and beam framing?
4. What is platform framing?
5. When a degree of fire resistance is required, what are the most commonly used interior finish materials?

References

1. Allen, Edward and Iano, Joseph *Fundamentals of Building Construction, Materials and Methods 4th ed.,* John Wiley and Sons.
2. International Code Council®, *International Building Code®, 2006 ed.*
3. National Fire Protection Association®, *Handbook of Fire Protection, 19th ed., Sect. 8, Chap 3 and 4.*
4. Stalnaker, Judith J. and Harris, Ernest C. *Structural Design in Wood, 2nd ed.,* Kluwer Academic Publications.
5. Society of Fire Protection Engineers, *Handbook of Fire Protection Engineering 2nd ed.*, Section 4, Chapter 11.
6. Strand, Renee, P.E., *The Evolution of Structural Composite Lumber,* Structure Magazine, August, 2007.
7. U.S. Department of Agriculture, *Wood Handbook,* Chapter 17.

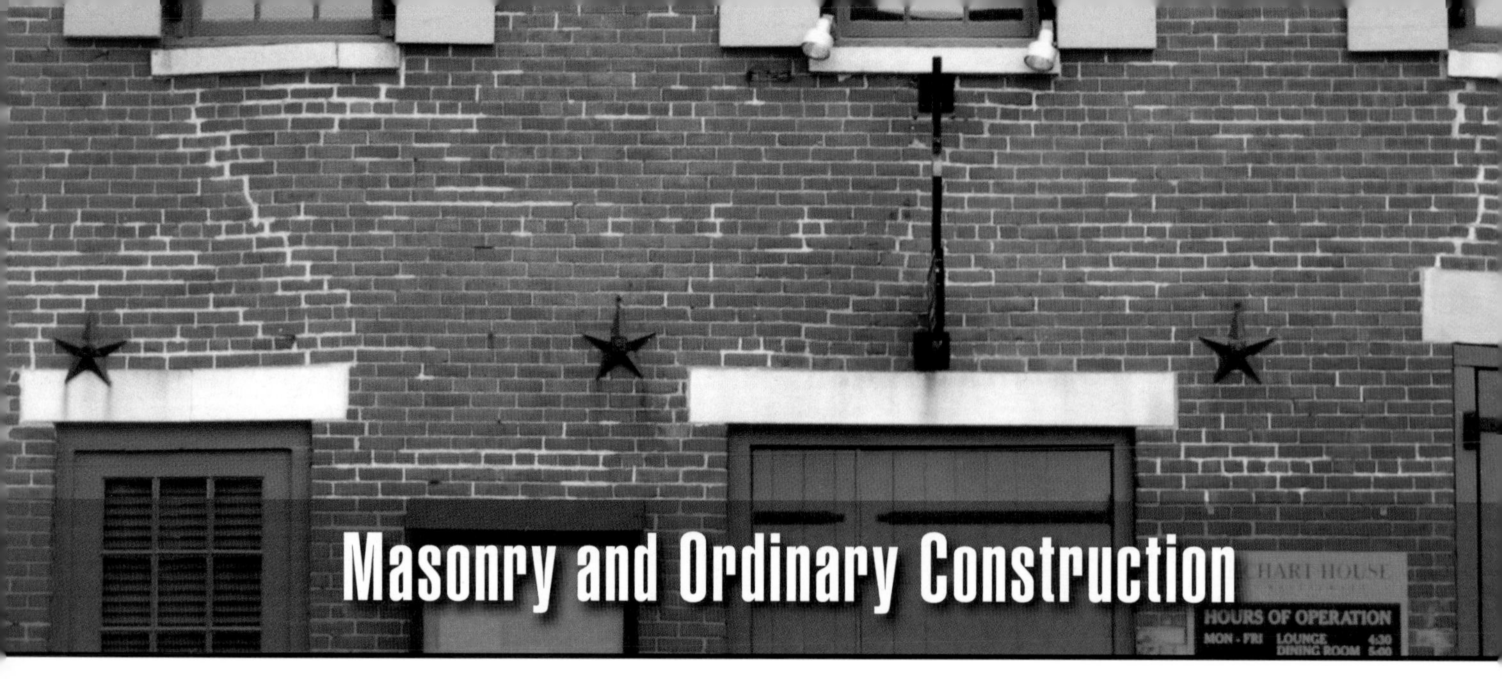

Masonry and Ordinary Construction

Chapter Contents

Divider page photo courtesy of McKinney (TX) Fire Department.

Key Terms

FESHE Objectives

Fire and Emergency Services Higher Education (FESHE) Objectives: *Building Construction for Fire Protection*

1. Demonstrate an understanding of building construction as it relates to firefighter safety, buildings codes, fire prevention, code inspection and firefighting strategy and tactics.
8. Identify the indicators of potential structural failure as they relate to firefighter safety.

Masonry and Ordinary Construction

Learning Objectives

After reading this chapter, students will be able to:

1. Describe the properties of the masonry products used as building material.
2. Describe the construction techniques and characteristics of masonry walls.
3. Describe the characteristics of the interior structural framing used in masonry buildings.
4. Discuss the factors that affect fire behavior in masonry structures.
5. Explain the differences between mill construction and ordinary masonry construction.

Chapter 8 Masonry and Ordinary Construction

Case History

Event Description: We responded to a reported fire in a residential structure. The first-arriving engine reported heavy smoke in an approximately 6,000 square foot, two-story house. There was an IC in place, accountability in place, and fireground communications on a dedicated TAC channel. There was a primary and secondary water supply established and two aerial trucks positioned for a defensive attack if needed. About ten minutes into the fire, the IC asked for a progress report. Interior crews reported some progress. They were operating two 1 3/4" handlines on the second floor at the top of the stairwell.

About 15 minutes into the fire, a significant structural collapse occurred. An emergency evacuation order was declared on the TAC and Dispatch channel, secondary emergency evacuation signal (apparatus horns sounded) was given, and radio confirmation was received. Crews were removed safely. The structural collapse was a large section of masonry chimney that extended past the roof line, but was supported by a 2x4 frame underneath the roof. The chimney was for aesthetics only and was not functional. The large piece fell into the garage where it smashed a late-model Cadillac like it was a beverage can. If the chimney had fallen in the other direction, it would have collapsed directly where the crews were operating. It is unlikely any of them would have survived the impact.

Lessons learned: Try to identify these structures in their construction phase. Our department is considering a visible marker to indicate structures with these features. It is difficult to identify real chimneys from ornamental ones post-construction. We are also working with plans review and the fire marshal's office to receive notification of these structures. This will be entered into our dispatch program. The owner of the home rebuilt after the fire; the new home does not have the chimneys.

Source: National Fire Fighter Near-Miss Reporting System.

Masonry is one of the oldest and simplest building materials; its use dates back thousands of years. Masonry has its origin in the simple piling up of available stones that were found in fields. The fundamental building technique that evolved over the centuries consisted of stacking individual masonry units on top of one another and bonding them into a solid mass. Masonry remains a commonly used construction material today and many different types of buildings are constructed using masonry. This chapter will describe the construction methods used in masonry buildings and the performance of masonry buildings under fire conditions.

Masonry as a Building Material

Masonry is a very durable building material. Masonry structures that have stood for centuries are not uncommon **(Figure 8.1)**. Although masonry units are inherently resistive to weather, fire, and insects, deterioration of mortar joints often occurs over a long period of time **(Figures 8.2 a and b)**. One drawback to the use of masonry is that the basic construction technique of laying individual units by hand is labor intensive.

Masonry units can be made of several materials, including the following:

- Brick
- Terra cotta
- Concrete block
- Stone
- Clay tile block
- Gypsum block
- Glass

Figure 8.1 This stone masonry building illustrates the durability of a masonry buildings. *Courtesy of Ed Prendergast.*

Figures 8.2 a and b The mortar used in masonry walls is subject to deterioration over time. Note the obvious repairs in both structures. *Photo a Courtesy of Ed Prendergast; photo b courtesy of Dave Coombs.*

Bricks

Bricks are produced from a variety of locally available clays and shales. The hardness of brick is dependent on the soil used in its composition. Bricks are manufactured by placing moist clay in molds and then drying the bricks after removing them from the molds. The clay can also be extruded through a rectangular die and sliced by a cutter into the desired size. The bricks are then fired in a kiln during which they are subjected to temperatures as high as 2,400° F (1,300° C). This intense heat converts them to a ceramic material. The firing process takes 40 to 150 hours.

Bricks are produced in a number of sizes. A typical brick size is 3⁵/₈ x 7⁵/₈ x 2¼ inches (90 mm x 190 mm x 57 mm). For most brick sizes, three courses (horizontal layers) of brick plus the intervening mortar joints equals a height of 8 inches (200 mm).

Concrete Blocks

Concrete blocks are also known as concrete masonry units (CMUs). The hollow concrete block is used most frequently. Hollow concrete blocks are produced in a number of sizes and shapes, but the most common is the nominal 8 x 8 x 16 inch (200 x 200 x 400 mm) block. Because a concrete block is larger than a brick, it is somewhat more economical to use because it takes the place of several bricks. In addition to the hollow block, concrete masonry units can be produced as either bricks or as solid blocks. The hollow blocks can be filled with cement or other material after a wall has been produced for added strength.

Concrete Block — Also known as concrete masonry units (CMU). The most commonly used concrete block is the hollow concrete block.

NOTE: For more information about concrete and its manufacture, see Chapter 10.

Stone

Stone masonry consists of pieces of rock that have been removed from a quarry and cut to the size and shape desired. The most common types of stone used in construction are granite, limestone, sandstone, slate, and marble. Firefighters in many urban areas are familiar with buildings constructed of brownstone, which is one form of sandstone.

Stone can be used in several ways in construction. It can be laid with or without mortar to form walls similar to brick or concrete block or it can be used as an exterior veneer attached by supports to the structural frame of the building **(Figure 8.3)**.

Figure 8.3 An example of stone construction. *Courtesy of Ed Prendergast.*

Other Masonry Units

Clay tile blocks and gypsum blocks were once widely used for the construction of interior partitions. Their use has diminished in modern practice although they can still be found in many buildings. Structural glazed tile is still frequently used where a smooth, easily cleaned surface is desired, such as in shower rooms, institutional kitchens, or corridors.

Glass block is available in many textures and is architecturally popular for both interior partitions and exterior applications **(Figure 8.4, p. 236)**. Glass block is also frequently used to fill in windows in existing buildings. In addition, fired clay tile known as structural terra cotta, which was once popular for decorative effects, may be found in older buildings.

Figure 8.4 Glass block used in an exterior application in a masonry wall. *Courtesy of Ed Prendergast.*

Compressive Strength of Masonry

Masonry units have no significant tensile strength. In their structural application they are used to support compressive loads. The ultimate compressive strengths for various masonry materials are shown in **Table 8.1**. In practice, lower allowable stresses than those shown are used to take into account the mortar joints and to provide a factor of safety. Thus, the allowable compressive strength of brick masonry would be 250 psi (1.72 kPa).

Density of Masonry Units

The density of masonry units varies depending on the specific type, as shown in **Table 8.2**. The mass of a masonry wall is significant to fire personnel because nonreinforced masonry walls can topple when a building becomes heavily involved in fire, thus posing a serious safety risk to emergency responders. A relatively small chunk of a brick wall having a thickness of 8 inches (200 mm) and measuring 4 x 4 feet (1.2 m x 1.2 m) would have a weight of approximately 1,267 pounds (575 kg). As noted previously, masonry is inherently resistive to fire. However, some deterioration can occur upon prolonged exposure to a fire.

Mortar

Mortar — Cement-like liquid material that hardens and bonds individual masonry units into a solid mass.

Mortar is an inherent part of masonry construction. The primary function of mortar is to bond the individual masonry units into a solid mass. Mortar also serves to cushion the rough surfaces of the masonry units permitting uniform transmission of the compressive load from unit to unit. Mortar provides a seal between masonry units and is important in the final appearance of a masonry wall. The mortar joints, however, can be the weakest part of the wall. Because the mortar must bear the same weight of the masonry wall, its compressive strength is as important as that of the masonry units. Mortar is available in five basic types with strengths varying from as low as 75 psi (0.52 kPa) to as high as 2,500 psi (18.3 kPa).

Table 8.1
Ultimate Compressive Strengths for Masonry

Type	Compressive Strengths
Brick	2,000 to 20,000 psi (14-140 Mpa)
Concrete Masonry Units	1500 to 6000 psi (10 - 41 Mpa)
Limestone	2,600 to 21,000 psi (18 – 147 Mpa)
Granite	15,600 to 30,800 psi (108 – 202 Mpa)
Sandstone	4,000 to 28,000 psi (28 – 195 Mpa)

Table 8.2
Densities of Masonry Materials

Type	Density
Brick	100 to 140 lb/cu ft (1,600 – 2,240 kg/cu m)
Concrete Masonry Unit	75 to 135 lb/cu ft (1,200 – 2,160 kg/cu m)
Limestone	130 to 170 lb/cu ft (2,080 – 2,720 kg/cu m)
Granite	165 to 170 lb/cu ft (2,640 – 2,720 kg/cu m)
Sandstone	140 to 165 lb/cu ft (2,240 – 2,640 kg/ cu m)

Portland Cement — Most commonly used cement consisting chiefly of calcium and aluminum silicates. Portland cement is mixed with water to form a paste that hardens and is, therefore, known as a hydraulic cement.

Most mortar is produced from a mixture of portland cement, hydrated lime, sand, and water. The portland cement functions as the bonding agent. At one time, mortar was produced without portland cement and the lime was the bonding agent.

Sand-lime mortar was commonly used in masonry construction until the 1890's. Today sand-lime mortar is found only in historic buildings. Because of the differences in properties between portland cement mortar and sand-lime mortar, older masonry construction that made use of sand-lime mortar must be repaired with sand-lime mortar.

Firefighters should be aware that use of master streams during a fire can weaken mortar, either from the pressure of the stream or from the flushing effect of the water.

Masonry Structures

Masonry can be used for a variety of purposes in architecture including primarily decorative functions such as a masonry fence or stonework trim. The main interest for the firefighter, however, is in its use for the construction of walls.

Construction of Masonry Walls

Bearing Wall — Wall that supports itself and the weight of the roof and/or other internal structural framing components such as the floor beams above it.

Masonry can be used to construct bearing walls that provide the basic structural support for a building. These walls are known as *bearing walls* **(Figure 8.5, p. 238)**. Masonry can also be used for non-load-bearing curtain walls or partitions. In modern practice, the most commonly encountered load-bearing masonry walls are constructed from brick, concrete block, or combination of

Figure 8.5 This older school was constructed with brick bearing walls. *Courtesy of McKinney (TX) Fire Department.*

brick and block. Other masonry materials, such as gypsum block and lightweight concrete block, are limited to non-load-bearing partition walls. Stone masonry is often used as an architectural veneer **(Figures 8.6 a and b)**.

Masonry exterior walls can be found with a variety of interior structural framing systems including unprotected steel, protected steel, and wood. Therefore, masonry walls are found in both fire-resistive and non-fire-resistive buildings.

The thickness of a masonry wall depends on the height of the building and the method of construction used. When used as a supporting wall, the thickness of masonry walls varies from a minimum of 6 inches (150 mm) to several feet. Walls that provide the structural support for multistory buildings must be greater in thickness than those supporting single-story buildings. This is because masonry units in the lower portions of a wall must support the weight of the upper portion of the wall.

Figures 8.6 a and b Stone masonry veneers are common. Notice that in the first photo, a crack has occurred between the stone veneer and the rest of the building. The building has also been braced at the second-floor level. *Photo a Courtesy of Ed Prendergast; Photo b courtesy of McKinney (TX) Fire Department.*

As the height of a wall is increased, the dead load that must be supported at the base of the wall increases. To keep the compressive stresses within acceptable limits and to provide greater lateral stability, the load-bearing area of the wall is increased by increasing its thickness **(Figure 8.7)**.

The increasing weight of a load-bearing wall with increased height makes very tall masonry structures largely impractical or more costly than alternative designs unless the masonry is reinforced with steel as illustrated in Figure 8.11. Nonreinforced masonry walls are usually limited to a maximum height of around six stories. These types of walls are more commonly found in structures built during the early part of the 20th century.

One noteworthy example of a load-bearing masonry wall structure is the Monadnock building in Chicago. This high-rise building was constructed to a height of 16 stories in 1893 using masonry exterior bearing walls. To support the weight of the building, the walls at the base are 6 feet (2 m) thick **(Figure 8.8)**.

In contemporary practice, when a building is to be more than three or four stories tall, the use of a steel or concrete structural frame is usually more economical than a erecting a nonreinforced masonry bearing wall. However, exterior masonry veneer walls may be used in combination with a steel-frame multistory design and will give the appearance of a masonry bearing wall **(Figure 8.9)**.

If a masonry wall is reinforced with steel, the required thickness can be reduced. By using reinforced masonry, it is possible to construct load-bearing masonry walls to a height of ten stories or more having a wall thickness of only 12 inches (300 mm).

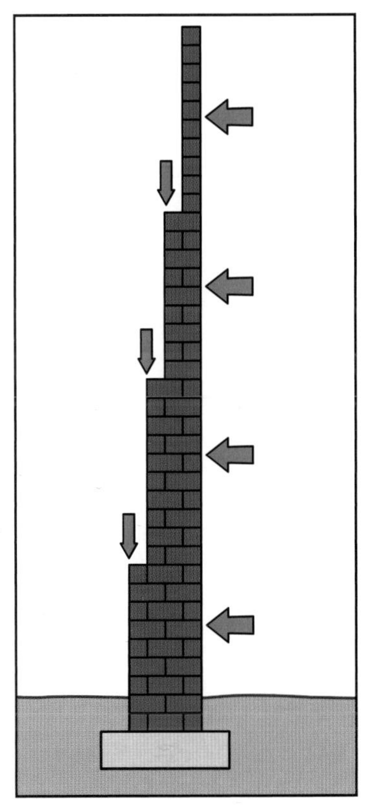

Figure 8.7 In older brick masonry buildings, the wall thickness increases at the base to handle the increasing load and to provide stability.

Nonreinforced Walls

When a masonry wall is constructed, the masonry units are laid side by side in a horizontal layer known as a *course*. The horizontal courses of brick are laid on top of each other in a vertical layer known as a *wythe* **(Figure 8.10, p. 240)**. Although the simplest brick wall consists of a single wythe, multiple

Course — Horizontal layer of individual masonry units.

Wythe — Single vertical row of multiple rows of masonry units in a wall, usually brick.

Figure 8.8 The Monadnock building in Chicago has masonry bearing walls 6 feet (2 m) thick at the base of the building. *Courtesy of Ed Prendergast.*

Figure 8.9 Masonry veneer is frequently applied over steel structures. The final result looks like a masonry structure but takes far less time to build. *Courtesy of McKinney (TX) Fire Department.*

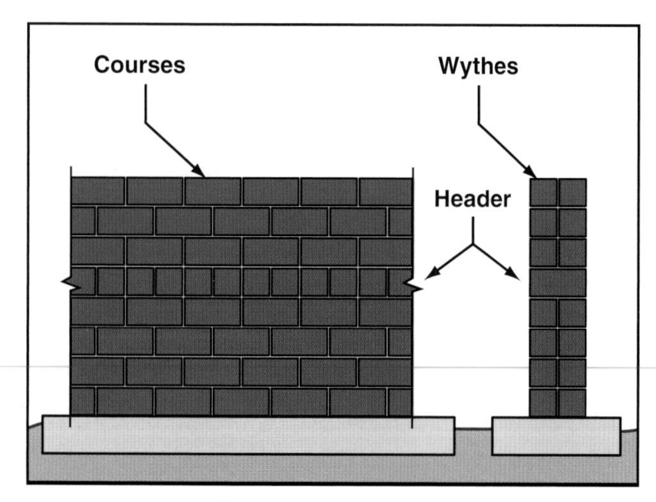

Figure 8.10 Parallel wythes of brick can be bonded using a header course every sixth course. A header course has the ends of the bricks facing outward.

wythes are normally provided to supply the necessary strength and stability in a masonry wall. A brick wythe is commonly used in combination with a concrete block wythe. Such a design is referred to as concrete block brick faced (CBBF).

In an ordinary nonreinforced wall, the strength and stability of the wall are derived from the weight of the masonry and horizontal bonding between adjacent wythes. When bricks are used to construct a masonry wall, the bricks can be placed in various positions for appearance or strength. When bricks are placed end-to-end, they create a *stretcher course.* If bricks are placed vertically on end, a *soldier course* is created.

One means of providing a horizontal bond between the wythes is to place a course of bricks across two wythes with the ends of the bricks facing out. A course of bricks laid in this manner is known as a *header course* (see Figure 8.10). The existence of header courses in a wall is one way to identify the method of construction of a masonry wall.

Header Course — Course of bricks with the ends of the bricks facing outward.

Horizontal bonding can also be accomplished through the use of different types of corrosion-resistant metal ties between the wythes. The metal ties are commonly used when the wall consists of brick and block (CBBF). In addition, metal ties are frequently attached directly to metal or wood studs for masonry curtain walls.

An exterior brick wall often is constructed with a vertical cavity between the exterior wythe and the interior wythes. Such a wall is known as a *cavity wall.* The cavity prevents water seepage through the mortar joints to the interior of the building and increases the thermal insulating value of the wall. In the case of a cavity wall, the placement of metal ties is especially important because the use of a brick header course usually is not practical.

Reinforced Walls

Masonry walls are reinforced to permit a taller building (as noted previously) or to provide lateral stability against horizontal forces such as seismic shock. Masonry walls can be reinforced by placing vertical steel rods in a cavity between two adjacent wythes of brick wall. Tie rods can also be placed between the brick wythe and the concrete block wythe in a concrete- block-brick-faced wall **(Figure 8.11)**. The cavity is then filled with grout, which is a mixture of cement, aggregate, and water. Concrete block walls can be reinforced by placing the steel rods in the openings in the individual blocks and filling the openings with grout.

The reinforcement of masonry walls can take other forms and can include such architectural features as buttresses, flying buttresses, and pilasters **(Figure 8.12)**. In older multistory buildings, interior load-bearing masonry walls were used to enclose stairwells or elevator shafts. Such walls also provide lateral support. In general, wherever masonry walls intersect, they will support and reinforce each other.

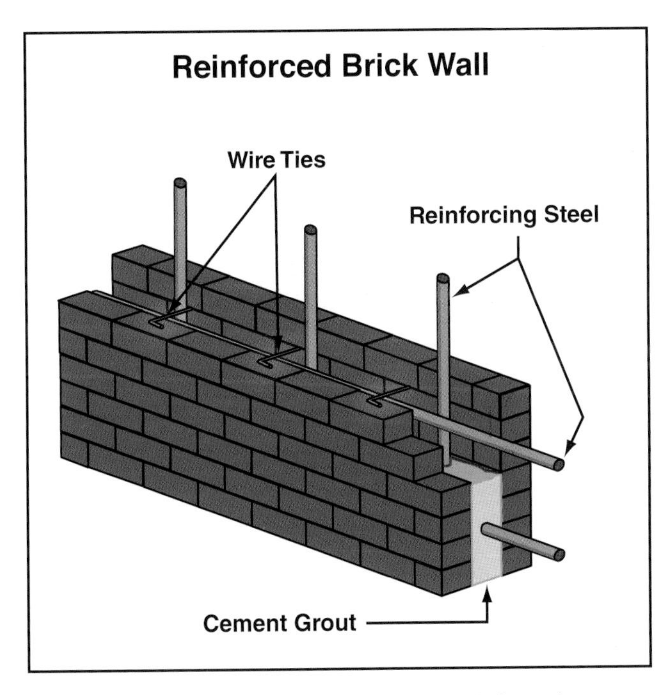

Figure 8.11 Details of reinforcement in a reinforced masonry wall.

Figure 8.12 An example of a buttressed wall used in a church. *Courtesy of Ed Prendergast.*

Openings in Masonry Walls

As with other types of walls, openings must be provided in masonry walls for doors and windows. Adequately supporting the weight of the masonry units over these openings poses a design problem because the mortar joints between the individual bricks or blocks provide little tensile support. Support of the masonry over an opening is accomplished by the use of a lintel, an arch, or corbelling **(Figure 8.13)**.

A *lintel* is a beam over an opening in a masonry wall. Lintels frequently are steel beams but also can be reinforced concrete or reinforced masonry. Wood lintels have been used in the past but are not often seen in modern construction because of shrinkage problems.

Lintel — Support for masonry over an opening; usually made of steel angles or other rolled shapes singularly or in combination.

In designing lintels, engineers assume that it is not necessary to support the complete weight of the masonry wall above the lintel. It is assumed that a lintel is required to support only the weight of a triangular section, as indicated in

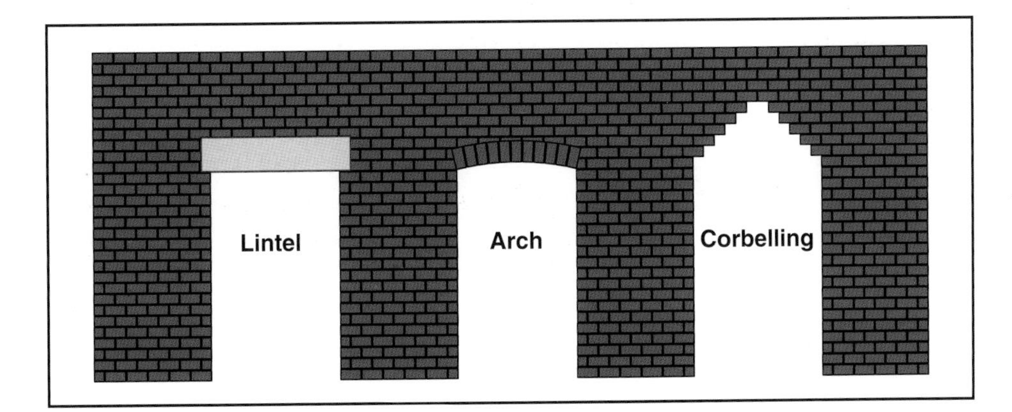

Figure 8.13 The wall portion over an opening in a masonry wall can be supported by a lintel, an arch, or a corbel.

Figure 8.14. This assumption is made because a certain amount of arching tends to occur between the masonry units above the units that reduces the load on the lintel. However, if the height of the wall above the opening is shorter than the height of the triangular section, it is assumed that the lintel must support the entire weight of the masonry above the opening. An example of this would be a lintel used over the opening for a garage door in a one-story building.

Lintels and, to a lesser extent arches, are the most common methods of supporting loads over openings in masonry walls. The corbelling shown in Figure 8.13 is used only where the architectural style makes it attractive.

In multistory buildings where large show windows are provided, the weight of two or more floors of masonry wall is often supported by a beam and column system at the grade level **(Figure 8.15)**. The weight of many tons of masonry is transmitted through the horizontal beam and columns to the foundation.

Figure 8.14 A lintel is designed to support the portion of the masonry wall indicated by the triangle.

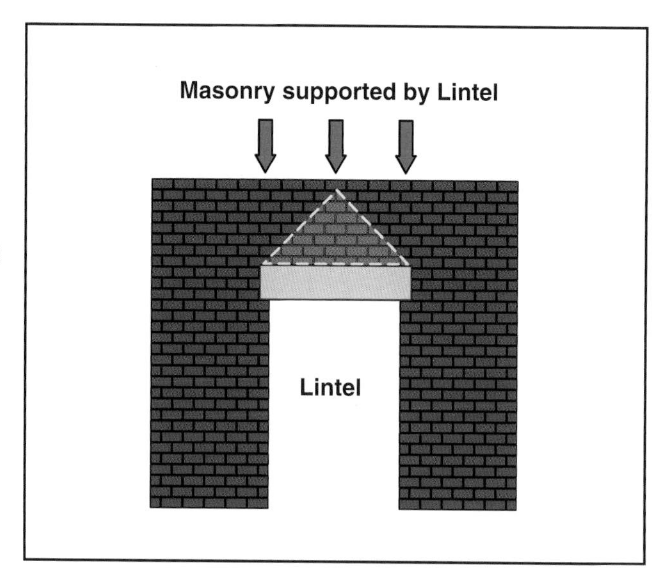

Figure 8.15 The masonry upper wall of this building is supported by the beams and columns at the first floor.

Parapets

A *parapet* is an extension of a masonry or steel wall that projects above the roof. Parapets are found on exterior masonry walls and fire walls of buildings with combustible roofs **(Figure 8.16)**. The purpose of a parapet on an exterior wall can be both architectural and functional. A masonry parapet may be used to enhance the architectural appearance of a building. A parapet may also be required by a building code to provide a barrier to the communication of fire between closely spaced buildings.

Parapet — Portion of the exterior walls of a building that extends above the roof. A low wall at the edge of a roof.

NOTE: Refer to Chapter 5 for further discussion of fire walls.

Parapets may pose a potential structural problem for firefighters, especially on the fronts of buildings. Parapets project from 1 to 3 feet (0.3 to 0.6 m) or more above the roof, usually without lateral support. They can collapse if exposed to high winds, particularly if erosion of the mortar joints or other deterioration has occurred. Furthermore, parapets can be dislodged if subjected to thrusts because of the collapse of a roof in the course of a fire **(Figure 8.17)**. Parapet walls are also susceptible to being knocked over by aerial ladders which are forcefully extended or retracted against them, as well as being struck by master streams.

NOTE: Designers have been known to occasionally make use of an artificial parapet made of Exterior Insulation Finishing Foam for esthetic purposes. These parapets offer no fire resistance but may be indistinguishable from masonry parapets.

Figure 8.16 a and b Parapets are found on structures of all types. They can collapse due to weather or become dislodged during fire fighting operations. *Both photos courtesy of McKinney (TX) Fire Department.*

Figure 8.17 During the course of a fire, the masonry parapet at the front wall of this building collapsed onto the sidewalk. *Courtesy of Ed Prendergast.*

Interior Structural Framing in Masonry Buildings

Bearing Walls

The traditional and most basic masonry structure consists of exterior load-bearing masonry walls that support the interior wood floors and roof that consists of wood joists and rafters **(Figure 8.18)**. The live loads of the building contents are transferred to the bearing walls and interior columns by joists and beams. This type of construction was so commonplace in the 19th and earlier part of the 20th century that it bears the designation "ordinary construction" (also known as Type III construction). Ordinary construction is also known as "masonry, wood-joisted" construction. In modern practice truss joists and wood roof trusses are often used in place of the traditional solid joists and rafters **(Figure 8.19)**.

Figure 8.18 In traditional masonry construction the masonry walls support wood floor and roof joists. *Courtesy of Ed Prendergast.*

Figure 8.19 Wood truss construction has largely taken the place of earlier construction methods. *Courtesy of McKinney (TX) Fire Department.*

Interior Framing

Masonry buildings can be found with interior framing systems other than wood, including masonry columns and interior bearing walls. Cast iron was frequently used for columns during the nineteenth century but it is not used in contemporary construction. In modern design the presence of an exterior masonry wall does not mean that a building is of "ordinary" construction. Both protected and unprotected steel interior framing can be used with masonry bearing walls; therefore, masonry walls can be encountered in both fire- resistive and non-fire-resistive buildings.

The interior framing of a wood-joisted building will be finished with plaster, drywall, or other interior finish materials to provide required fire resistance. Therefore, wood-joisted masonry construction will have concealed combustible voids similar to wood-frame construction **(Figure 8.20)**. Travel of fire through floor and ceiling spaces should be anticipated just as with wood-frame construction, especially where truss joists are used.

Firefighters must always be on the alert for early collapse involving any trusses.

In many applications such as residential and small commercial buildings, wood joists or beams simply rest on the masonry wall in an indentation known as a *beam pocket*. The beam pocket is several inches deep to provide an adequate bearing surface for the beam. A metal strap may be provided to function as a horizontal tie between the masonry and the end of the beam.

Fire Cut — Angled cut made at the end of a wood joist or wood beam that rests in a masonry wall to allow the beam to fall away freely from the wall in case of failure of the beam. This helps prevent the beam acting as a lever to push against the masonry.

The end of a wood joist or beam will be cut at a slight angle. This angle is known as a *fire cut* **(Figure 8.21)**. The purpose of a fire cut is to allow the beam to fall away freely from a wall in the case of structural collapse without acting as a lever to push against the masonry. However, fire cuts in joists do not totally preclude the collapse of a masonry wall.

When a beam transmits a large vertical load to a masonry wall, the wall may be increased in thickness at the point of support with a pilaster to reduce the compressive stresses in the masonry **(Figure 8.22)**. Wood roof trusses in commercial buildings, for example, are frequently supported on pilasters.

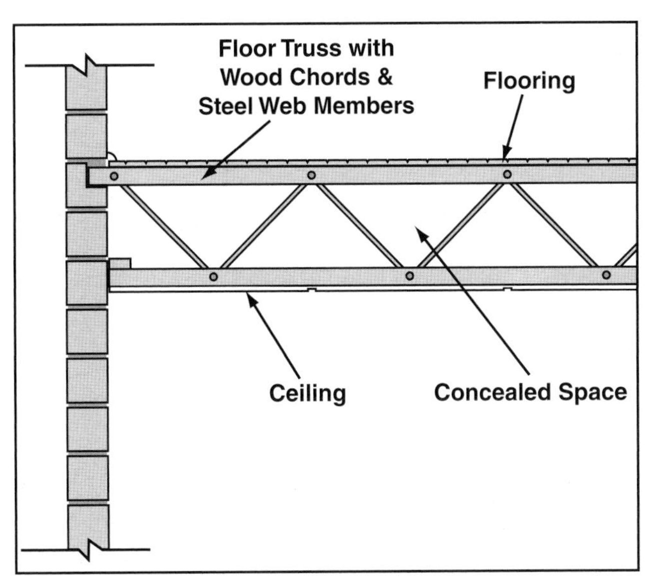

Figure 8.20 Combustible concealed space in a masonry building using a floor truss.

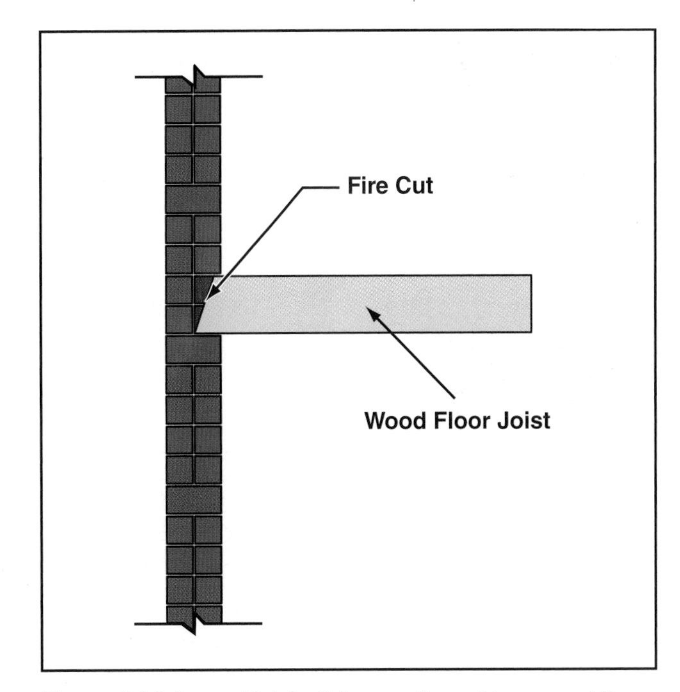

Figure 8.21 A wood joist will have a fire cut to prevent it from acting as a lever in case of collapse and exerting force on the entire wall.

Figure 8.22 A pilaster may be used to provide additional support where beams exert additional force.

Figure 8.23 Steel roof beam supported by a masonry bearing wall. *Courtesy of Ed Prendergast.*

Steel beams or trusses are supported on a masonry bearing walls in a manner similar to wooden structural members **(Figure 8.23)**. Precast concrete slabs are sometimes used for a floor or roof system with masonry walls. In this case the applied load can be distributed uniformly along the masonry wall.

The interior structural framing of a masonry building is often a combination of materials. It is not uncommon to have a wood or concrete floor deck supported by steel beams, which in turn are supported by a masonry wall.

Code Classification of Masonry Buildings

Buildings of ordinary construction, such as those with masonry exterior walls and wood joisted interior framing, are classified as Type III construction in the building codes. They may be Type III-A or Type III-B depending on the degree of fire resistance provided to the structural members. (See Chapter 2).

Buildings with masonry exterior walls can also be Type I or Type II construction. A multistory building with exterior masonry walls and noncombustible interior framing can be a Type I-A, I-B, or II-B building if the interior framing members are protected with fire-retardant materials as required by the building code. If the interior framing members are noncombustible and unprotected, the building would be a Type II-B (noncombustible) building.

Figure 8.24 The masonry exterior wall of this building remained despite the total burnout of the interior wood framing. *Courtesy of Ed Prendergast.*

Fire Behavior of Masonry Construction

Fire Resistance of Masonry Walls

The fire resistance of a masonry wall depends on the type of masonry units used and the thickness of the wall. Non-fire-rated hollow concrete blocks may have little fire resistance and may spall and crumble when exposed to a fire. By contrast, walls constructed with fire-rated concrete masonry units or bricks can have fire-resistance ratings of two to four hours or more. A massive masonry wall, 18 inches (450 mm) or more in thickness, will have an inherently high degree of fire resistance. A well-constructed masonry wall that has not been undermined or weakened is usually the last structural component to fail in a wood-joisted building **(Figure 8.24)**.

NOTE: Refer to Chapter 2 for a discussion of Type IV construction.

Masonry walls do collapse under fire conditions and, like all collapsing walls, pose a danger to firefighters. The likely fire behavior of a building with a masonry structural system is determined by several factors. Some of these factors are obvious and can be detected by visual observation. Others are subtle and not likely to be detected during the course of a fire.

Masonry walls are inherently difficult to breach, but newer reinforcing and security measures can make it even more difficult to force entry into these types of structures.

Over time, masonry walls can deteriorate from several causes. Deterioration can result from erosion of the mortar as a result of exposure to the elements. Moisture from rain, for example, can seep into cracks in the mortar. If the moisture then freezes, it expands on a microscopic basis, which subjects the mortar to tensile stresses. The stresses will cause the mortar to flake off.

The formation of cracks and the misalignment of a wall can occur because of the shifting of the foundation **(Figure 8.25)**. Wooden interior members can rot if they have been exposed to moisture from roof leaks or leaking plumbing. The resulting sagging of the interior members affects the stability of the masonry walls by a shifting of the forces to which they are subjected.

Deterioration of masonry walls is normally a slow process that takes place over many years. Old buildings have been known to collapse suddenly for no apparent reason. Any structural deterioration that may have occurred before a fire will contribute to structural failure under fire conditions.

It is possible for masonry walls to be repaired when deterioration has occurred **(Figure 8.26)**. If repairs are carried out by competent contractors, the stability of a structure will not be compromised. If the repairs are not made properly, however, the walls may remain in an unstable condition.

WARNING!
A fire officer needs to monitor changes or growth in masonry wall cracks during emergency operations. Changes in cracks or alignment can be indicators of imminent collapse.

Figure 8.25 Cracks and misalignment of a masonry wall can occur from shifting of the foundation. Note that this wall has been repaired.

Figure 8.26 A large section of repaired masonry wall. If repairs are made correctly, the wall will remain stable. *Courtesy of Ed Prendergast.*

If a masonry wall begins to bulge or lean outward it will tend to pull away from the interior framing. One way to stabilize a masonry structure is through the use of steel tie rods extended through the masonry walls, parallel to the joists, and attached to bearing plates or structural washers on the outside **(Figure 8.27)**. The tie rod usually has a turnbuckle to adjust the tension in the rod. The bearing plates are usually easily visible on the outside of a building.

The presence of tie rods does not necessarily mean that a building has undergone repair. The tie rods may have been part of the original construction. In the early 19th century, when masonry buildings began to be constructed more than one story in height and needed to support greater loads, designers used tie rods to reinforce the structures.

Bearing plates can take several forms. Some are intended to be architecturally decorative and can be in the shape of stars. They are, in fact, sometimes referred to as stars. Occasionally stars or similar devices are placed on masonry walls for purely ornamental purposes.

One potential problem with the use of tie rods is that the steel rods can become heated under fire conditions and stretch, resulting in weakening of the walls. The situation of greatest concern for firefighters is where tie rods have been used to repair a wall rather than where they are part of the original construction. It is usually possible to differentiate between repairs and original construction. The bearing plates used where repairs have been undertaken are likely to be:

- Less compatible with the architecture of the building. They may simply be in the form of steel channels **(Figure 8.28)**.
- Not uniformly positioned on a wall.
- Show other indications of repair such as repairs to the mortar joints in the immediate area of the bearing plate.

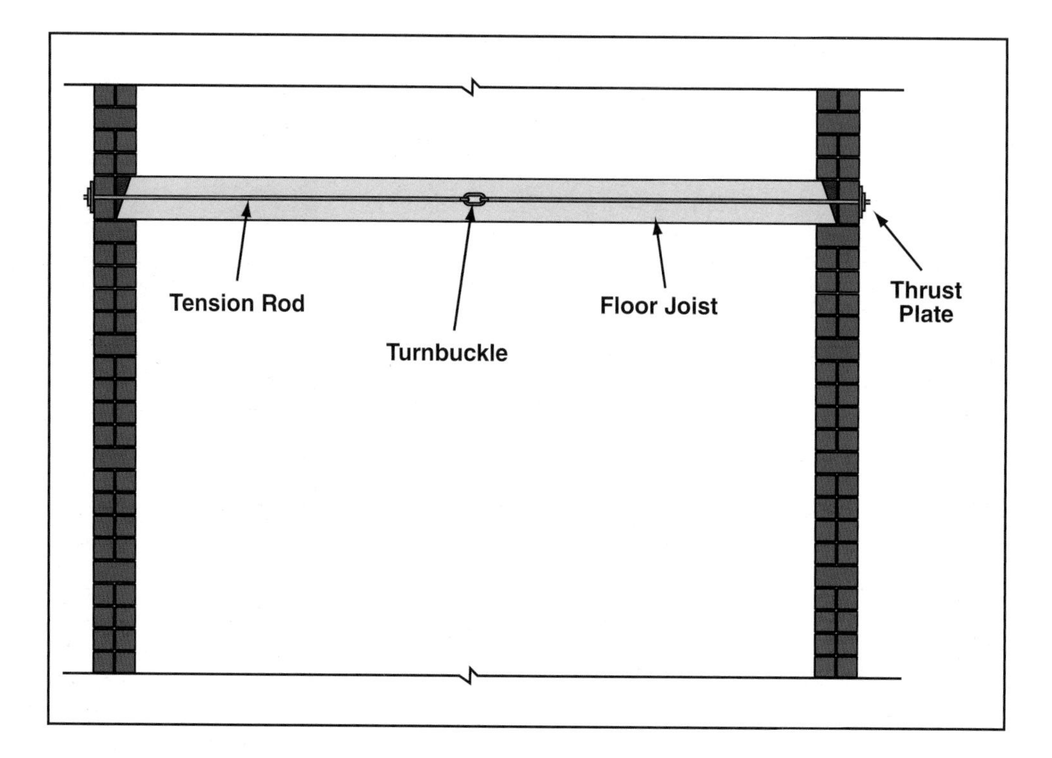

Figure 8.27 Use of a tie rod to reinforce a masonry wall.

Despite the basic structural qualities of masonry construction, total collapse of a masonry building is possible if it becomes heavily involved in fire. Masonry walls usually collapse as a result of the collapse of the interior wood framing. Collapsing interior floor or roof members can exert horizontal forces against a wall and push the wall outward **(Figure 8.29)**. These horizontal forces create tensile forces at the inner face of the wall that the mortar joints cannot resist. Because the collapse of interior framing also removes interior bracing for the wall, the wall is simply pushed out from the building. If a wall is of reinforced construction, the reinforcing steel can withstand the tensile force and collapse is less likely. The challenge for firefighters is that determining whether or not a masonry wall is reinforced cannot be readily made from the outside.

There is no sure way to predict when or how a masonry wall may collapse if a building is heavily involved in fire. A masonry wall may collapse partially or completely. A collapsing roof may dislodge only a parapet or an entire wall. Firefighters should always assume that if a wall collapses it will fall out from the building a distance *at least* equal to the height of the wall. Because intersecting masonry walls tend to support each other, the corners of the building or other points of intersection, such as stairwells or elevator shafts, will be the strongest points in a masonry structure.

The exterior fire-resistive walls of masonry construction do more than provide structural support; they also tend to reduce the communication of fire from structure to structure. Building codes usually require less clearance between buildings with masonry or other fire-resistive exterior walls than between buildings with combustible exteriors. Of course, large window openings, unprotected soffits, or cantilevers in a fire-resistive wall can still create an exposure problem.

Figure 8.28 Bearing plates on a building that has undergone repairs.

Figure 8.29 Masonry walls are frequently pushed out due to forces exerted by collapsing interior components. *Courtesy of Ed Prendergast.*

False Fronts and Voids

A masonry building is sometimes provided with a decorative false front or *fascia*. This may occur as a result of a renovation or it may be part of the original design.

Fascia — (1) Flat horizontal or vertical board located at the outer face of a cornice. (2) Broad flat surface over a storefront or below a cornice.

A false front can create a combustible void that is not normally encountered with masonry walls. One common example of this a mansard style *fascia* **(Figure 8.30)**. A mansard-style fascia forms a projection beyond the building wall that creates a concealed space through which a fire can communicate. Any construction can create unprotected void spaces in an otherwise sprinklered building.

Figure 8.30 Mansard-style fascia added to update a masonry building. *Courtesy of Ed Prendergast.*

Mill (Heavy-Timber) Construction

Mill construction is a type of masonry construction in which the exterior bearing walls are constructed of brick and the interior structural framing is of solid or laminated wood. Mill construction is also commonly known as heavy-timber construction; however, it should not be confused with the heavy timber *wood-frame* construction discussed in Chapter 7. In building codes mill construction is classified as Type IV construction.

There are two characteristics that distinguish mill construction from Type III masonry construction. They are the dimensions of the wood structural members and the fact that combustible concealed spaces are not permitted **(Figure 8.31)**. The wood structural members used in mill construction have larger minimum dimensions than those permitted in Type III construction. **Table 8.3** shows the fire resistance requirements for the structural elements of mill construction in the *International Building Code*. In addition:

- Floors are required to be 3-inch (75 mm) tongue and groove with 1-inch (25 mm) nominal tongue and groove flooring.
- Roofs are required to be 2-inch (50 mm) tongue and grove or $1^1/_8$ (32 mm) structural panel or planks.

NOTE: These dimensions are for sawn lumber, namely, a solid piece of lumber. Laminated members may also be used and would have slightly different dimensions.

- Nonbearing interior partitions and walls are required to be not less than two layers of 1-inch (25 mm) matched boards or laminated construction 4 inches thick (100 mm) thick or 1-hour fire rated construction.

Figure 8.31 Typical structural members in a masonry heavy-timber building. No concealed spaces are permitted in this type of construction. *Courtesy of Ed Prendergast.*

Table 8.3
Fire-Resistance Requirements for Mill Construction

Building Element	Fire Resistance in Hours
Structural Frame (Columns and beams, girders, trusses, and spandrels connected to the columns.)	HT
Bearing Walls Exterior Interior	 2 1 or HT
Nonbearing exterior walls and partitions	0 to 3 depending on occupancy and separation distance
Nonbearing interior walls and partitions	See below
Floor construction	HT
Roof Construction	HT

Members designated HT (Heavy Timber) are required to have minimum dimensions not less than the following:

Nominal Dimensions for Sawn Lumber

	Width, inches (mm)	Depth
Columns		
Supporting Floors	8 (203)	8 (203)
Supporting Roofs	6 (152)	8 (203)
Floor Framing	6 (152)	10 (254)
Roof Framing	4 (102)	6 (152)

Source: International Building Code

It is noteworthy that in Table 8.3 many of the structural components, such as columns and beams, are not given an actual fire resistance rating. Instead, minimum dimensions are specified. Type IV construction relies on the inherent fire resistance of lumber with larger cross-sectional dimensions.

Mill construction was used extensively in factories, mills, and warehouses in the 19th and early 20th centuries. It is not commonly used in new construction for multistory buildings although many old buildings of this type remain in use. Many old mill-constructed warehouse and industrial buildings have been converted to residential use.

The absence of concealed spaces is a major advantage in mill construction from a fire fighting standpoint:

- Firefighters can more easily "get at" a fire.
- It is not necessary to "pull ceiling" in mill construction because there are no suspended ceilings.

The advantageous features of mill construction, however, can be significantly lessened when a mill building undergoes a change of occupancy. Many older mill buildings have been converted to residential occupancies, which results in the subdivision of floor space into residential units. These divisions can lead to the creation of concealed spaces when partitions and ceilings are installed in individual units.

The firefighter must be concerned with the integrity of the timber connections under fire conditions and the likelihood of collapse of roof and floor decks. In light-frame wood construction, nails, staples, or screws may be adequate. In the case of heavy timber framing, the loads carried are greater and the connections between members will incorporate through bolts, special brackets, and the bearing of one member directly on another **(Figure 8.32)**. **Figure 8.33** illustrates a timber column and beam connection after a fire.

NOTE: Refer to Chapter 7 for illustrations depicting specific methods for joining heavy timber structural members.

Figure 8.32 An example of a heavy-timber joint. *Courtesy of Ed Prendergast.*

Figure 8.33 Timber column and beam connection after exposure to a fire. *Courtesy of Ed Prendergast.*

Summary

Masonry construction is encountered in several applications. The oldest form of masonry construction is ordinary construction with masonry bearing walls and interior wood framing. These structures are classified as Type III in the building codes. Mill buildings, also known as heavy-timber buildings, are masonry buildings constructed with heavy-timber interior framing and are classified as Type IV construction by the codes. Masonry veneer construction can be used with other framing systems. A building with a masonry exterior can be Type I, Type II or Type V construction.

Masonry itself is noncombustible and inherently fire resistive, so masonry walls reduce the communication of fire between structures. Combustible interior framing can collapse, however, and masonry walls can and will fail under heavy fire conditions. The interior combustible framing also contributes to the fuel load. Masonry walls can deteriorate with age from several causes. Any signs of deterioration of masonry must be viewed very cautiously by firefighters.

Review Questions

1. What is a drawback to the use of masonry?
2. What is a course?
3. How can parapets be affected by high winds?
4. How can the stability of a masonry wall be affected by repairs?
5. What is a mortise and tenon joint?

References

1. Allen, Edward and Iano, Joseph *Fundamentals of Building Construction. Materials and Methods 4th edition,* John Wiley and Sons.
2. Ambrose, James, *Building Structures 2nd edition,* John Wiley and Sons.
3. International Code Council®, *International Building Code®, 2006 edition.*

Chapter Contents

Divider page photo courtesy of McKinney (TX) Fire Department.

Key Terms

FESHE Objectives

Fire and Emergency Services Higher Education (FESHE) Objectives: *Building Construction for Fire Protection*

1. Demonstrate an understanding of building construction as it relates to firefighter safety, buildings codes, fire prevention, code inspection and firefighting strategy and tactics.
8. Identify the indicators of potential structural failure as they relate to firefighter safety.

Steel Construction

Learning Objectives

After reading this chapter, students will be able to:

1. Describe and differentiate the properties of steel and iron used as building material.
2. Describe the types of steel frame structures and their applications.
3. Identify the types and uses of steel frames in flooring systems.
4. Discuss how connections and lighter weight construction affect the potential for collapse of steel structures.
5. Describe the materials used to provide fire resistance to steel members and their effectiveness.
6. Discuss the importance of code modifications as they relate to firefighting.

Chapter 9
Steel Construction

Case History

Event Description: Firefighters were dispatched to a fire in a shop located in a single-story 3,000 sq. ft. stand-alone building with an awning in front. Arriving units found fire issuing from the front doors of the building. Crews began to perform an offensive fire attack in the front of the building using handlines and forcible entry tools. Crews reported difficulty gaining entry due to metal security gates in front of the building and at two other entrances. Heavy smoke was seen issuing from the upper part of the building.

Approximately 20 minutes into the response, the awning collapsed, striking seven firefighters. Five firefighters were covered by debris and two needed assistance and extrication. Fire was seen throughout the attic space after the collapse. All on-scene crew members were directed to the front of the structure to assist with extrication and extinguishment. Several firefighters were transported to a hospital for treatment but fortunately no one was killed.

Lessons Learned:

- The awning that collapsed had clay tiles and an estimated weight of 1,850 lbs. The roof and awning structure was constructed of pin-connected open-web trusses consisting of lightweight wooden top and bottom cords and tubular steel webbing connected with steel pins. This same type of truss contributed to an awning collapse that killed another firefighter in 2006. Personnel should consider the added weight of this type of roof covering.
- On-scene personnel reported visibility inside the structure; in reality, there was significant fire in the attic space and awning, causing heavy smoke production. More coordinated on-scene assessment would have helped officers determine the status of the fire.
- The security gates that were difficult to breach were later determined to be made of aluminum, which could have been easily cut with a bolt cutter, permitting firefighters to gain entrance earlier when the fire was smaller.

Source: National Fire Fighter Near-Miss Reporting System.

Steel is a strong and noncombustible building material. The development of steel structural framing at the end of the 19th century permitted the construction of high-rise buildings in the 20th century. Because of strength issues, steel beams and trusses have largely replaced wood beams and trusses in commercial structures. Steel framing is found in buildings of all heights and is used in both fire-resistive and non-fire resistive buildings. Steel is used in applications varying from heavy beams and columns to door frames and nails.

The firefighter must understand that the behavior of steel under fire conditions depends on the mass of the steel and the degree of fire resistance provided. This chapter discusses the properties, use, and protection of steel. Firefighters should understand that the protection of steel depends on the integrity of the installation, protection methods, maintenance, and failure points.

Properties of Steel and Iron

The basic properties of steel are as follows:

- Strongest of the building materials
- Non-rotting, resistant to aging, and dimensionally stable
- Consistent quality due to controlled industrial process used in its manufacture
- Relatively expensive, but strength and variety of forms enable it to be used in smaller quantities than other materials

Steel is basically an alloy of iron and carbon. Common structural steel has less than three tenths of one percent carbon. Cast iron, by contrast, has a carbon content of three to four percent. The higher carbon content of cast iron produces a material that is hard but brittle.

The composition of steel can be altered by adding other materials. Molybdenum, for example, can be added to increase strength. Vanadium increases strength and toughness. Manganese increases the resistance of steel to abrasion.

Ductility of Steel

Ductile — Capable of being shaped, bent, or drawn out.

The lower carbon content of steel compared to that of cast iron results in a material that is *ductile* rather than brittle. This attribute is important because it enables steel to be rolled into a variety of shapes **(Figures 9.1 a and b)**. The rolling process consists of repeatedly passing ingots of steel heated to 2,200°F (1,200°C) between large rollers until the desired shape is achieved.

The ductility of steel also allows it to be rolled in a cold state. Cold-rolled steel (known as CRS) is used for members that have a thin cross-section, such as floor and roof decking and wall studs. The CRS studs can be used either for interior non-load bearing partitions or for exterior bearing walls.

Figure 9.1 Steel can be formed into many shapes. *Both photos Courtesy of McKinney (TX) Fire Department.*

A very commonly used steel for structural purposes is designated ASTM A36. The ductility of this steel is illustrated by comparing the stress exerted on the steel with the resulting deformation known as the strain. In **Figure 9.2** the stress and strain for different types of steel are plotted. The resulting curve for A36 steel is initially steep and straight. However, when the "yield point stress" is reached (approximately 36,000 psi [248,220 kPa]), the steel undergoes a pronounced deformation. Finally, the steel breaks when the "ultimate stress" is reached. Because excessive deformation of steel is undesirable, the maximum stress that develops within steel members under conditions of loading must be kept below the yield point.

In examining Figure 9.2 it can be seen that the steels with higher yield points have less ductility. Steels for special application, such as those used in bridge strands, have strengths as high as 300,000 psi (2,068,500 kPa) but have very little ductility.

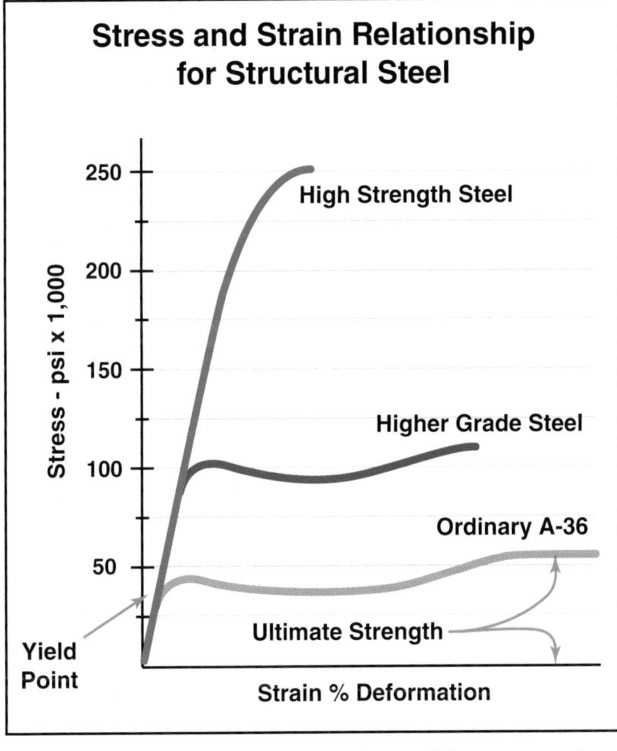

Figure 9.2 Different types of steel have different strength characteristics. The maximum allowable stress for A-36 steel must be kept below its yield point of approximately 36,000 psi (248 220 kPa) -- the point at which it begins to deform.

Deterioration of Steel Strength at Elevated Temperatures

Steel possesses two inherent disadvantages:

- Tendency to rust when exposed to air and moisture
- Loss of strength when exposed to the heat of a fire

Steel can be protected in several ways from the formation of rust. Methods include painting the surface with a rust-inhibiting paint and coating the material with zinc and aluminum. Steel also can be produced using ingredients that resist rust, as in the case of stainless steel.

To the fire service, the deterioration of the strength of steel at elevated temperatures is its most significant characteristic. The fires typically encountered by firefighters do not create temperatures hot enough to melt steel. However, they do create enough heat to greatly weaken steel, resulting in structural failure. **Figure 9.3** illustrates the effect of temperature on the yield point of steel. Note that at a temperature of around 1,000°F (538°C), the yield point of the steel has dropped from 36,000 psi (248,220 kPa) to approximately 18,000 psi (124,110 kPa). At 1,200°F (650°C), the yield point has dropped to approximately 10,000 psi (68,950 kPa) - a loss in strength of approximately 72 percent.

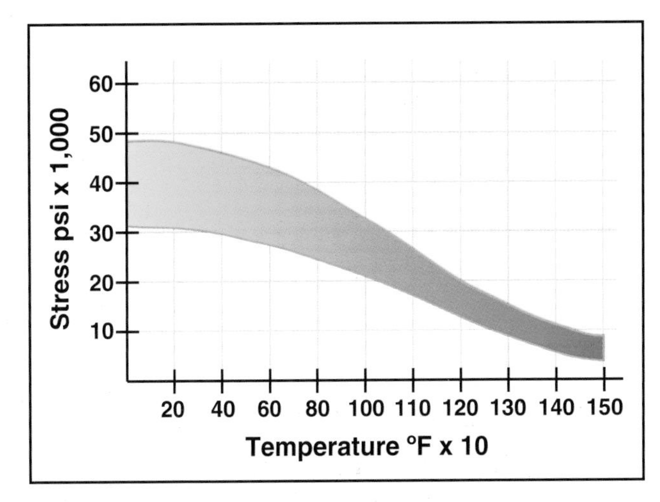

Figure 9.3 Strength characteristics of steel at different temperatures.

NOTE: The temperatures in Figure 9.3 are the temperatures of the *steel*, not the temperatures of the exposing fire.

Because temperatures in excess of 1,200°F (650°C) are regularly encountered in fires, failure of unprotected steel to a greater or lesser degree can be anticipated. The loss of strength because of increased temperature is not a sudden occurrence; rather, the steel loses its strength gradually as its temperature increases. The speed with which unprotected steel fails when it is exposed to a fire depends on several factors, including the following:

- Mass of the steel members
- Intensity of the exposing fire
- Load supported by the steel
- Type of connections used to join the steel members **(Figures 9.4 a and b)**
- Type of steel

Although steel is a good conductor of heat, it is also a very heavy material, having a density of around 490 pounds per cubic feet (7 840 kg/ cubic m). A great deal of heat is required to raise the temperature of heavy steel structural members. A wide flange steel beam 14 inches (350 mm) in depth with a 10-inch (250 mm) wide flange would have a weight of 74 pounds per foot of length (110 kg/m). A beam 20 feet (6.1 m) long would weigh 1,480 pounds (671 kg).

To raise the temperature of this entire beam would require a minimum of 137,000 Btu of thermal energy. However, this number is a theoretical approximation. In a real fire, a large amount of energy is radiated away from the fire and lost in the upward flow of heat. Therefore, the fire would have to generate a large amount of heat above the theoretical amount to raise the temperature of the beam to 1,000°F (538°C).

Figure 9.4 a and b The way steel members are attached affects their behavior during a fire. *Courtesy of McKinney (TX) Fire Department.*

Unprotected steel structural members that have less mass require less heat to reach the temperature at which they begin to fail. Members such as bar joists or slender trusses can be expected to fail early when exposed to a fire **(Figure 9.5)**. By contrast, massive steel beams and girders frequently remain in place under severe fire exposure.

Because the mass of steel requires a large amount of heat to raise its temperature, the intensity of a fire directly affects the behavior of steel. In a structure with a light fuel load, unprotected steel may not fail if the fire does not supply enough thermal energy.

The load that steel members are supporting also affects the behavior of steel because the loads produce the stresses in the steel. When lower stresses exist in the steel, it must be heated to a higher temperature for the yield point to be reached (See Figure 9.3).

When individual steel members are rigidly welded or bolted into a large structural system, they are better able to resist failure than if they are simply supported. The end restraint provided by rigid connections exerts a resistance to the deformation of individual members.

Steel expands as it is heated. The amount of expansion for slender members, such as beams and columns, can be determined through a property known as the *linear coefficient of thermal expansion.* If an unrestrained steel beam 20 feet (6.1 m) long were heated from 70°F (21°C)

Figure 9.5 Lightweight steel trusses can be expected to fail early during a fire. *Courtesy of McKinney (TX) Fire Department.*

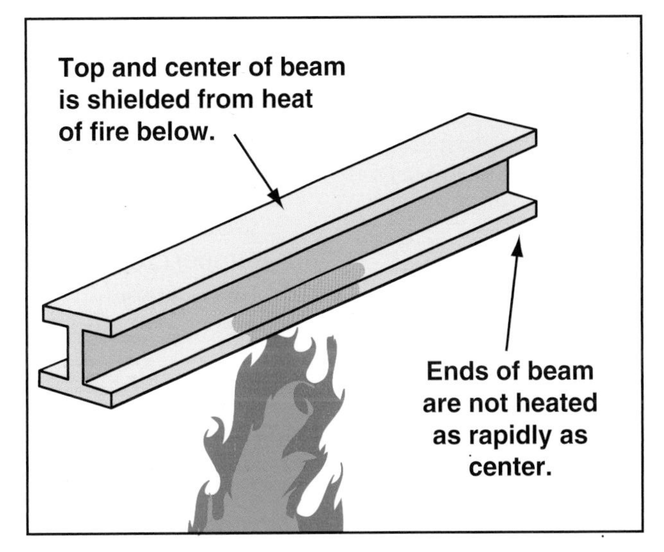

Figure 9.6 The uneven heating of a steel beam.

to a uniform temperature of 1,000° F (538°C), it would expand 1.4 inches (35 mm). If the beam were restrained, as in a rigid framework, the expansion of the steel would be restricted and large forces would be exerted against adjacent members, greatly altering their internal stresses.

In fire situations, the heating of steel or other materials does not occur in a uniform manner. In the beam shown in **Figure 9.6,** the bottom of the beam at the center is heated more by the fire than it is at the top or the ends. Therefore, the resulting yielding and deformation is not uniform over the length of the beam.

The combination of all the previously discussed variables accounts for the variations in the distortion or failure of steel that may be observed under different fire scenarios. Both yielding and thermal expansion take place simultaneously; however, the tendency for steel to yield and, therefore, to bend or buckle is the more significant concern in most fire situations.

Figure 9.7 Cast-iron fascia typical of buildings constructed at the turn of the last century and found in many older cities. *Courtesy of Ed Prendergast.*

Cast Iron

Cast iron was used in buildings in structural framing before the turn of the 20th century. Cast iron columns and staircases can still be found in older buildings. A few structures were built with complete cast iron fronts **(Figure 9.7)**. Today cast iron has been completely displaced by steel. Because cast iron is a brittle material, it tends to fail by fracturing from impact loading rather than by yielding as in the case of steel. The firefighter needs to be concerned with the way the cast iron fronts are attached to the structures. It may be a failure of these attachment points that will lead to the collapse of the cast iron front.

Steel-Framed Structures

Girder — Large, horizontal structural member used to support joists and beams at isolated points along their length.

The basic method by which steel is used in the design of buildings is the construction of a structural framework that supports the floors, roof, and exterior walls. Several different techniques can be used to construct a steel frame. Steel structural shapes can be used to construct a frame of columns, beams, and girders **(Figure 9.8)**. Steel also can be used in heavy or lightweight trusses to support roofs and floors **(Figures 9.9 a and b)**. Rigid frames and arches can be constructed from steel. Steel cables or rods can be used to support roofs. Cold-rolled steel studs are being used to construct exterior walls.

Because steel is a strong but very dense material, it is not efficient to use it in the form of solid slabs or panels as is done with other materials such as wood or concrete. Steel in sheet form, however, is used for applications such as floor decking and exterior curtain walls. The exterior envelope of a steel-frame building can consist of concrete, masonry, or glass **(Figures 9.10 a and b)**.

Beam and Girder Frames

Beams and columns in steel-frame buildings are connected by one of two methods: bolting or welding. Riveting was used in the first half of the 20th century but is not practical to use today.

Figure 9.8 Steel frame-buildings are a common construction sight. When they are completed they can appear as steel structures or receive veneers of brick or stone. *Courtesy of McKinney (TX) Fire Department.*

Figures 9.9 a and b Steel trusses stacked for use and installed as roof support in a steel-framed structure. *Both photos courtesy of McKinney (TX) Fire Department.*

The design of the connections in steel-frame buildings is extremely important. The connection of a beam to a column not only transfers the loads between members but also determines the rigidity of the basic structure. In the framework shown in **Figures 9.11 a and b**, some structural rigidity must be provided to resist wind load and other lateral forces that would tend to cause the distortion of the building. Beam and girder steel frames can be classified as rigid, simple, or semi-rigid.

Figures 9.10 a (a) A steel-frame building being constructed with a masonry exterior wall. (b) A steel-frame building with a concrete exterior. Photo a *courtesy of Ed Prendergast; photo b courtesy of McKinney (TX) Fire Department.*

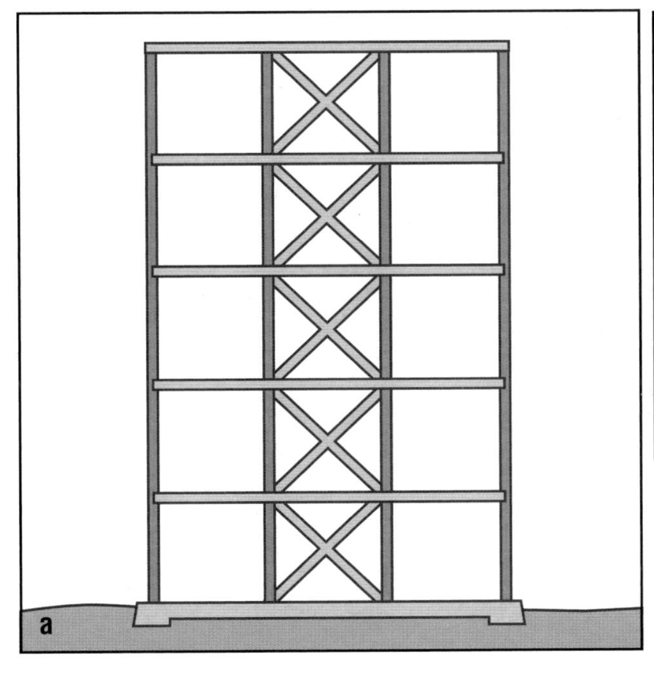

Figures 9.11 a and b When necessary, diagonal bracing is provided to meet the structural stability requirements of steel-frame buildings. *Photo courtesy of Ed Prendergast.*

Rigid Frame

When a framing system is classified as a rigid frame, the connections between the beams and the columns are designed to resist the bending forces resulting from the supported loads and lateral forces **(Figure 9.12)**. In a rigid-frame connection, sufficient rigidity exists between the beam and the column so that no change occurs in the angle between the beam and the column as the loads are applied.

Simple Frame

In the case of a simple frame, the joints are designed primarily to support a vertical force. A degree of angular change between beams and columns can occur if some form of diagonal bracing is not provided **(Figure 9.13)**. Steel beams and trusses are frequently supported by a masonry wall. These designs are also examples of simply supported systems.

Semi-Rigid Frame

In a semi-rigid frame, the connections are not completely rigid but possess enough rigidity to provide some diagonal support to the structure. When rigid connections are not used, lateral stability for a frame must be provided through the use of diagonal bracing or shear panels **(Figure 9.14)**. Shear panels are reinforced walls located between columns and beams to brace them laterally. Ideally a shear wall should be continuous from the foundation of a building to the highest story at which it is needed.

Steel Trusses

Steel trusses provide a structural member that can carry loads across greater spans more economically than beams can. Steel trusses can be fabricated in a variety of shapes to meet specific applications **(Figure 9.15)**. They are

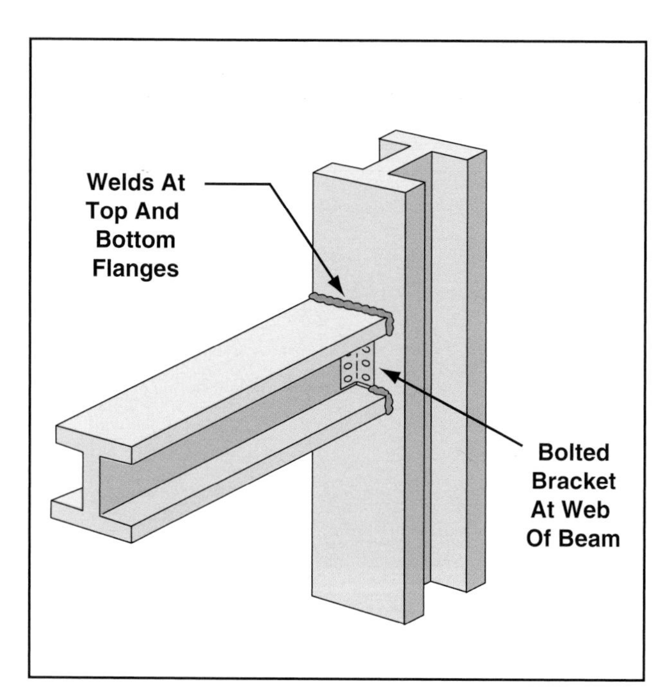

Figure 9.12 A rigidly designed beam and column connection with both bolts and weld joints.

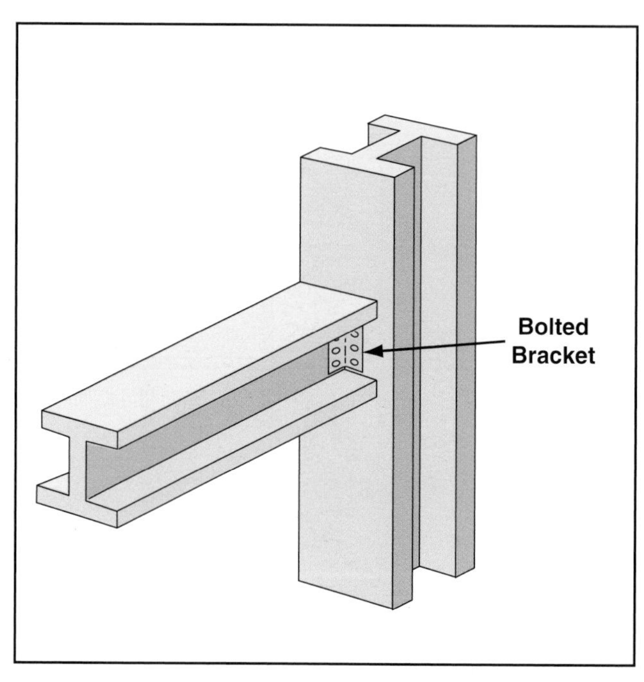

Figure 9.13 A simple connection using a bolted bracket capable of withstanding a vertical force.

frequently used in three-dimensional space frames, in which case they are known as delta trusses because the cross-section resembles the Greek letter Delta **(Figure 9.16)**. Two commonly encountered applications of the basic steel truss are the open web joist and the joist girder.

Open web joists are mass produced and are available with depths of up to 6 feet (2 m) and span up to 144 feet (44 m). However, they are more frequently found with depths less than 2 feet (0.6 m) and spans of 40 feet (13 m). The top and bottom chords of a web joist can be made from two angles, two bars, or a T-shaped member. The diagonal members can be made from flat bars welded

Bar Joist — Open web truss constructed entirely of steel, with steel bars used as the web members.

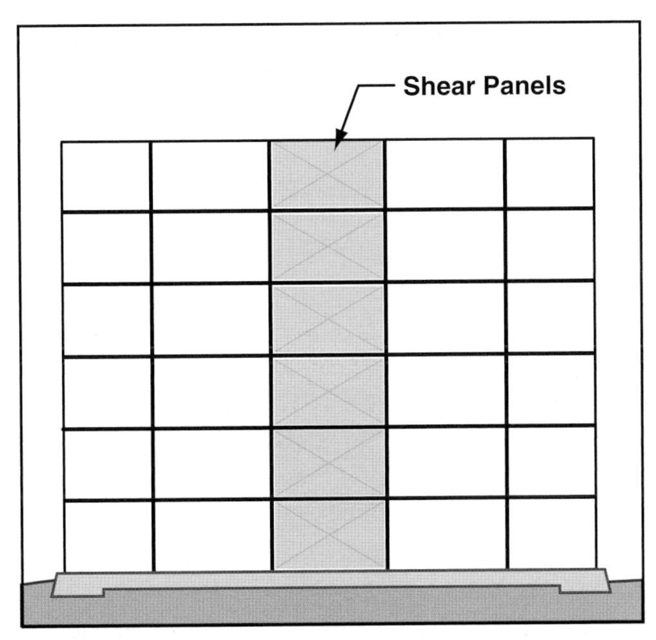

Figure 9.14 A steel-frame building can be strengthened against lateral forces with shear panels, providing stiffness to the frame.

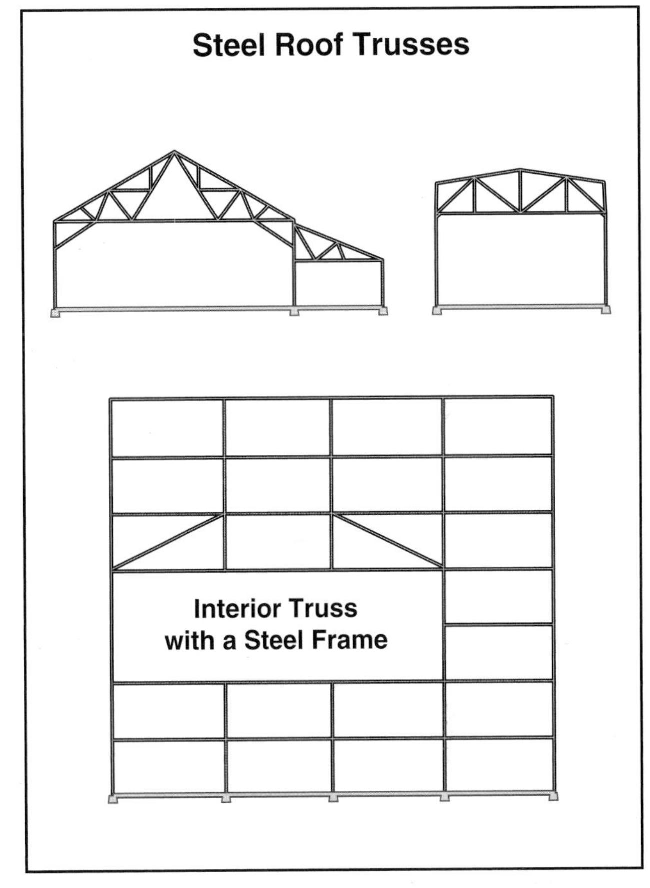

Figure 9.15 Steel trusses are made in a variety of shapes.

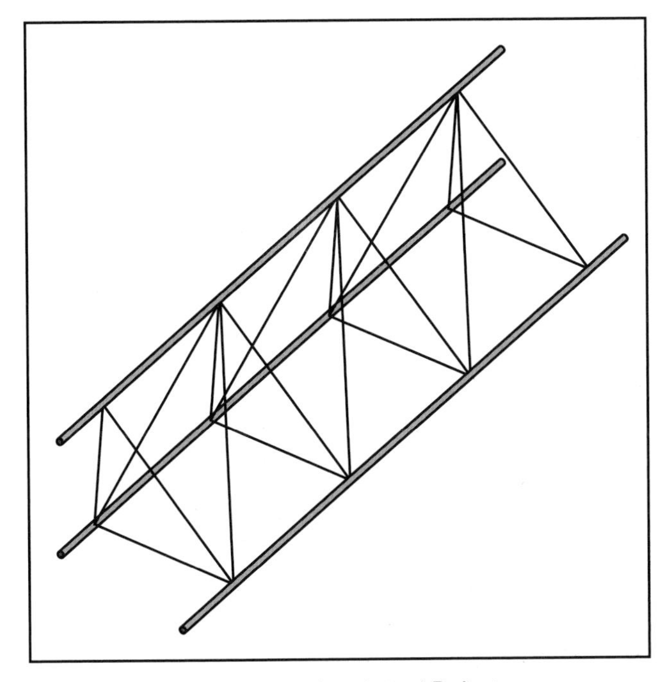

Figure 9.16 A three-dimensional steel Delta truss.

to the top and bottom chords or they can be a continuous round bar bent back and forth and welded to the chords. When round bars are used for the diagonal members, the open web truss is known as a *bar joist* **(Figure 9.17)**.

Bar joists are frequently used in closely spaced configurations for the support of floors or roof decks **(Figures 9.18 a and b)**. Bar joists are frequently supported on a masonry wall to support a roof. In multistory buildings, they are supported by the steel framing beams and are used for the support of the floor decks. Joist girders are heavy steel trusses used to take the place of steel beams as part of the primary structural frame **(Figure 9.19)**.

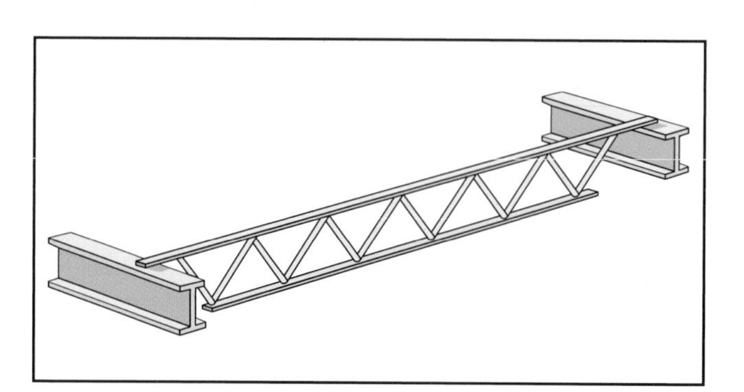

Figure 9.17 The bar joist is a common type of open web steel joist.

Figures 9.18 a and b Bar joists used to support a metal deck floor and a roof. *photo a courtesy of Ed Prendergast; photo b Courtesy of McKinney (TX) Fire Department.*

Figure 9.19 An example of a steel joist girder used to support a roof.

Gabled Rigid Frames

Steel rigid-frame buildings with inclined (or gabled) roofs are widely used for the construction of one-story industrial buildings, farm buildings, and a variety of other applications **(Figures 9.20 a and b)**. The inclined top members of the one-story rigid frame configuration allow an increase in interior clear space. Steel rigid frames usually are used for spans from 40 to 200 feet (13 m to 66 m) and are fabricated by welding or bolting together steel shapes and plates.

The top of the rigid frame is known as the *crown* and the points where the inclined members intersect the vertical members are known as the *knees*. The crown and the knees are designed as rigid joints with no rotation between members. The vertical members may or may not be rigidly connected to the foundations depending on anticipated wind loads.

Gable roof rigid-frame structures must be braced diagonally to prevent deflections in the direction perpendicular to the plane of the frame sections. This is accomplished by providing diagonal cross-members in the plane of the roof and in the vertical plane of the walls between the rigid frame sections.

Figure 9.20 a A steel rigid-frame building with a gabled roof under construction. *Courtesy of Ed Prendergast.*

Figure 9.20 b A completed structure of the same type. *Courtesy of Ed Prendergast.*

Steel Arches

Steel arches are used to support roofs on buildings where large unobstructed floors are needed. These include occupancies such as gymnasiums and convention halls. Steel arches can be constructed to span distances in excess of 300 feet (100 m).

Steel arches can be designed as either girder arches or trussed arches. A *girder arch* is constructed as a solid arch that may be built up from angles and webs with a cross section similar to that of a beam. A *trussed arch* is built using truss shapes as shown in **Figure 9.21**, which shows a trussed arch with pin connections. The pin connections allow for slight movement between the two halves due to settling or temperature change.

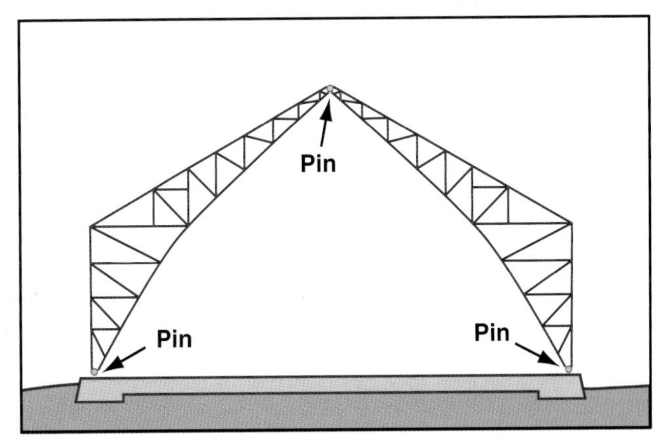

Figure 9.21 A hinged trussed steel arch.

NOTE: Refer back to Figure 3.39 for an illustration of a hinge in an arch connection.

Steel Suspension Systems

The strength of steel is such that it can be used in very slender forms such as rods and cables. Drawing steel bars through a die to produce wire greatly increases the strength of the steel. It is possible to produce wire for use in bridge cables with strengths as high as 300,000 psi (2,100,000 kPa). Such slender shapes are subject to buckling and therefore are limited to the support of tension forces.

Steel rods and cables are sometimes used in suspension systems to support roofs **(Figure 9.22)**. Suspension roof systems can provide large unobstructed areas similar to arches without the reduction in vertical clearance at the sides of a building that occurs with an arch. As with arches, applications include sports complexes and convention halls.

Steel suspension systems make some unique designs possible. One useful application of a suspension system cantilever roof is shown in **Figure 9.23**. Note that the rods supporting the roof are supported by vertical masts which, in turn, are balanced by other steel rods anchored to a support.

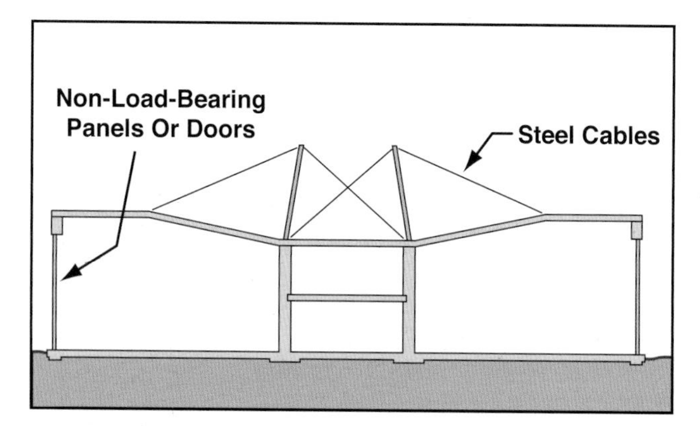

Figure 9.22 Steel cables can be used to support overhanging roofs.

Figure 9.23 Cantilever roof supported by steel rods. *Courtesy of Ed Prendergast.*

Steel Columns

Because of the high compressive strength of steel, the cross-section of steel columns can be very small compared to their length. Because of this slenderness, the possibility of buckling is greater with steel columns than with columns made of other materials. In the design of a steel column, engineers must evaluate this possibility and take steps to prevent failure by buckling. Furthermore, heat from a fire will reduce the yield point of the steel and an unprotected slender column can easily buckle.

Steel columns can vary from such simple single-piece members as cylindrical pipes to complex tower assemblies. The most common column cross-sections are the hollow cylinder, the rectangular tube, and the wide flange shape similar to the cross section of an I-beam **(Figures 9.24 a and b)**.

The possibility of buckling in an individual column is a function of its length, its cross-section, and the method by which the column is supported at its top and bottom. A property of a given column, known as its "slenderness ratio," is used in combination with the condition of the column end to determine the load that can be safely supported without buckling.

The slenderness ratio is a number that compares the unbraced length of a column to the shape and area of its cross-section. The higher the numerical value of the slenderness ratio, the more likely it is that buckling will occur. In general, columns used for structural support in buildings should not have a slenderness ratio greater than 120.

In evaluating a given design, the slenderness ratio is modified by the manner in which the ends are attached to the rest of the structure. Columns that are erected so they cannot rotate at their ends have less tendency to buckle than columns that are free to rotate at their ends **(Figure 9.25)**. For example, a round steel column that is simply resting on a concrete footing has no resistance to rotation at its ends. In contrast, a wide flange column with rigid connections at each end has a large resistance to buckling.

The stability of columns is critical to the structural integrity of buildings under circumstances such as earthquakes, impact, and/or the shifting of a foundation. The investigation of a structural collapse under any circumstance must include an evaluation of the columns and their means of support. It is possible for a beam that is simply supported by a column to become dislodged and fall off the supporting column if the column shifts or buckles.

Figures 9.24 a and b Rectangular and circular steel columns. *Both photos courtesy of McKinney (TX) Fire Department.*

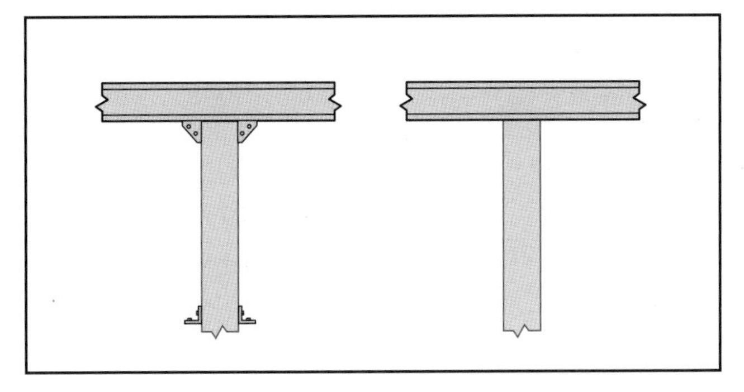

Figure 9.25 The column on the left has rigid connections and is able to withstand buckling better than the column on the right. Without rigid connections the column is more prone to rotate at the ends.

Floor Systems in Steel-Framed Buildings

There are three methods by which steel structural members can be used to support floors in multistory buildings. These include open web joists (bar joists) or trusses, steel beams, and light-gauge steel joists.

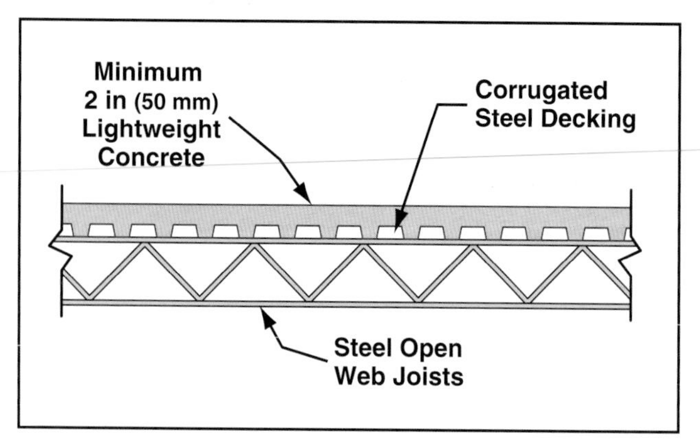

Figure 9.26 A common lightweight floor system.

Open-Web Joists

A very common floor design in steel-frame buildings uses a lightweight concrete with a minimum thickness of 2 inches (50 mm) supported by corrugated steel decking. The corrugated steel is, in turn, supported by open-web steel joists **(Figure 9.26)**. The steel joists can be supported by steel beams or directly supported on a masonry wall. The open web joists can also be used to support precast concrete panels or wood decking.

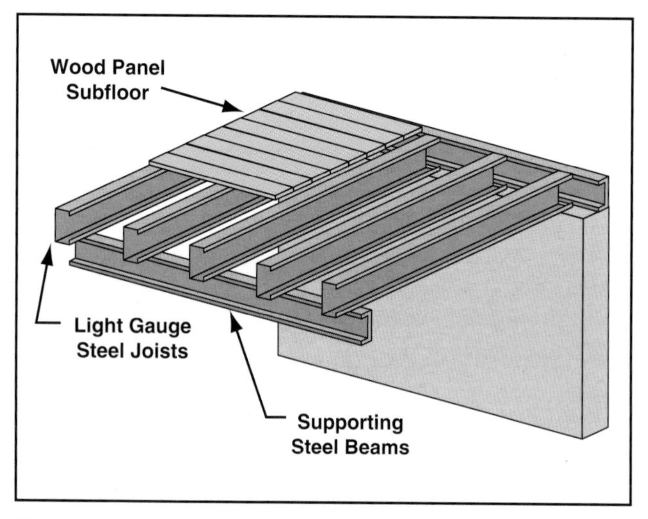

Figure 9.27 Light-gauge steel joist system.

Steel Beams and Light-Gauge Steel Joists

Where floor loads or spans dictate, steel beams are used to support flooring instead of the lighter open-web joists. Light-gauge steel joists are also sometimes used to support flooring **(Figure 9.27)**. The light-gauge joists are produced from cold-rolled steel and are available in several cross-sections. Like the open-web steel joists, the light-gauge joists can be used to support metal decks or wood panel flooring systems. The steel joists are produced with depths of 6 to 12 inches (150 mm to 300 mm) and can be spaced 16 to 48 inches apart (400 mm to 1 200 mm) depending on the span and the load to be supported.

Collapse of Steel Structures

Earlier in this chapter the characteristic decrease of the strength of steel as a result of an increase in temperature was discussed. The factors that affect the possible failure of steel were also discussed. Two of these factors — the mass of the steel members and the type of structural connection — result from the structural design used. In general, although unprotected steel is not fire-resistive, the greater the mass of a steel member, the less likely it is to fail in a fire. Of particular importance to firefighters are the additional strength provided by certain connections and the weakness of lighter-weight construction.

Connections and Structural Strength

Mass of Steel. The rigid connections used in the beam and girder type of frame have a greater mass of steel at the point of connection than do simple connections. Therefore, it takes much more heat to produce failure in rigid connections than it does in the less-massive simple connections. Rigid connections are frequently found intact after a fire even after other parts of a frame have failed. In contrast, a simply supported beam may fail under fire conditions as loads shift **(Figure 9.28)**.

Gusset plates. Steel connections, both in the case of rigid connections used with beam and girder frames and heavy trusses, frequently make use of a steel web known as a *gusset plate* **(Figures 9.29 a and b)**. Although the primary purpose of a gusset plate is to strengthen the connection, the gusset plate also increases the steel mass at the connection, thereby decreasing its possibility of failure.

Gusset Plates — Metal or wooden plates used to connect and strengthen the intersections of metal or wooden truss components roof or floor components into a load-bearing unit.

Knee joints. In gabled rigid-frame structures the knee joint between the roof and the wall will be the strongest part of the frame and the last part to fail **(Figure 9.30)**.

Redundancy. In a structure with a large beam and girder frame made of repeating sections, the adjacent sections of the frame tend to be mutually supporting. These adjacent sections provide a degree of redundancy to the overall system that reinforces the structure's strength.

Figure 9.28 A simply supported beam collapsed as a result of fire.

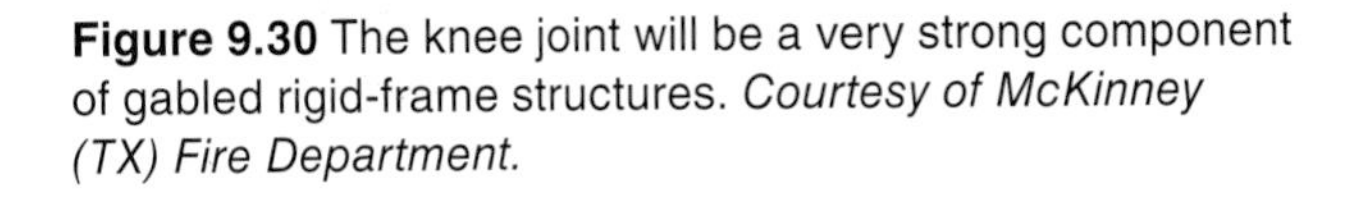

Figure 9.30 The knee joint will be a very strong component of gabled rigid-frame structures. *Courtesy of McKinney (TX) Fire Department.*

Figures 9.29 a and b Steel gusset plates provide strength and additional mass at the joints. *Photo b courtesy of McKinney (TX) Fire Department.*

Lighter Construction

If steel trusses are constructed with slender chords and diagonal members and are simply supported on a bearing wall or beam, they can easily fail and collapse under fire conditions. As in the case of steel beams, failure will not occur unless the trusses are exposed to a large amount of heat.

The light-gauge steel sheeting used in floor systems and in roofs has a large surface area compared to its mass (large surface area to mass ratio). It will heat rapidly under fire conditions. Therefore, unprotected, light-gauge steel sheeting may fail structurally although it will not melt.

Fire Protection of Steel

Steel, no matter what its mass, ultimately is not inherently fire-resistive. If exposed to enough heat long enough, steel will fail. In order to be used in fire-resistive buildings, steel must be made more fire-resistive. The usual way to accomplish this is with an insulating material to protect the steel from the heat of a fire.

In older buildings, the steel framework was encased in brick or ordinary concrete. A steel column encased in 3 inches (75 mm) of concrete with a siliceous aggregate would have a fire resistance of four hours. This method is effective but increases the weight and therefore the cost of a structure. Structural designers prefer to use lighter-weight materials for the protection of steel.

In contemporary practice the most commonly used insulating materials are gypsum, spray-applied materials, and intumescent coatings. Lightweight concrete can also be used because it is a durable material. Some of these insulating materials are applied by spraying and are known collectively as *spray-applied fire resistive materials* (SFRMs).

Gypsum

Gypsum Board — Widely used interior finish material. Consists of a core of calcined gypsum, starch, water, and other additives that are sandwiched between two paper faces. Also known as gypsum wallboard, plasterboard, and drywall.

Gypsum can be used as an insulating material either in the form of flat boards or a plaster **(Figure 9.31)**. Gypsum board consists of a core of calcined gypsum, starch, water, and other additives that are sandwiched between two paper faces. Gypsum board is available as regular or type X. Regular gypsum board has no special additives to enhance its fire resistance, although it will provide some degree of fire protection. Type X gypsum board contains additives to increase its fire resistance and is usually required where a specified fire resistance is desired.

The value of gypsum as an insulating material arises in part from the water that is chemically combined within the material. Gypsum consists of approximately 21 percent entrapped water. This water content enhances gypsum's performance as a fire-insulating material. The water turns to steam upon exposure to fire and, in doing so, absorbs the heat. This process is known as *calcination*. Once the moisture has been driven off, the remaining gypsum will act as an insulating material.

Gypsum can be used to protect both columns and beams, providing fire resistance ratings from one to four hours. Gypsum board is used in multiple layers to attain higher fire resistance ratings.

NOTE: For more information about construction systems and materials and their ratings, consult the *UL Fire Resistance Directory*.

Figure 9.31 The gypsum covering the interior of this steel-frame store provides a measure of fire resistance. *Courtesy of McKinney (TX) Fire Department.*

Figure 9.32 Sprayed-on insulating material for steel.

Spray-Applied Materials

Spray-applied fire-resistive materials (known as SFRMs) are efficient and inexpensive. The most commonly used SFRMs are mineral fiber or expanded aggregate coatings such as vermiculite and perlite. The degree of fire resistance provided will depend not only on the material but also on the thickness of the application.

Most fire-insulating materials can be applied in varying thicknesses to achieve different fire-resistance ratings. **Figure 9.32** illustrates a vermiculite-based SFRM applied to a steel beam. The applied fireproofing can vary from 7/8 to 1 7/8 inches (22 mm to 47 mm) to produce a fire-resistive rating of 1 to 4 hours.

Density. Low-density mineral fiber materials are relatively soft and can be easily dislodged from the steel. Low-density fiber materials are not suitable for exterior use. More durable mineral fiber products with densities greater than 20 lb/cu ft. can endure conditions of limited weather exposure and higher humidity such as might exist in parking facilities.

High-density SFRMs that use magnesium oxychloride have densities ranging from 40 to 80 lb/cu ft.

NOTE: Asbestos has not been used in SRFMs since the 1970's.

Cementitious — Containing or composed of cement. Has cementlike characteristics.

Cementitious. Cementitious materials are produced in various formulations. Ingredients can include Portland cement, gypsum, perlite, and vermiculite. Some manufacturers use magnesium oxychloride, or oxysulfate, calcium aluminate, phosphate, or ammonium sulfate. The cementitious materials have densities varying from 15 to 50 lb/cu ft.

Because of the variety of SFRMs available, it can be difficult to ensure that the applied material meets the specifications of the design documents. Furthermore, for an SFRM to be effective, proper installation procedures must be followed. Application of SFRMs must be in accordance with the manufacturer's listing and recommendations. The surface to which the SFRM is applied must be clean of oil, dirt, loose paint, and any other substance that would prevent good adhesion. Hangers and supports are frequently attached to the steel members and workers will sometimes remove some of the fireproofing material to attach these supports. Because this action can lessen the effectiveness of the fireproofing material, it is very important that supports and hangers be installed *before* application of the SFRM.

Intumescent and Mastic Coatings

Intumescent materials undergo a chemical reaction when exposed to the heat of a fire. An intumescent coating will char, foam, and expand when heated. The coating material will expand to 15 to 30 times its original volume. The expanded coating then acts as an insulating material to protect the steel.

Intumescent Coating — Coating or paintlike product that expands when exposed to the heat of a fire to create an insulating barrier that protects the material underneath.

Mastic coatings function in a manner similar to intumescent coatings except they are based on more complex organic materials and their reaction to heat is more complex.

Intumescent coatings are applied as a paint. They have an applied thickness of 0.03 to 0.4 inches (0.76 mm to 10.1 mm) which is less than the thickness of the spray-applied materials. Because they have the appearance of paint, it can be difficult to visually establish that a fire-resistive coating has been applied to steel members.

Both intumescent and mastic coatings are relatively expensive. Their advantages include lighter weight, durable surfaces, and good adhesion. In addition they are frequently the most aesthetically pleasing.

Membrane Ceilings

A very commonly used method of protecting a steel floor or roof assembly is the *membrane ceiling*. A membrane ceiling consists of a ceiling material suspended from the supports for the floor or ceiling above. The most common method is to use mineral tiles in a steel framework suspended by wires **(Figure 9.33)**. The mineral tiles are a lightweight insulating material and usually contain perforations for acoustical purposes. Gypsum panels are also used for membrane ceilings.

Membrane Ceiling — Usually refers to a suspended, insulating ceiling tile system.

The ceiling material acts as a thermal barrier to protect the steel that supports the floor or ceiling above **(Figure 9.34)**. They are frequently used in steel framing systems that make use of the open web joists described earlier. The use of a membrane ceiling can provide a floor and ceiling assembly or a ceiling and roof assembly with a fire rating of one to three hours depending on the specific details of the installation.

Ceiling materials are never rated independently. A ceiling is always rated as part of a floor and ceiling assembly, so it is not accurate to speak of the fire resistance of a ceiling alone. Because the membrane ceiling forms an integral part of a fire-rated assembly, it follows that any removal or penetration of the ceiling material reduces or even eliminates the fire resistance of the total assembly.

Membrane ceilings are popular, in part, because building services such as electrical wiring, automatic sprinkler piping, and ventilation ducts can be concealed above the ceiling. Penetration of membrane ceilings is frequently necessary for lighting fixtures and ventilation diffusers. When a floor and ceiling assembly is rated, any such penetrations must be provided for in the

Figure 9.33 Membrane ceiling. *Courtesy of McKinney (TX) Fire Department.*

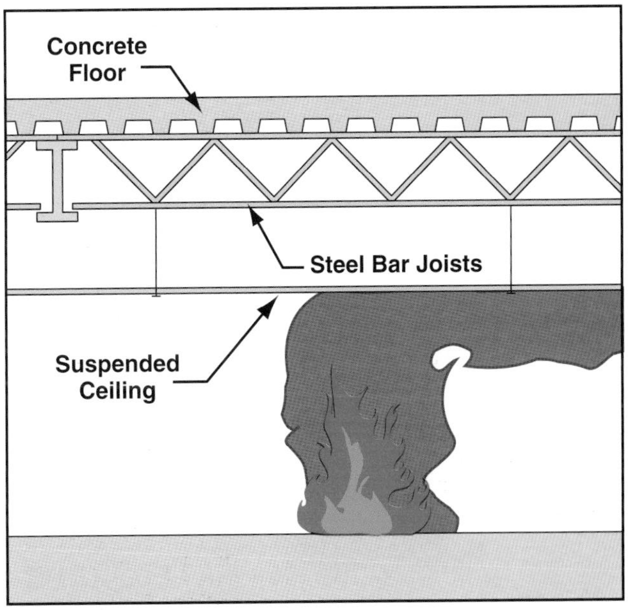

Concrete Floor
Steel Bar Joists
Suspended Ceiling

Figure 9.34 A suspended ceiling provides protection for the building structure above.

testing. It may be necessary to provide additional insulation on the back of lighting fixtures and to equip ventilation ducts with fire dampers so fire does not penetrate through the opening.

The fire-rated membrane floor and ceiling assemblies are listed by the testing laboratories as a total assembly. All the specific details of the assembly must be adhered to in its installation. Deviation from the laboratory specifications will affect the fire rating of the assembly.

Code Modifications

Protected steel is one of the two common methods of providing fire-resistive construction, such as Type IA and IB and Type IIA. The other method is reinforced concrete (See Chapter 10, Concrete Construction). Although building codes specify the degree of fire resistance required for various structural members, they do permit reductions under certain circumstances. Several examples are as follows:

- Eliminating the fire-resistance rating for roof construction located more than 20 feet (6.6 m) above the floor below for some occupancies
- Allowing a reduction of the required fire resistance when an automatic sprinkler system is provided that is not otherwise required by the provisions of the code

The consequence of these code provisions is that firefighters may encounter unprotected steel structural members in buildings that are classified under the building code as fire-resistive. Proper preincident planning is necessary to help avoid making tactical decisions based on incomplete information.

Summary

Steel is a durable and noncombustible building material. Its strength permits it to be used for the framework in high-rise buildings. Steel can be used in columns, beams, and trusses. Steel trusses and beams can span greater distances than other materials. Steel, however, is affected by the heat of a fire. Steel members will lose a large percentage of their strength at the temperatures encountered in fires and can fail. The extent of failure that may occur will depend on the mass of the steel member. If steel is to be used in a fire-resistive building it must be protected and several methods are available to achieve this protection.

Review Questions

1. What are two disadvantages of steel?
2. Why is cast iron no longer used in structural framing?
3. How is a trussed arch built?
4. What is a gusset plate?
5. How does an intumescent coating protect steel under fire conditions?

References

1. Ambrose, James, *Building Structures, 2nd edition,* John Wiley and Sons.
2. Harper, Charles A., *Handbook of Building Materials for Fire Protection,* McGraw-Hill.
3. Parker, Arthur J., Beitel Jesse J., and Iwankiw, Nestor R. *Fire Protection Materials for Architecturally Exposed Steel,* Structures Magazine, February 2005.

Concrete Construction

Chapter Contents

Divider page photo courtesy of McKinney (TX) Fire Department.

Key Terms

FESHE Objectives

Fire and Emergency Services Higher Education (FESHE) Objectives: *Building Construction for Fire Protection*

1. Demonstrate an understanding of building construction as it relates to firefighter safety, buildings codes, fire prevention, code inspection and firefighting strategy and tactics.
8. Identify the indicators of potential structural failure as they relate to firefighter safety.

Concrete Construction

Learning Objectives

After reading this chapter, students will be able to:

1. Describe the production process of concrete.
2. Describe the methods used to reinforce concrete used in building structures.
3. Discuss the methods of ensuring the quality of concrete.
4. Describe the concrete framing systems used in building structures.
5. Discuss the factors that affect the performance of concrete under fire conditions.

Chapter 10
Concrete Construction

Case History

Event Description: Two firefighters were working in a defensive posture at a structure fire in a single-family dwelling that had undergone a partial roof collapse. Two firefighters were operating near a window and were struck by concrete roofing tile as the remainder of the roof collapsed. The two were part of a division supervised by a single officer, who was monitoring another pair of firefighters. They were rapidly extricated by nearby firefighters and assessed for injuries. One was uninjured and the other suffered a minor back strain.

Lessons Learned: Situational awareness is key to survival on the fireground. If a team of firefighters is not directly supervised by an officer, it is essential that the back-up person remain aware of the current surroundings because the nozzle person may be focusing on the fire. Just because a roof has collapsed once does not mean the threat is over. A 360-degree situational assessment must be made and constantly updated as the operation continues.

Source: National Fire Fighter Near-Miss Reporting System.

Concrete has many applications in building construction. It is used for pavement, foundations, columns, floors, walls, and concrete masonry units **(Figures 10.1 a - c, p. 282)**. Its advantages are that it can be produced from raw materials that are usually locally available and are low in cost. Like masonry, concrete does not burn. It also resists insects and the effects of contact with soil. It can be placed in forms to create a variety of architectural shapes.

Because concrete is noncombustible, it is widely used in fire-resistive (Type I) construction. In almost all structural applications concrete is reinforced with steel. Only in a few applications such as sidewalks or driveways would concrete be used without reinforcement.

Concrete

Concrete is produced from Portland cement, coarse and fine aggregates, and water. The aggregates used in concrete are inert mineral ingredients that reduce the amount of cement that would otherwise be needed. The coarse aggregates consist of gravel or stone and the fine aggregate is sand. The cement combines with the water to form a paste that coats and bonds the pieces of fine and coarse aggregate. The aggregates make up a large percentage of the total volume of concrete.

Aggregate — Gravel, stone, sand, or other inert materials used in concrete. These materials may be fine or coarse.

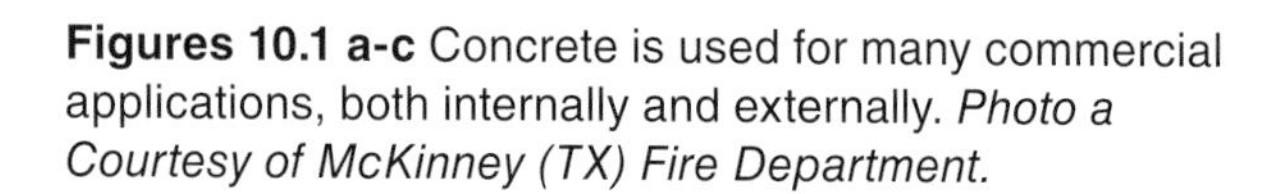

Figures 10.1 a-c Concrete is used for many commercial applications, both internally and externally. *Photo a Courtesy of McKinney (TX) Fire Department.*

Curing

Curing — Maintaining conditions to achieve proper strength during the hardening of concrete.

Concrete must be cured in order to reach its proper strength. Curing fresh concrete requires correct hydration and temperature control. Improper curing methods will negatively affect the finished surface of concrete as well as its strength.

Hydration

Heat of Hydration — During the hardening of concrete, heat is given off by the chemical process of hydration.

Hardening of concrete involves a chemical process known as *hydration*. In hydration, water combines with the particles of cement to form a microscopic gel. As the concrete hardens, this gel gives off heat, which is known as the *heat of hydration*.

Concrete initially hardens fairly quickly but then begins to harden more slowly. Because hydration involves water, proper curing requires that the concrete be kept moist until it reaches its desired strength. **Figure 10.2** shows the hardness of concrete as a function of time in days for concrete that is moist cured and concrete that is cured in air. It can be seen that the moist curing of concrete produces a stronger concrete.

Concrete can be kept moist by several techniques including sprinkling it with water, ponding, or covering it with a plastic film. Sealing compounds are also available that form a membrane on the concrete to slow evaporation.

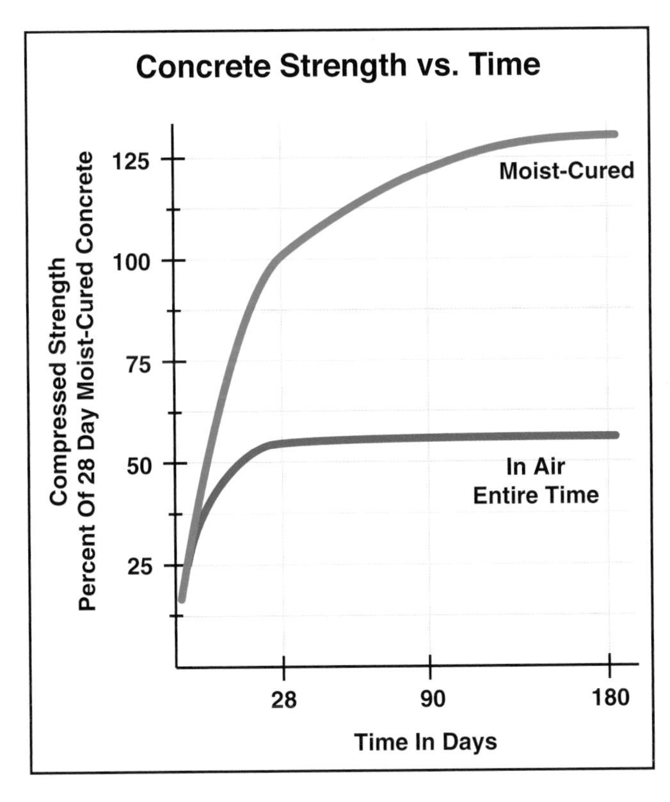

Figure 10.2 These curves illustrate the positive impact of keeping fresh concrete well-hydrated while it cures.

Temperature Control

In addition to maintaining proper moisture, concrete that is curing must be maintained at the correct temperature, ideally between 50°-70°F (10°-21°C). Concrete that is cured at or above 100°F (37°C) will not reach its proper strength; concrete cured near freezing temperatures will harden more slowly. In massive concrete structures, the heat of hydration generated can adversely affect the final strength of the concrete. Concrete can be cooled by using cold water during curing or using chilled water in the mixing process. If concrete is to be placed near freezing temperature it can be protected with a heated enclosure.

During the curing process concrete will shrink slightly. Theoretically, properly cured concrete continues to harden indefinitely at a gradual rate; however, normal design strength is reached after 28 days.

Admixtures

Different types of concrete can be produced for specific purposes. This is accomplished by varying the ingredients or adding chemicals to the concrete mixture; these are generally known as *admixtures*. For example, the density of concrete can be reduced by using a lighter-weight aggregate such as shale or clinker. An admixture known as a superplasticizer can be used to produce a mixture that flows more freely.

Admixture — Ingredients or chemicals added to concrete mix to produce concrete with specific characteristics.

Concrete types include the following:

- Ordinary stone concrete
- Structural lightweight concrete
- Insulating lightweight concrete
- Gypsum concrete
- High early-strength concrete
- Expansive concrete
- Water-permeable concrete

Coloring can also be added to concrete for aesthetic or safety reasons, such as coloring concrete when it is placed over buried electrical cables.

Reinforced Concrete

Like masonry, concrete is strong in compression but weaker in tension. The ultimate compressive strength can be varied from 2,500 psi to 6,000 psi (17,237 kPa to 41,370 kPa) with the allowable stress used in design reduced by a factor of safety.

Because concrete is weak in tension, it cannot be used alone where tensile forces occur in a structure. To resist the tensile forces, concrete is reinforced with steel reinforcing bars placed within the concrete before it hardens.

NOTE: In some structural designs, such as an arch, the forces are primarily compressive. In these cases concrete could, theoretically, be used without reinforcing. However, in most structural designs, tensile forces exist and the concrete must be reinforced to resist them.

- The techniques that are used to reinforce concrete are:
- Ordinary reinforcing
- Prestressing reinforcing (pretensioning or posttensioning)

Ordinary Reinforcing

With ordinary reinforcing, steel bars are placed in the formwork and the wet concrete is placed in the formwork around the bars. The concrete must be properly compacted as it is placed in the forms to completely surround the reinforcing bars (rebar) and to avoid cavities in the hardened concrete. Mechanical vibrators are used to ensure that the wet concrete fills all the spaces within the formwork **(Figure 10.3)**. The concrete then is permitted to cure. When the concrete has hardened, it adheres to the reinforcing bars because of deformations on the surface of the bars.

NOTE: For more information about formwork, see Chapter 13.

It is the job of the design engineer to specify the number of reinforcing bars to be used, their size (diameter), and the depth of concrete cover around the bars. For example, the diameters of standard-size reinforcing bars vary from .375 inches to 2.257 inches (9.5 to 57 mm).

The actual tensile and compressive forces developed within structural components are complex. For example, the fundamental tensile and compressive forces are not uniform throughout a beam. In addition, diagonal tension forces occur that must also be resisted. In **Figure 10.4** additional reinforcing steel is shown in a vertical position. The vertical reinforcing bars are known as stirrups and are provided to resist the diagonal tension. **Figure 10.5** illustrates the extensive placement of reinforcing steel needed in a concrete rigid frame.

Concrete beams are frequently cast in the shape of a tee **(Figure 10.6)**. The wider cross-sectional area at the top of the tee beam permits the concrete to support a greater load. At the same time, it reduces the dead load that would result with a simple rectangular beam. Reinforcing steel is placed in the bottom of the tee to resist the tensile force.

Although the primary function of placing reinforcing steel in concrete is to resist tensile forces, the steel can also be used to support some of the compressive forces **(Figure 10.7)**. The steel bars support some of the compressive load and also resist bending forces in the column from such sources as wind load and settling.

Figure 10.3 Concrete is smoothed to make sure it fills all the spaces in the formwork. *Courtesy of McKinney (TX) Fire Department.*

Figure 10.4 Additional rebar in a vertical position.

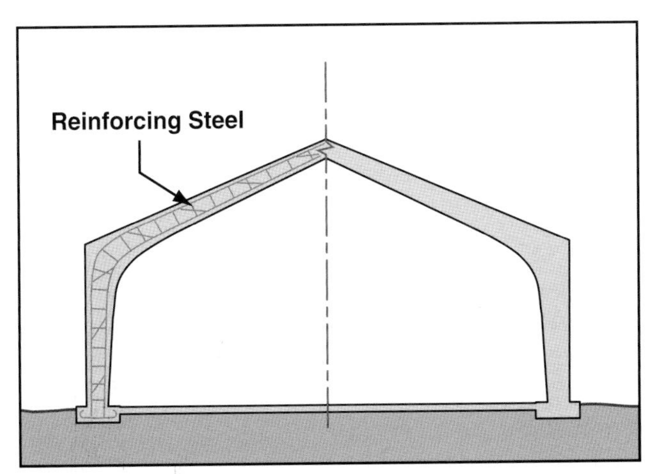

Figure 10.5 Reinforcing steel pattern in a concrete rigid frame.

Figure 10.7 Vertical reinforcing bars in a concrete column. *Courtesy of Ed Prendergast.*

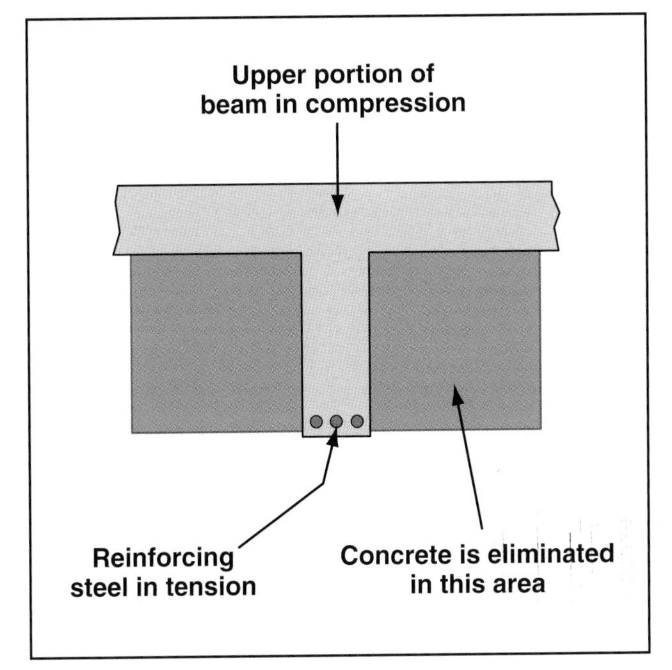

Figure 10.6 A tee-shaped concrete beam is a more efficient and lightweight design because the concrete in the shaded area is eliminated.

The compressive forces in a column could be great enough to cause the steel reinforcing bars to buckle even though they are imbedded in the concrete. To avoid possible buckling, lateral reinforcing is provided around the vertical bars.

Prestressing Reinforcing

Ordinary reinforcing is a fundamental and useful construction method, but it has inherent limitations. When a concrete beam or floor slab supports a load, the concrete in the part of the beam in tension is essentially doing no work. More efficient use of concrete is made using a technique known as *prestressing*. In prestressing, a compressive force is induced in the concrete before the load is applied. This "pre" stress is applied by tightening or "pre" loading the reinforcing steel. The preloading of the steel creates compressive stresses in the concrete that counteract the tensile stresses that result when the loads are applied **(Figure 10.8)**.

The prestressing process requires large loads to be applied to the concrete along the axis of the beam. These loads can result in the shortening of the concrete over time. The steel may also slowly stretch in length. This would result in a reduction of the compressive force and a possible loss of load-carrying capacity. To compensate for this loss, the forces that are initially applied in the prestressing process are slightly higher than the forces theoretically needed to support the concrete and the applied loads.

Prestressing is widely used in concrete structures. There are two methods of prestressing concrete. These are pretensioning and postensioning. Both processes use the same basic materials. In pretensioning the cables are tensioned before the concrete is poured; in posttensioning the cables are tensioned after the concrete has reached a certain strength.

Pretensioned Reinforcing — Used in pretensioned concrete. Steel strands are stretched between anchors producing a tensile force in the steel. Concrete is then placed around the steel strands and allowed to harden.

Pretensioning. In pretensioned concrete, steel strands are stretched between anchors producing a tensile force in the steel. Concrete is then placed around the steel strands and allowed to harden. After the concrete has hardened sufficiently, the force applied to the steel strands is released. As the force on the strands is released, the strands exert a compressive force in the concrete.

When the steel strands are released, the concrete member usually takes on a slight upward deflection. As loads are applied to the pretensioned member, the deflection usually disappears and the member becomes flat.

Post-Tensioned Reinforcing — Technique used in post-tensioned concrete. Reinforcing steel in the concrete is tensioned after the concrete has hardened.

Posttensioning. When concrete is posttensioned, the reinforcing steel is not tensioned until after the concrete has hardened. The reinforcing strands are placed in formwork and covered with grease or a plastic tubing to prevent binding with the concrete **(Figures 10.9 a and b)**. When the concrete has hardened, the strands are anchored against one end of the concrete member and a jack is positioned at the other end. The jack is used to apply a large tensile force to the steel that stretches the steel and results in a compressive force in the concrete. The pulled end of the reinforcing strand is anchored to the concrete and the jack is removed. The reinforcing cables are trimmed to the edge of the concrete and grouted.

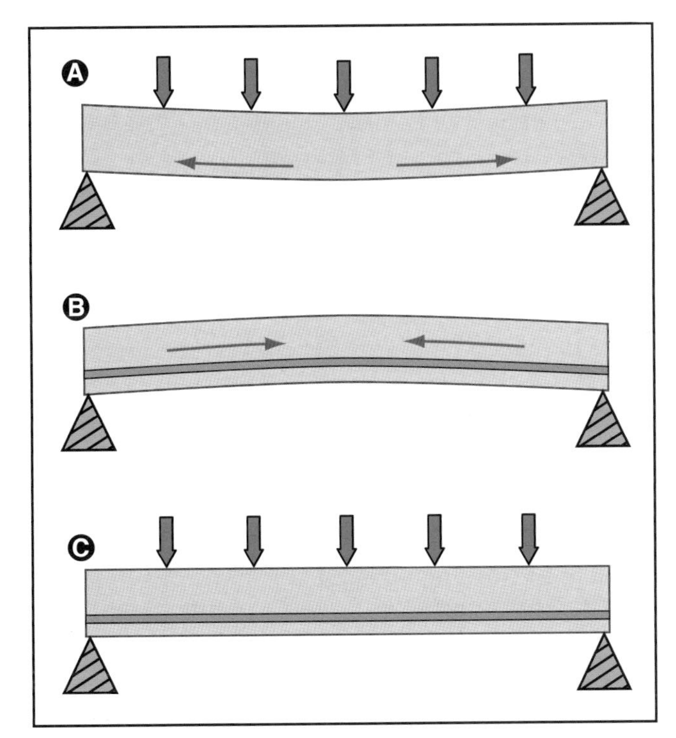

Figure 10.8 "A" illustrates a load on a beam, which causes downward deflection and tensile stresses in the lower part of the beam. "B" shows how prestressing the beam with reinforcing steel causes an upward deformation and creates compression forces in the beam. "C" illustrates that when a load is applied to the beam, the tensile and compressive forces equalize and the slight upward deflection disappears.

Figures 10.9 a and b Posttensioning strands in a concrete slab. *Photo a courtesy of Ed Prendergast; photo b Courtesy of McKinney (TX) Fire Department.*

The reinforcing steel used in concrete is essential to the strength of a reinforced concrete member. The reinforcing steel bars support large loads. Furthermore, the forces that are produced in the prestressing process remain locked in the steel for the life of the assembly. As a general rule, reinforcing steel should not be cut in the course of fire fighting operations unless it is necessary to rescue trapped victims.

Cutting through reinforcing steel with a saw or torch is particularly dangerous in posttensioned concrete because the steel is not bonded to the concrete. The steel strands are stretched like giant rubber bands. If they are cut, they are likely to spring out of the concrete, injuring emergency responders. Releasing the posttensioned element may also lead to the failure of the concrete structural element, resulting in a collapse or partial collapse.

Concrete Structures

Concrete can be either cast in place or precast. Cast-in-place concrete is placed into forms at the building site as a wet mass and hardens in the forms. Precast concrete is placed in forms and cured at a precasting plant away from the job site. Precast concrete structural members are then transported to the job site.

Cast-in-Place Concrete

Cast-in-Place Concrete — A common type of concrete construction. Refers to concrete that is poured into forms as a liquid and assumes the shape of the form in the position and location it will be used.

Cast-in-place concrete permits the designer to cast the concrete in a wide variety of shapes. This type of concrete does not develop its design strength until after it has been placed in the location where it will be used. Most cast-in-place concrete is proportioned at central bulk plants and then mixed in a mixing truck en route to the job site. There the wet concrete is transported from the truck to the formwork either by hoisting it in large buckets or by pumping.

Great care must be exercised in the mixing, placing, and curing of concrete to ensure good quality **(Figure 10.10)**. Poor-quality concrete will not attain its desired strength upon hardening or will otherwise be defective. For example, using sand from an ocean beach as a shortcut introduces salt into the concrete. Eventually the salt will cause corrosion and deterioration of the reinforcing steel. If the concrete is vibrated excessively as it is placed in the forms, segregation of the aggregate results. The heavy coarse aggregate settles at the bottom of the mixture and the water and cement rise to the top.

The quality of concrete is of significance to firefighters. Concrete that is of poor structural quality will behave poorly under fire conditions by spalling and even breaking apart.

Figure 10.10 Cast-in-place concrete must be poured and cured correctly. *Courtesy of McKinney (TX) Fire Department.*

Water-to-Cement Ratio

The single most important factor in determining the ultimate strength of concrete is the water-to-cement ratio. Water is a necessary ingredient in concrete because it reacts with the cement powder in the hydration process. An amount of water greater than required for curing is added to the concrete mix to increase its workability as it is placed in the forms. Some of this excess moisture evaporates and leaves microscopic voids in the hardened concrete. A portion of the excess moisture remains locked in the concrete.

If too much water has been used in the mix, the final product will not achieve its desired strength. The presence of excess moisture in the concrete also produces spalling in the concrete due to freezing conditions or the heat of a fire.

Quality Control of Concrete

Slump Test

When the concrete arrives at the job site, its quality is checked by administering a test known as a *slump test.* The slump test is used to check the moisture content of concrete by measuring the amount that a small, cone-shaped sample of the concrete settles or "slumps" after it is removed from a standard-sized test mold **(Figure 10.11)**. Concrete with a high moisture content has a more liquid consistency and a greater slump. Concrete with excessive slump will fail the test and may be rejected by the structural engineer.

Compression Test

Another method of testing concrete is to make small test cylinders of a concrete batch and subject them to compression testing. This method is accurate but has the disadvantage of requiring that the concrete be permitted to harden before the results are known. This test would be very costly if the concrete were ultimately found to be unsatisfactory.

Concrete Framing Systems

Large cast-in-place structures cannot be cast in one operation. Construction joints unavoidably occur between successive pours. To provide for transfer of loads and forces from one placement to the next, the reinforcement steel will overlap the joints **(Figure 10.12)**.

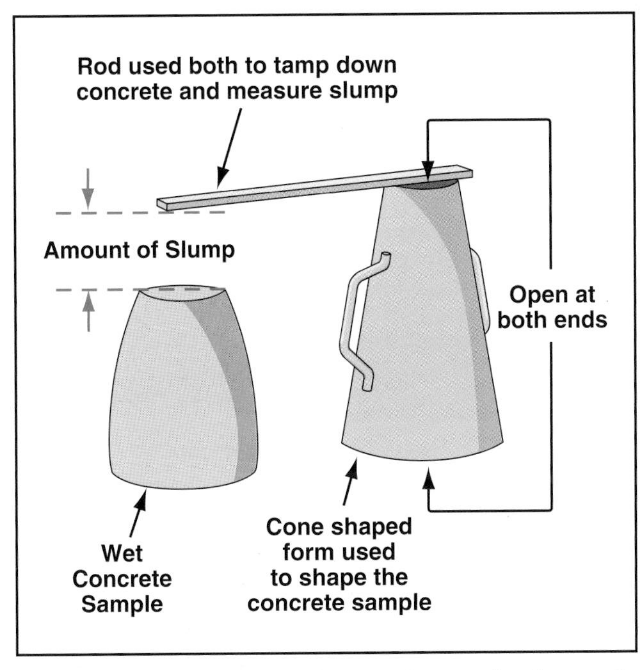

Figure 10.11 Measuring the 'slump' of a sample of concrete gives an indication of the amount of moisture in the concrete mix.

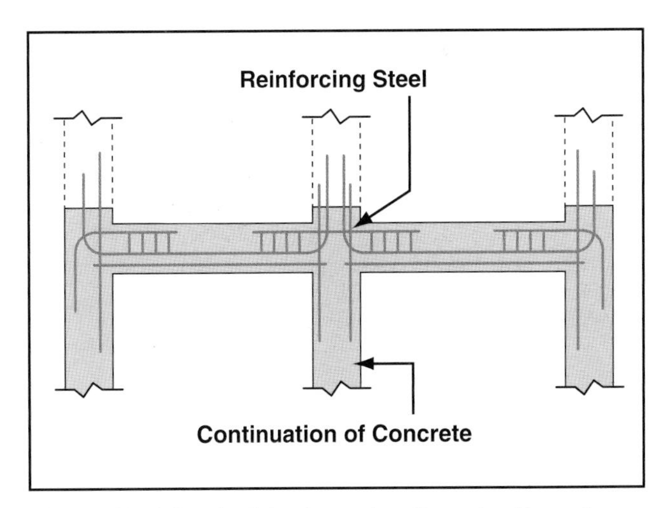

Figure 10.12 Steel reinforcing rods will overlap through joints in the concrete to provide continuity of strength between successive pours.

NOTE: See Precast Connections section later in this chapter.

Concrete buildings are constructed with structural systems that use bearing walls formed from cast-in-place concrete. However, a more typical design is to construct a concrete frame. The following are common cast-in-place structural systems:

- Flat slab
- Slab and beam
- Waffle construction

Flat-Slab Concrete Frames

The flat-slab concrete frame is a simple system that consists of a concrete slab supported by concrete columns **(Figure 10.13)**. The slab of concrete varies in thickness from 6 to 12 inches (150 mm to 300 mm). Shear stresses develop in the concrete where the slab intersects the supporting columns. In a building that will have heavy live loads, the area around the columns is reinforced with additional concrete in the form of drop panels or mushroom capitals. If the building will support light loads, this additional reinforcing is not necessary. The system then is known as a *flat plate.*

Slab and Beam Framing

A slab and beam frame consists of a concrete slab supported by concrete beams **(Figure 10.14)**. This framing system is extremely light weight and is best suited for buildings with light floor loads. Slabs in this type of construction can sometimes be as thin as 2 inches (50 mm). Due to the thin slab, the concrete beams must be closely spaced in order to provide adequate support. This spacing often gives an appearance similar to wood joists and is sometimes referred to as concrete joist construction.

Figure 10.13 Concrete slabs supported by concrete columns. *Courtesy of McKinney (TX) Fire Department.*

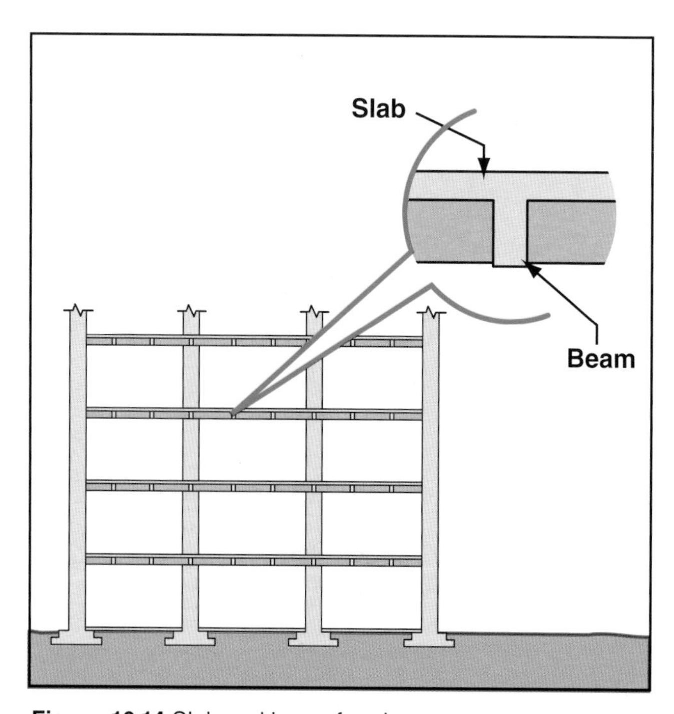

Figure 10.14 Slab and beam framing.

When the concrete beams run mainly in one direction, the framing is known as a one-way reinforced slab. Slab-and-beam concrete framing can also feature concrete beams running in two directions. Such a system is known as a two-way slab construction. The two-way framing system is used where spans are short and floor loadings are high.

Buildings using slab and beam systems are often highly susceptible to fire damage due to their thin nature. Fireproofing of some sort is often necessary, especially when specific fire ratings must be achieved.

Waffle Construction

Waffle construction derives its name from the waffle-like pattern of the bottom of the concrete slab **(Figure 10.15)**. The pattern results from the placement of square forms over which the wet concrete is placed. This design provides a thicker slab while eliminating the weight of unnecessary concrete in the bottom half of the slab. Reinforcing steel placed in the bottom of the formwork provides reinforcement in two directions. Slabs of this type, therefore, are also known as two-way slabs.

Figure 10.15 Waffle construction represents a type of two-way slab. This method of forming concrete adds strength without extra weight.

Concrete Plus Structural Steel

A poured concrete slab can be supported by structural steel beams instead of concrete beams. If the assembly is intended for a Type I (fire-resistive) building the steel must be provided with some form of fireproofing as described in Chapter 9.

Precast Concrete

As was noted earlier, precast concrete is placed in forms and cured at a location other than the construction site. Precast concrete may be produced at a precasting plant some distance from the work site. The precast structural shapes, including slabs, wall panels, and columns are transported to the job site and hoisted into position **(Figures 10.16 a and b)**.

Precast Concrete — Method of building construction where the concrete building member is poured and set according to specification in a controlled environment and is then shipped to the construction site for use.

NOTE: Some precast components, such as tilt-up panels, are cast at the site and moved into position.

Figure 10.16 a and b Precast concrete sections being hoisted into place. *Both photos courtesy of McKinney (TX) Fire Department.*

Advantages and Disadvantages

There are several advantages to using precast concrete:

- Higher degree of quality control possible than with cast-in-place concrete:
  - — Precasting forms can be located in a sheltered environment not exposed to the weather
  - — A high degree of quality control can be exercised over the ingredients
  - — Mixing and pouring the concrete can be more mechanized and efficient.
- Work can proceed more quickly at the job site:
  - — Precasting is faster because there is no need to construct formwork at the job site.
  - — No need to wait for concrete to harden before work can proceed, as with cast-in-place concrete.
  - — No need to construct and remove forms, which may result in additional costs.
- Precast concrete sandwich panels can be produced using a polystyrene core, which improves the insulating properties of the precast concrete (Figures 10.17 a - c).

Figures 10.17 a - c Foam insulation in precast concrete.
Courtesy of Gregory Havel, Burlington, WI.

A major disadvantage to using precast concrete is the need to transport the finished components to the job site. Transportation increases costs and limits the size of the shapes that can be precast.

Precast concrete buildings can be built using whole precast modular units but it is more common to assemble precast parts into a framework for a building. Therefore, from a construction standpoint, precast concrete structures have more in common with steel-framed buildings than with cast-in-place concrete buildings.

NOTE: Because precasting concrete did not become common practice until after World War II, it is not generally found in older buildings.

Precast concrete slabs for floor systems can be cast in standard shapes that include solid slabs, hollow-core slabs, single tee slabs, and double tee slabs **(Figure 10.18)**. Solid slabs are used for short spans up to approximately 30 feet (10 m), while the tee slabs can be used for spans up to 120 feet (40 m).

A common form of construction used with precast concrete is known as *tilt-up* construction. In tilt-up construction, reinforced wall panels are cast at the job site in horizontal casting beds. After the concrete has cured, the wall panels are tilted up into the vertical position by a crane. Temporary bracing is provided until the roof supports or other permanent horizontal bracing is provided **(Figure 10.19)**.

Tilt-Up Construction — Type of construction in which concrete wall sections (slabs) are cast on the concrete floor of the building and are then tilted up into the vertical position. Also known as Tilt-Slab Construction.

NOTE: Cast-in-place concrete is not moved after it has hardened.

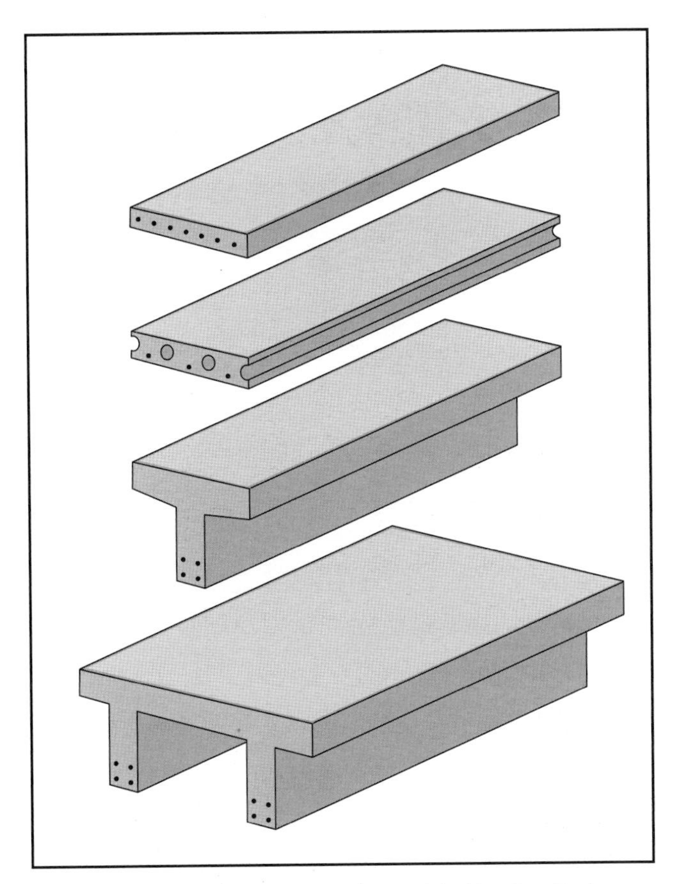

Figure 10.18 Precast concrete is provided in standard shapes such as slabs, hollow core slabs, single tees, and double tees.

Figure 10.19 Temporary bracing for tilt-up construction. *Courtesy of Dave Coombs.*

Figure 10.20 This tilt-up structure is supporting a steel roof. *Courtesy of McKinney (TX) Fire Department.*

Tilt-up walls can be several stories in height. Tilt-up walls can support several types of roof including timber beams, precast slabs, steel beams, and steel trusses **(Figure 10.20)**. The integrity of the roof is critical under fire conditions. Failure of a roof can result in outward horizontal forces against the wall resulting in collapse.

Precast Connections

Precast structural elements are usually lighter than corresponding cast-in-place components. However, the use of precast concrete results in a loss of continuity that is inherent with cast-in-place concrete frames. In a precast concrete structure, the connections between the individual components are a critical aspect of structural engineering.

A variety of techniques, such as bolting, welding, and posttensioning can be used to connect precast structural elements **(Figures 10.21 a and b)**. In the simplest of precast designs, precast slabs simply rest on a bearing wall or column. Simple designs of this type are not inherently rigid and the slabs need to be laterally tied together to resist horizontal forces.

When precast beams are to be supported by columns, the beams may be supported by *corbels* cast into the column. A corbel, also known as a bracket, is a ledge that projects from the column and supports the beam. Precast beams can also be supported by a short steel beam that is cast into the precast column. The precast beams are secured to the column through the use of steel angles cast into the columns or through the use of posttensioned steel cables.

Figures 10.21 a and b These precast sections have been bolted together. *Courtesy of McKinney (TX) Fire Department.*

Garages

A very common application of precast concrete is in the construction of parking garages. The floor loads and the span lengths make a precast structural system a practical choice. The precast columns and beams are typically left exposed **(Figure 10.22)**. Precast parking facilities perform well structurally under fire conditions. For example, a fire in a parking garage in Chicago involved 11 automobiles with only minor damage to the structure.

Identification of Concrete Systems

Buildings supported by a concrete frame are usually enclosed by a nonbearing curtain wall. The curtain wall is the building's exterior enclosure and can be made of such materials as aluminum, glass, steel panels, and masonry. The choice of material is determined by architectural style, thermal insulation properties, and cost. A curtain wall tends to conceal the structural details of a building and makes it difficult to accurately identify the structural system by observation alone.

It is difficult to know with certainty if a concrete frame building was constructed with ordinary reinforcing or posttensioned reinforcing. It may also be impossible to distinguish between cast-in-place concrete and precast concrete after a building is completed. In addition, some building systems, such as stucco and exterior insulation finish systems (EIFS), may appear to be concrete when they are not **(Figure 10.23)**. Frequently all fire personnel can do is to be familiar with prevailing construction methods in their jurisdictions and conduct preincident surveys.

Curtain Wall — Nonbearing exterior wall attached to the outside of a building with a rigid steel frame. Usually the front exterior wall of a building intended to provide a certain appearance.

Figure 10.22 Concrete is commonly used in garages. *Courtesy of McKinney (TX) Fire Department.*

Figure 10.23 Exterior finish insulation system resembling concrete applied to the exterior of a wood-frame building. *Courtesy of Ed Prendergast.*

Fire Behavior of Concrete Structures

Concrete is a fundamentally fire-resistive material that performs well under fire conditions. It is noncombustible and has good insulating properties. Concrete structural elements usually retain their integrity and concrete slabs and walls act as good fire barriers. Although the steel used in reinforced concrete is not fire resistive, the concrete surrounding the steel acts as insulation to protect it from the heat of the fire. The greater the depth of the concrete over the reinforcing steel, the greater it will protect the steel.

Prestressed concrete systems may be somewhat more vulnerable to failure than ordinary reinforced concrete. This increased vulnerability is due to the fact that the reinforcing cables and rods used in prestressed systems are made of high-strength steels that have lower yield-point temperatures. They can yield at a temperature of around 752°F (400°C). Therefore for the same depth of cover a prestressed assembly will fail sooner than a conventional reinforced assembly.

Concrete structural systems can have fire-resistance ratings from one to four hours. The fire resistance of a concrete assembly is affected by such variables as the following:

- Concrete density
- Concrete thickness
- Concrete quality
- Load supported by the concrete
- Depth of concrete cover over the reinforcing bars

Structural lightweight concrete has a lower density than ordinary concrete and has a lower thermal conductivity. Therefore, it acts as a better insulator against the heat of a fire than does ordinary concrete of comparable thickness. However, structural lightweight concrete is not used for load-bearing members.

Both cast-in-place and precast concrete buildings can achieve the fire-resistance ratings required by building codes. Cast-in-place concrete buildings have an advantage over precast buildings because the continuity of the assembly provides an inherent restraint to movement of the intersecting members such as columns and beams. This results in a fundamentally greater structural rigidity for cast-in-place buildings.

The fire resistance of concrete assemblies can be compromised in several ways. If concrete floor slabs or wall panels are supported by non-fire-resistive members, the overall construction would not be fire-resistive. In **Figure 10.24**, for example, the precast wall panels are horizontally braced by exposed steel roof beams. This is a very common design used for one-story mercantile and industrial buildings. When openings exist in concrete slabs or walls, the ability of the concrete to act as a barrier is lost unless the opening is protected by an appropriately rated assembly such as a fire door or shutter.

Just as with other types of construction, concrete buildings are not designed to withstand the force of an explosion **(Figure 10.25)**.

Figure 10.24 Precast walls supported by steel beams. *Courtesy of Dave Coombs.*

Figure 10.25 San Juan, Puerto Rico, November 22, 1996: Task Force members began performing search and rescue operations inside the Humberto Vidal Building following a gas mainline explosion. *FEMA News Photo*

Spalling

Spalling of concrete is caused primarily by the expansion of the excess moisture within the concrete when it is either heated or when it freezes. The expansion of the water creates tensile forces within the concrete. Because concrete has little resistance to tension, small pieces of the concrete break off. This spalling weakens the concrete structurally and exposes the reinforcing steel to the fire **(Figure 10.26)**. As a general observation, concrete that is of poor quality structurally will also perform poorly when exposed to a fire.

Spalling — Expansion of excess moisture within concrete due to exposure to the heat of a fire resulting in tensile forces within the concrete, causing it to break apart. The expansion causes sections of the concrete surface to violently disintegrate, resulting in explosive pitting or chipping destruction of the material's surface.

The extent to which concrete undergoes spalling depends on the amount of excess moisture in the concrete and the length of time that has passed the concrete was placed. New concrete that is not completely cured is subject to more severe spalling when exposed to a fire. The severity of spalling also depends on the duration and severity of the structure's exposure to the fire.

Figure 10.26 Spalling of concrete has exposed the reinforcing steel. Note the collapsed concrete slab. *Courtesy of Ed Prendergast.*

Heat Sink Effect

Concrete absorbs heat when it is exposed to a fire. Because concrete has relatively good insulating properties, it tends to retain the heat of an exposing fire and release it slowly, similar to the manner in which a masonry oven releases heat. This general effect is referred to as the *heat sink* effect. Firefighters performing overhaul in a concrete structure will experience the gradual release of the heat from the surrounding concrete for some time after the fire has been extinguished. The heat released from the concrete is not enough to reignite combustibles, but is enough to make the overhaul operation uncomfortable.

Summary

Concrete is a noncombustible building material that is strong in compression but weak in tension. Therefore it must almost always be combined with reinforcing materials. There are several methods of reinforcing concrete, including ordinary reinforcing and prestressing by pretensioning or posttensioning. The specific method by which a concrete structure has been reinforced cannot be identified visually. In any case, reinforcing bars should not be cut unless it is necessary for rescue purposes.

Concrete structures are built from either cast-in-place or precast concrete. Precast concrete buildings are assembled in a manner similar to steel structures. Cast-in-place structures have greater inherent rigidity than precast structures.

Reinforced concrete structures generally perform well under fire conditions. Their fire resistance is affected by the quality of the concrete. Spalling of concrete will result in failure of the reinforcing bars and failure of the structural assembly.

Review Questions

1. What is ordinary reinforcement of concrete?
2. How is a slump test performed?
3. What is tilt-up construction?
4. How is the choice of material for a curtain wall determined?
5. What is the heat sink effect?

References

1. Ching, Francis D. K., and Adams, Cassandra, *Building Construction Illustrated*, 3rd edition, John Wiley and Sons.
2. *Connections For Tilt-up Wall Construction*, Portland Cement Association, 1987.
3. NFPA® *Fire Protection Handbook 19th edition.* National Fire Protection Association®, Section 12, Chapter 11.
4. Allen, Edward and Iano, Joseph, *Fundamentals of Building Construction, Materials and Methods,* 4th edition., John Wiley and Sons.
5. Nilson, Arthur H., *Design of Concrete Structures, 12th edition, McGraw Hill.*

Roofs

Chapter Contents

Divider page photo courtesy of McKinney (TX) Fire Department.

chapter 11

Key Terms

FESHE Objectives

Fire and Emergency Services Higher Education (FESHE) Objectives: *Building Construction for Fire Protection*

1. Demonstrate an understanding of building construction as it relates to firefighter safety, buildings codes, fire prevention, code inspection and firefighting strategy and tactics.

Roofs

Learning Objectives

After reading this chapter, students will be able to:

1. Identify the ways roofs can affect structural firefighting.
2. Describe the characteristics of the different architectural styles of roofs.
3. Describe the systems used to support roofs.
4. Explain the functions of the roof deck and describe the materials used to construct it.
5. Describe the types and materials used as roof coverings for the different types of roofs.
6. Describe the testing process used to determine the fire rating of roof coverings.
7. Describe the characteristics of roofs installed for specific purposes.
8. Discuss the purpose of penthouses and skylights and their impact on firefighting tactics.
9. Describe the impact ceilings have on fire spread to roofs.

Chapter 11
Roofs

Case History

Event Description: Our fire department was dispatched to a commercial structure fire in an old automobile repair shop. Upon arrival of the first-in engine company, heavy fire was seen on the second story of the structure. A 2½-inch (65 mm) hoseline was stretched to the fire area. Upon the arrival of the truck company, the roof crew opened up the exterior wall of the main garage just below the main roof. After the exterior wall was opened, the roof crew was sent to the peak of the roof to cut an additional vent hole, which they accomplished.

As the crew was about to head down from the roof, one of the firefighters fell through. He caught himself with his arms and was dangling about 25 feet from the ground. Two firefighters on the roof immediately grabbed his arms and another firefighter wrapped a rope around him. Interior crews could see the firefighter's legs and set up a 24-foot extension ladder underneath him. With the ladder braced against an exposed rafter a firefighter climbed up the ladder and used his shoulders to push the firefighter up and out of the hole to safety.

The roof was made of metal decking and the area where the firefighter went though was not over the fire area. Later it was discovered that eight roof sections near the peak of the roof had been replaced with fiberglass decking. This fiberglass was painted over with a metallic paint and looked the same as the metal portion of the roof, especially in a smoke-filled environment.

Lessons learned: 1. Stay on the roof ladder as much as possible. If the firefighter had been on the roof ladder the entire time, he wouldn't have gone through the roof. 2. If at all possible, tie off to something or wear a ladder belt and hook to the roof ladder. 3. Pre-incident plans of commercial structures in the district may have alerted us to the fiberglass sections.

Source: National Fire Fighter Near-Miss Reporting System.

Roofs function as the primary sheltering element for the interior of a building. The significance of roofs in fire protection and fire fighting operations has long been recognized. A number of critical fire fighting operations are conducted from roofs. If roofs are constructed of combustible materials, however, they can become the fire fighting problem. For example, the combustibility of thatched roofs was one of the first fire hazards addressed in the earliest fire regulations adopted in colonial America. In modern times, fires that communicate from building to building by way of combustible wood shake shingles continue to plague some communities. Today, more roofs are being used as part of 'green design' with solar panels and even gardens. Solar panels in particular represent a new element of safety hazard during emergency operations.

This chapter will describe the details of different types of roof construction, roof support systems, fire ratings of roofs, advantages and disadvantages of different types of roofs, and their role in fire control.

Roofs and Fire Fighting

Roofs play an important role in structural fire fighting. Firefighters have to work on roofs either to combat a fire involving a roof or to ventilate the products of combustion from within the structure **(Figure 11.1)**. Sometimes it is desirable to use a roof as a vantage point for attacking a fire involving a neighboring structure. Finally, it should be pointed out that firefighters work beneath a roof whenever they are inside a building. For these reasons firefighters must be familiar with all aspects of roof construction, including materials, means of support, architectural styles, and such functional aspects as the use of a roof to support ventilation equipment.

There are several fundamental safety points that must always be kept in mind regarding roofs:

- Roofs are usually not as strong as floors because they are typically designed to support lighter live loads.
- Many types of roof construction have inherent concealed spaces between the ceiling and the roof deck, making it difficult to determine the extent to which a fire has developed overhead.
- Over time, loads may be added to roofs for which they were not originally designed **(Figure 11.2)**.
- Roofs are subject to wear and deterioration from the elements **(Figure 11.3)**. They are often repaired or renovated along with other parts of a building. A roof can wind up with several layers of roofing materials, again making it difficult for firefighters to determine the extent of fire or to perform ventilation.

Figure 11.1 When the roof is too dangerous to work on, firefighters have to work as close as they can from a safe position to assist with ventilation. *Courtesy of Ron Jeffers.*

Figure 11.2 Roofs can become overloaded over time if more and larger air-conditioning units and equipment are added.

Figure 11.3 This deteriorated roof collapsed after a snowstorm. The heavy timbers fell onto a gas line and the electrical panels were still energized. *Courtesy of West Allis (WI) Fire Department.*

The fact that a roof is a waterproof covering for a building can frustrate efforts to control a fire from above. The waterproof nature of a roof tends to limit the penetration of streams from ladder pipes and platform apparatus into the seat of the fire. Furthermore, if the roof collapses into a structure, it will form a waterproof covering over buried fire.

Architectural Styles of Roofs

From a fire fighting standpoint, the roofs of buildings can be classified into three styles: flat, pitched, and curved.

Flat Roofs

Flat roofs are found on all types of buildings including large-area warehouses, factories, shopping centers, schools, and numerous other applications **(Figure 11.4, p. 306)**. It is possible to construct a roof that is completely horizontal, but this design presents a drainage problem. If the roof is not constructed with a method for drainage, pools of water will form, which leads to early deteriora-

Figure 11.4 A typical roof on a large-area store.

tion. Therefore, many flat roofs are provided with a slight slope, typically from front to rear to facilitate drainage. Many large buildings have roofs sloped toward drains in the center of the roof. These roof drains may have the capacity to handle a heavy rainfall but not the water from a master stream. Firefighters must remember that these streams may overload the roof structure and cause it to collapse.

Flat roofs are the easiest roofs on which firefighters can work. However, they can provide a false sense of security and are not without potential dangers. In darkness, a firefighter can step off the edge of a roof or stumble over a low parapet. Firefighters can also to fall through an opening in a roof such as an opened roof hatch or skylight, or hole that has been cut for ventilation.

Pitched Roofs

Pitched roofs have inclined surfaces. They may be categorized into low slope roofs and medium to high slope roofs. Low slope roofs have a slope of up to 3 to 12, meaning for each 12 units of horizontal dimension the roof slopes upward 3 units. Medium to high slope roofs have slopes of 4/12 to 12/12. (A slope of 12/12 equates to 45-degree angle.)

Some structures, such as certain churches and mansions, have roofs with slopes of 18/12 or greater — too steep to work from a roof ladder.

Pitched roofs are designed in a number of styles that are determined by climate, function, and aesthetic considerations. In fact, architects can design roofs in an almost endless variety **(Figures 11.5 a and b)**.

Shed Roof — Pitched roof with a single sloping aspect, resembling half of a gabled roof.

Several commonly encountered pitched roof styles include the gable, hip, gambrel, mansard, butterfly, monitor, and sawtooth **(Figure 11.6)**. The simplest pitched roof is the *shed roof* that slopes in only one direction.

Figures 11.5 a and b Roofs can come in many shapes. Their design, pitch, and materials may make them very difficult to work on. *Photo a courtesy of Chris E. Mickal, NOFD Photo Unit; photo b courtesy of McKinney (TX) Fire Department.*

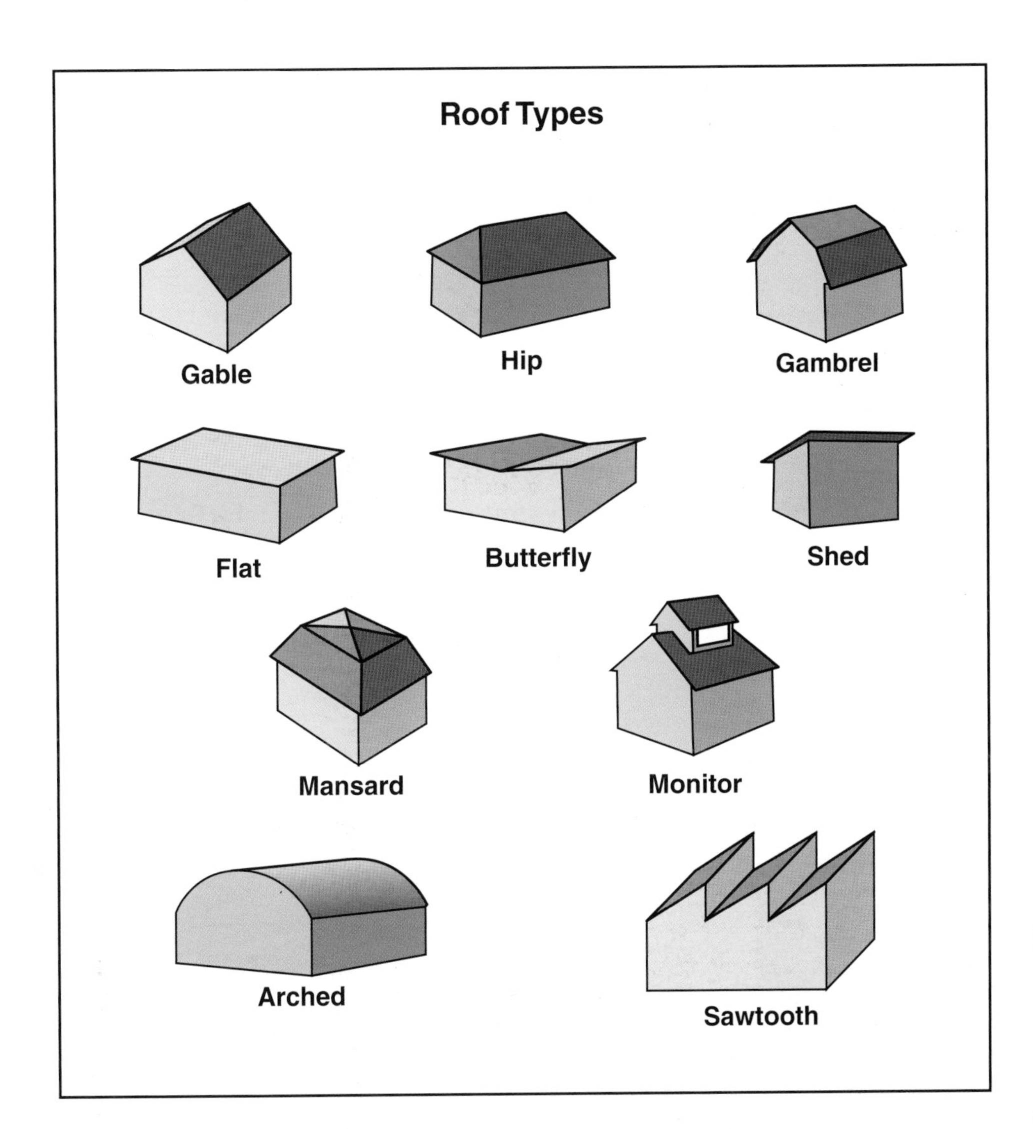

Figure 11.6 Examples of roof types.

- The *gable roof* is a very common roof style that consists of two inclined surfaces that meet at their high side to form a "ridge."
- The *hip roof* slopes in four directions and has a degree of slope similar to the gable roof.
- *Gambrel roofs* slope in two directions, but there will be a break in the slope on each side. Gambrel roofs are functional because the space created by the roof can be used as an attic or living space.
- A *mansard roof* has the break in the slope of the roof on all four sides. A mansard roof can also be constructed with a flat deck, in which case it is sometimes known as a modern mansard or deck roof. The mansard style roof forms a projection beyond the building wall that creates a concealed space through which a fire can communicate. A false mansard front is sometimes added to the front of a flat-roofed building as an architectural detail **(Figure 11.7, p. 38)**. Firefighters may be exposed to danger while working under these structures, which can collapse in large sections.
- The *butterfly roof* is a roof style that slopes in two directions — basically two shed roofs that meet at their low eaves.

Gabled Roof — Style of pitched roof with square ends in which the end walls of the building form triangular areas beneath the roof.

Hip Roof — Pitched roof that has no gables. All facets of the roof slope down from the peak to an outside wall.

Gambrel Roof — Style of gabled roof on which each side slopes at two different angles; often used on barns and similar structures.

Mansard Roof — Roof style with characteristics similar to both gambrel and hip roofs. Mansard roofs have slopes of two different angles, and all sides slope down to an outside wall.

Butterfly Roof — V-shaped roof style resembling two opposing shed roofs joined along their lower edges.

Monitor Roof — Roof style similar to an exaggerated lantern roof having a raised section along the ridge line, providing additional natural light and ventilation.

Sawtooth Roof — Roof style characterized by a series of alternating vertical walls and sloping roofs that resembles the teeth of a saw. This type of roof is most often found on older industrial buildings to provide light and ventilation.

- A *monitor roof* is designed to provide light and ventilation. Monitor roofs were once very commonly used on factory buildings. A raised central section of the roof extends several feet above the surrounding roof surface. The vertical sides of this monitor section, which are normally openable windows, are known as "clerestories."
- *Sawtooth roofs* were also once commonly used on industrial buildings for light and ventilation. Ideally the glass vertical sections should face north because the northern light is more constant during the day and the glare of the sun can be avoided.

NOTE: Modern ventilation and lighting systems have largely eliminated the need for monitor and sawtooth roofs.

A pitched roof is designed to shed water and snow. The pitch of a roof presents a major hazard to firefighters because the steepness of the roof results in a lack of secure footing. This hazard is increased when the roof is wet or covered with ice but also exists when the roof is dry because of the loose or granular texture of some roof coverings **(Figure 11.8)**. Loose or broken pieces of roof tiles can also slide off a pitched roof and create a hazard to firefighters on the ground.

Figure 11.7 Mansard-style fascia.

Figure 11.8 Some roofs are pitched too steeply to permit firefighters to work on them safely. *Courtesy of McKinney (TX) Fire Department.*

Curved Roofs

Curved roof surfaces take their form from the structural system used to support them. Curved roofs are most frequently supported by arches and bowstring trusses (see also Chapter 3).

When the area to be enclosed by the roof is circular, a dome roof can be used **(Figure 11.9)**. A dome can be thought of as an arch rotated 360 degrees. A dome roof produces structural forces similar to those of an arch. That is, horizontal thrusts exist at the base and a compressive force exists at the top. However, the forces in a dome are exerted around a complete circle instead of just one plane. More architecturally spectacular curved roofs can be created using geodesic domes or "lamella" arches.

Lamella Arch

The lamella arch is a special form of arched roof. It is constructed from short pieces of wood known as *lamellas*. Lamellas vary from 2 x 8 inches (50 mm x 200 mm) to 3 x 16 inches (75 mm to 400 mm) and in lengths varying from 8 to 14 feet (2.5 m to 4.3 m). The short lamellas are bolted together in a diagonal pattern with a special plate known as a "lamella washer" **(Figure 11.10)**. The curvature of the lamella arch results from the beveling (inclining) of the ends of the individual lamellas. The lamella technique can be adopted to form a dome as well as an arch. Roofs of this type have been used for occupancies such as gymnasiums, exhibitions halls, and auditoriums.

Lamella Arch — A special type of arch constructed of short pieces of wood called lamellas.

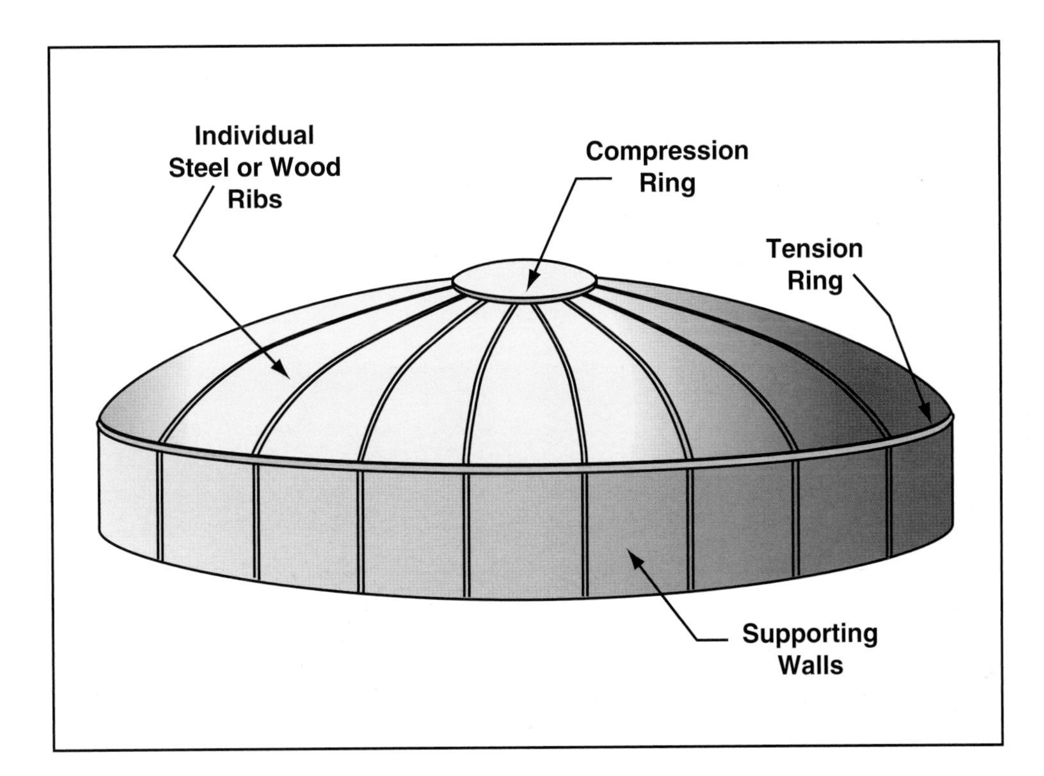

Figure 11.9 A circular dome roof.

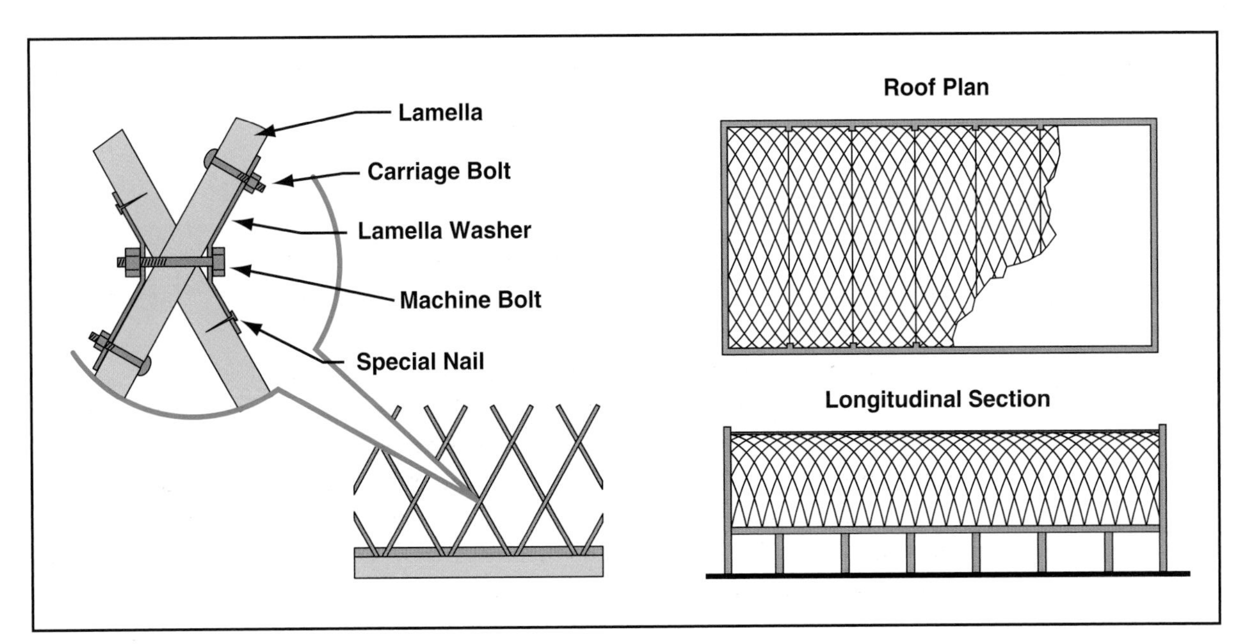

Figure 11.10 Lamella construction details.

Geodesic Domes

A geodesic dome is created using spherical triangulation. That is, triangles are arranged in three dimensions to form a nearly spherical surface **(Figures 11.11 a and b)**. A geodesic dome can be constructed from wood, steel, or concrete as well as plywood, bamboo, and aluminum.

Dormers

A dormer is frequently provided in buildings with pitched roofs to increase the usable space in an attic by increasing the light and ventilation. Dormers often become living space **(Figures 11.12 a and b)**. False dormers are also used as decorative elements.

Figures 11.11 a and b Geodesic domes are created using triangles.

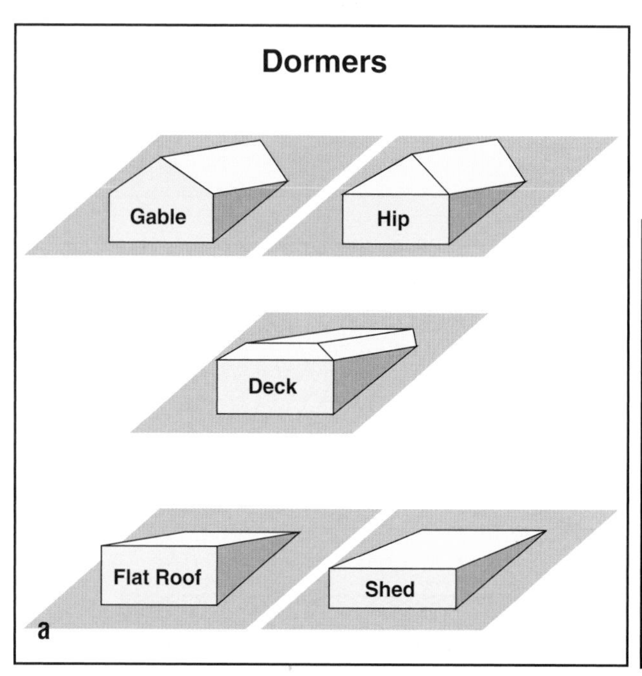

Figures 11.12 a and b Common styles of dormers. Dormers are designed to create extra living space and allow additional light into a structure.

Roof Support Systems

Flat Roof Support

Several common methods are used to support flat roofs. The simplest system uses ordinary wood joists supported at either end by a load-bearing wall **(Figure 11.13)**. The wood joists function as beams to support the roof deck just as floor beams support a floor system. Solid or laminated beams and columns may be used to support the wood roof joists.

The traditional wood-joisted roof uses solid wood joists that tend to lose their strength gradually as they burn. This loss of strength results in the roofs becoming soft or "spongy" before failure, especially with a wood plank roof deck.

Although the softening or sagging of a roof is an obvious indication of structural failure, it should not be looked upon as the only sign of imminent collapse. The relatively thin plywood or oriented strand board used for roof sheathing can fail quickly without prior warning. Therefore, in general, firefighters should view any indication of advanced or heavy fire development as a warning sign that the roof is weakening.

In modern practice box beams and I-beams manufactured from plywood and wood truss joists are often used to support flat roofs. Although these beams provide adequate strength, the thin web portion of plywood I-beams renders them susceptible to early failure in a fire (see also Chapter 7). The relatively slender members of which truss joists are manufactured are also susceptible to early failure during a fire. In addition, the open web design of truss joists also permits the rapid spread of fire in directions perpendicular to the truss joist instead of simply along the long dimension of the member.

Flat roofs are also supported by open-web steel joists and steel beams **(Figures 11.14 a and b, p. 312)**. (See also Chapter 9). Depending on the fuel load within an occupancy, unprotected lightweight open-web joists can be expected to fail quickly in a fire.

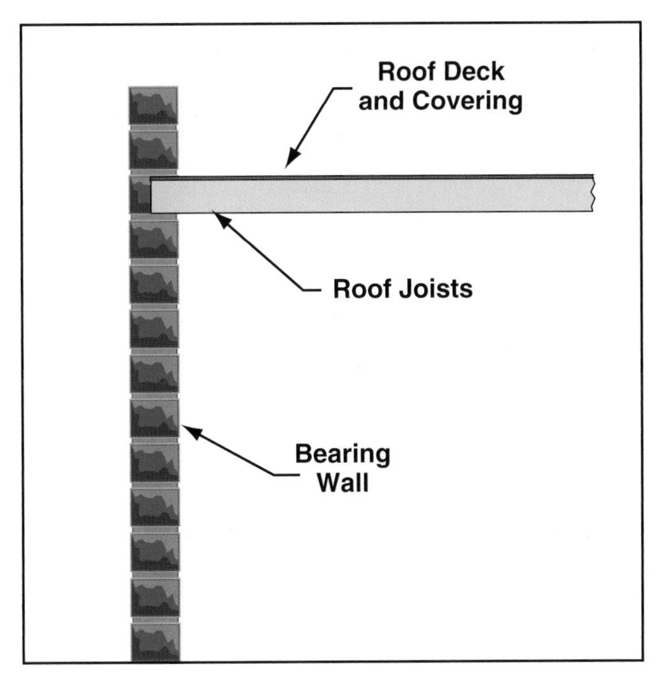

Figure 11.13 Flat joisted roof details.

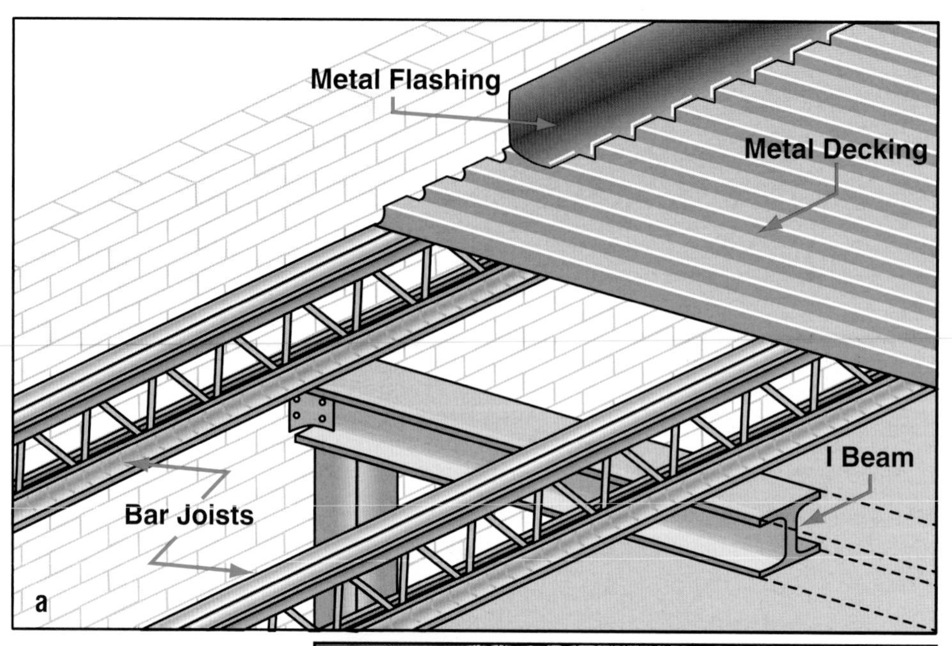

Figures 11.14 a and b Flat roof supported by bar joists and steel beams. *Photo courtesy of McKinney (TX) Fire Department.*

Firefighters should be on the alert when responding to structures with unprotected lightweight open-web joists of any material. These joists can be expected to fail quickly in a fire.

Under certain conditions, building codes will allow the omission of structural fire proofing from roof supports in Type I construction. For example, fireproofing can be omitted from roof supports when the roof is located more than 20 feet (6.6 m) above the floor in an assembly occupancy. Therefore, unprotected steel roof supports may be encountered in a building in which the main structural supports are fire-resistive.

Because roofs are designed for lighter live loads than floors are, it is not unusual for modern flat roofs to deflect or vibrate noticeably as personnel walk across them. Nonetheless, flat roofs usually must be designed to support the weight of at least a few workers so they can be accessed safely for maintenance purposes. Therefore, deflection or vibration under the weight of firefighters may not signal imminent failure. However, the deflection and vibration *are* an indication of lightweight roof construction, and firefighters should view such construction cautiously.

A variation of the flat roof is a type known as the "inverted roof." Inverted roofs differ from conventional roofs primarily in the location of their main roof beams. In a conventional roof system the main joists are located at the final roof level directly supporting the roof deck. A ceiling is attached to the underside of the joists or, more commonly, suspended below the joists. With the inverted roof, the main joists are located at the level of the ceiling and a framework is constructed above the main joists to support the roof deck.

From the outside, the inverted roof looks like any other flat roof. The design of the inverted roof creates a concealed space that may be several feet in height between the ceiling and the roof deck.

Raftered Roof

Rafters are the inclined joists used to support some types of pitched roofs **(Figures 11.15 a and b)**. Rafters are the standard supports used in shed, gable, hip, gambrel, and mansard style roofs, although trusses can also be used for these roof types.

The basic design of a raftered roof results in an outward thrust against the walls, similar to the action of an arch. The outward thrust of the rafters is resisted by ceiling or attic floor joists or collar beams that are in tension. If

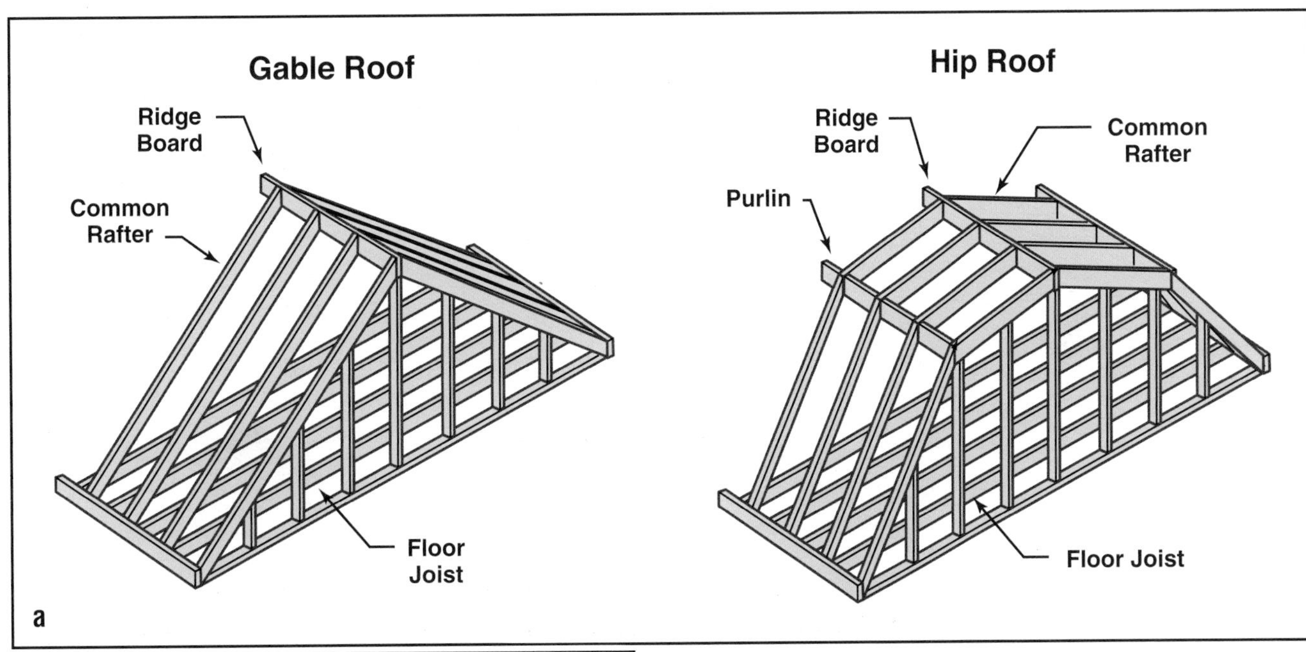

Figures 11.15 a and b (a) The arrangement of rafters for a simple gable and hip roofs. (b) A gable roof under construction. *Courtesy of Wil Dane.*

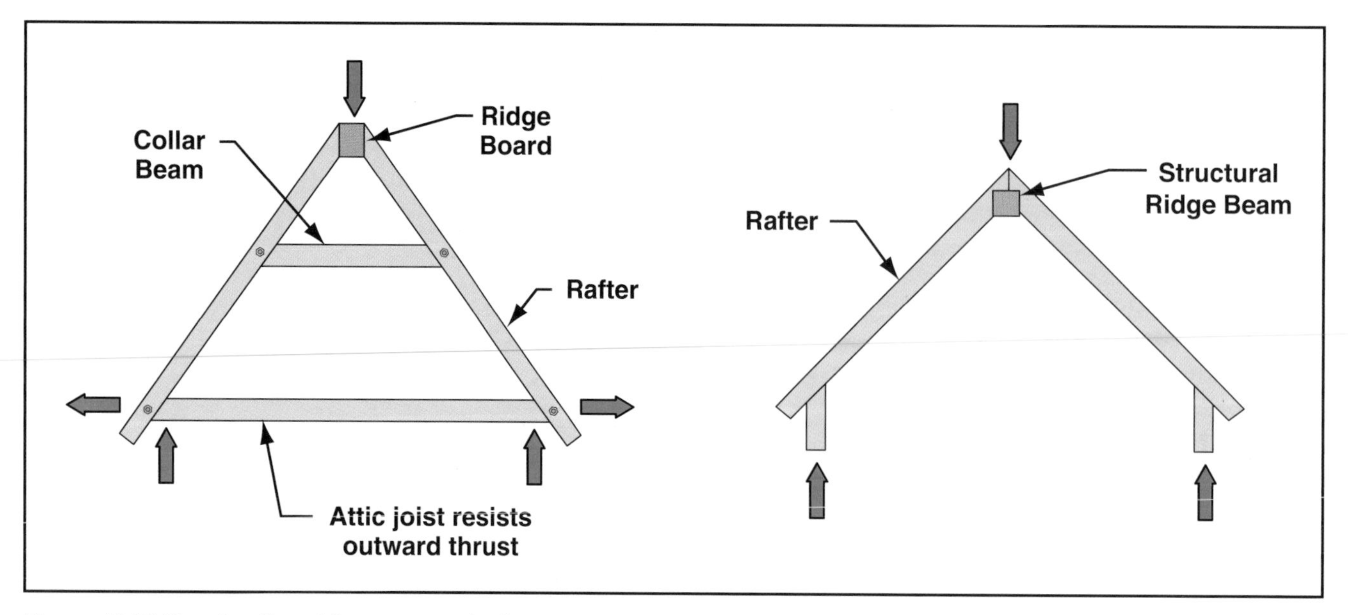

Figure 11.16 The direction of forces on roof rafters.

these joists are damaged or destroyed in a fire, the roof can push out against the walls. If the architect desires to leave the underside of the roof exposed without joists, a structural ridge beam must be used to support the rafters **(Figure 11.16)**.

Rafter — Inclined beam that supports a roof, runs parallel to the slope of the roof, and to which the roof decking is attached.

Rafters are commonly made of wood, although steel beams and steel trusses can be found in contemporary commercial construction. Wood rafters vary in size from 2 x 4 inches (50 mm x 100 mm) to 2 x 14 inches (50 mm x 350 mm). They can be spaced from 12 to 24 inches apart (400 mm to 600 mm), depending on the span and design load.

Trusses

Trusses are a very common roof support system **(Figures 11.17 a and b)**. As noted in Chapter 3, trusses use less material and are lighter than a comparable beam or joist for an equal span. However, the reduced mass of their components

Figures 11.17 a and b Trusses are a very common sight in roof construction. Notice how much lighter in weight the roof structural components are in the second photo compared to the walls. *Photo a courtesy of McKinney (TX) Fire Department; photo b courtesy of Ed Prendergast.*

and the interdependence of those components make them vulnerable to early failure under fire conditions. Furthermore, ceilings are often suspended from a roof truss, creating concealed spaces between the top and bottom chords and throughout the ceiling to the roof.

Several types of roof trusses may be used to support roofs. The details of a Fink truss that could be used in a light wood-frame structure are illustrated in **Figure 11.18.** The truss in **Figure 11.19** is a "monoplane" truss: that is, all of its chords and diagonal members lie in the same plane. This configuration is typical of lightweight trusses.

NOTE: Refer to Figure 3.43 for an illustration of common types of trusses.

Figure 11.20, p. 318, illustrates the use of lightweight wood trusses. Of note is the height of the concealed space created by the truss. Note also that the joints used in the trusses are connected using gang nail-type gusset plates. In trusses of this type, the individual members would be wooden 2 x 4s (50 mm x 100 mm) or 2 x 6s (50 mm x 150 mm) and individual trusses would be spaced 2 to 4 feet (0.61 m to 1.22 m) apart center to center. A heavier wood truss made up of multiple members is illustrated in **Figure 11.21, p. 316**.

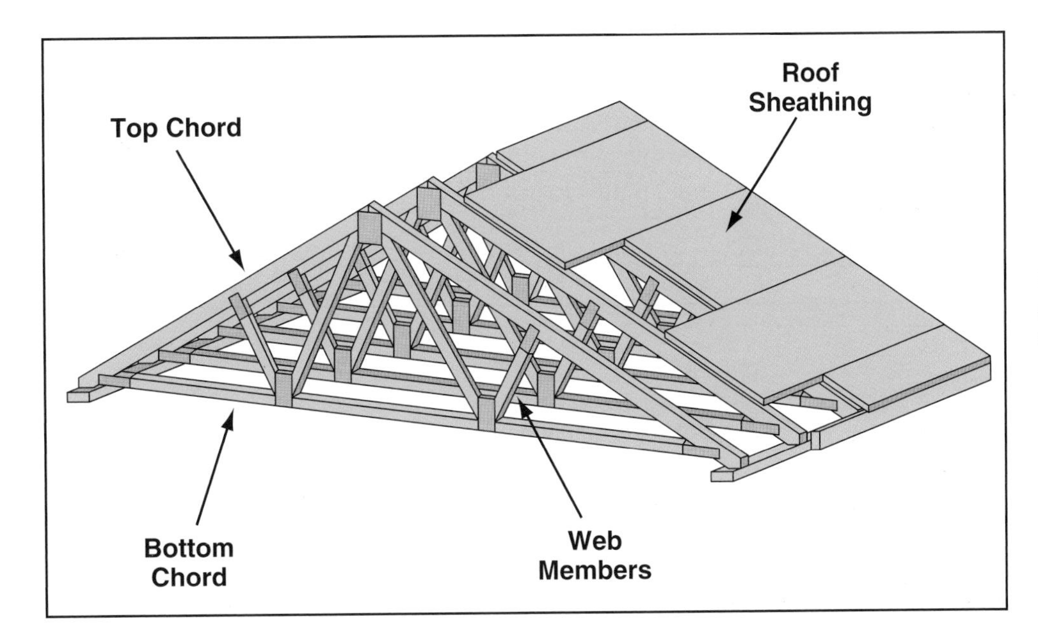

Figure 11.18 A Fink truss is common in residential-type construction.

Figure 11.19 All the chords and diagonal members lie in the same plane in a monoplane truss. *Courtesy of McKinney (TX) Fire Department.*

Figure 11.20 A typical lightweight wood truss. *Courtesy of Dave Coombs.*

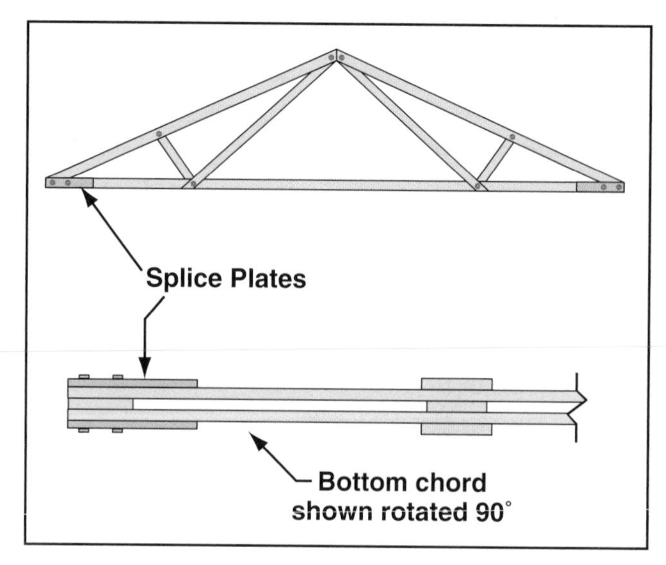

Figure 11.21 A multiple-member wood truss.

Figure 11.22 The bowstring truss is distinctive because of its curved top chord. *Courtesy of McKinney (TX) Fire Department.*

Bowstring Trusses

Bowstring trusses, which use a curved top chord, were once commonly used for roofs and many remain in use **(Figure 11.22)**. **Figure 11.23** illustrates the details of a bowstring truss with a laminated top chord. This assembly uses split-ring connectors at all joints except the heel plates located at the ends of the truss.

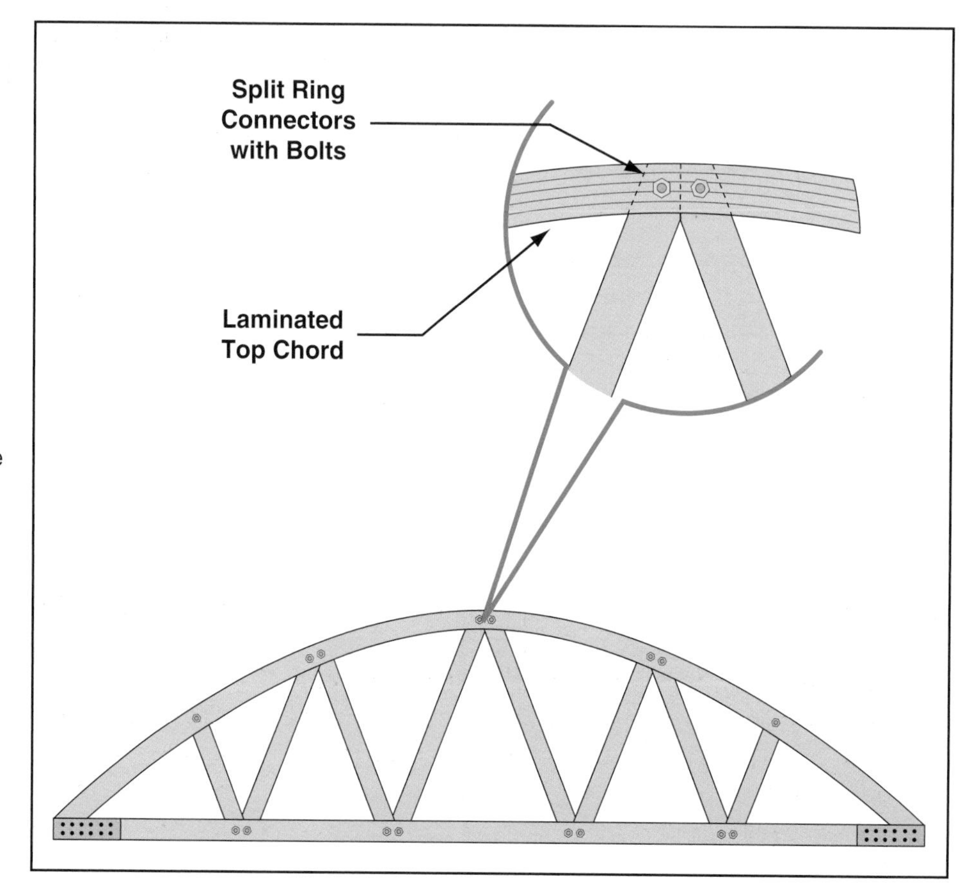

Figure 11.23 Components of the typical bowstring truss.

Figure 11.24 Although this truss has obviously failed, it can still exert pressure against the wall. *Courtesy of Ed Prendergast.*

Figure 11.24 illustrates a failed wood bowstring truss. Note that the heel plates of the bowstring truss have remained in place. Notice also that the truss is leaning against the wall. This action creates an outward horizontal force against the wall that can result in collapse of the wall.

Bowstring Truss — Lightweight truss design noted by the bow shape, or curve, of the top chord.

Wood and Steel Trusses

Roof trusses are fabricated from steel as well as wood. The Fink and Pratt-style trusses are the most common types used for pitched roofs. Both wood and steel trusses are usually fabricated elsewhere and shipped to the job site. If a truss is too large to be transported as one unit, it can be moved in sections and connected in the field.

Arches

The basic structural principles of an arch were described in Chapter 3. Because arches may be constructed from masonry, laminated wood, or steel, the behavior of a roof supported by an arch is basically determined by the material from which the arch is constructed. Thus, laminated wood arches will behave much like laminated beams and steel truss arches will react to a fire in a manner similar to a steel truss.

There is one characteristic of some arch-supported roofs that can cause a catastrophic failure under fire conditions. Some arch roofs use a steel tie rod between the two ends of the arc to resist the outward thrust of the arch. The tie rods extend through the interior of the building and are usually unprotected. Failure of the tie rods will permit the arches to spread outward and the roof will collapse.

Roof Decks

Function of the Roof Deck

The deck of a roof is the portion of roof construction to which the roof covering or "roofing" is applied. Through the deck, the loads on the roof are transmitted to the roof supporting members. The components of roof decks include sheathing, roof planks or slabs, and purlins. Sometimes, as in concrete deck roofs, the roof deck serves as the roof support. In other cases, the roof covering and the deck are the same. Corrugated steel decking is frequently used in applications where it serves as both the deck and the exterior roof covering **(Figure 11.25, p. 318)**.

Figure 11.25 Corrugated steel may serve as both decking and roof. *Courtesy of McKinney (TX) Fire Department.*

A roof deck must be stiff enough that it does not deflect excessively under anticipated loads. From a construction standpoint, the deck material should be clean and smooth so that any insulation or roof covering can be attached.

Roof Deck Materials

Roof decks can be constructed of plywood, wood planks, corrugated steel, precast gypsum or concrete planks, poured gypsum, poured concrete, and cement planks containing wood fiber. Wood panel decking may have a thickness of ½ inch (13 mm) on supports 24 inches (600 mm) on center **(Figures 11.26 a and b)**. Wood planks will have a minimum 1-inch (25 mm) nominal thickness.

Corrugated steel used in roof decking ranges from 29 gauge, the thinnest, to 12 gauge, the thickest. The overall depth varies from ¾ to 2 inches (19 mm to 50 mm). Corrugated steel decking can be used with a sheet of flat steel welded to the bottom to form cellar decking. The attached flat steel increases the stiffness of the deck.

Gypsum has the advantage of being "nailable," which means it is possible to nail into the material. Precast concrete can also be made nailable by choosing an appropriate aggregate. Cast-in-place concrete decks are not nailable. When cast-in-place concrete is used as a roof deck some provision must be made for attaching the roof to the deck. This is sometimes accomplished with

Figures 11.26 a and b Examples of wooden decking. *Photo b courtesy of McKinney (TX) Fire Department.*

Figures 11.27 a and b Flat roofs are often constructed from the same materials as the rest of the building. *Both photos courtesy of McKinney (TX) Fire Department.*

Figure 11.28 Buildings that are being remodeled may have roofs of different materials. *Courtesy of McKinney (TX) Fire Department.*

wood nailing strips that are either imbedded in the concrete at intervals of 3 feet (0.9 m), drilled and anchored, or placed between rigid insulation panels if rigid insulation is used.

If a multistory building is to have a flat roof, the usual practice is to use the same structural system for the roof and the floors because it is more economical **(Figures 11.27 a and b)**. Therefore, a building with wood-joisted floors usually will have a wood-joisted roof system and a steel-framed building will have a steel roof. It is possible to encounter exceptions to this general rule, especially where an additional story has been added to an older building **(Figure 11.28)**. If a building has been fire damaged and the roof is replaced, a roof structure different from the original may be installed. As a result, firefighters may encounter unusual combinations of floor and roof construction. An example would be steel bar joists supporting a steel deck roof in a building with wood-joisted or heavy-timber floor construction.

Roof Coverings

The roof covering provides the water-resistant barrier for the roof system. The type of roof covering used depends on the form of the roof structure, the slope of the roof, the local climate, and the appearance desired. Some other factors that affect the choice of roof covering include maintenance requirements,

Roof Covering — Final outside cover that is placed on top of a roof deck assembly. Common roof coverings include composition or wood shake shingles, tile, slate, tin, or asphaltic tar paper.

durability, required wind resistance, and fire resistance. For example, hail can puncture asphalt shingles and roll roofing. Fog, salt air, smoke, and other pollutants tend to corrode metal roofing. In some regions, roofs are subjected to summer temperatures well over 100°F (38°C) and winter temperatures below 0°F (-18°C) with resulting expansion and contraction of the roof covering.

Given the previous considerations, it is not surprising that a variety of roof covering materials and systems are in use. Furthermore, over time roofs are repaired and resurfaced, meaning that firefighters may encounter more than one layer of roof covering on a given roof.

Flat Roof Coverings

A roof covering can consist of a single layer of material as in a corrugated steel roof. More typically, it can consist of several layers of material used in combination. Because flat roofs drain more slowly than pitched roofs they usually require more complex roof covering assemblies than pitched roofs.

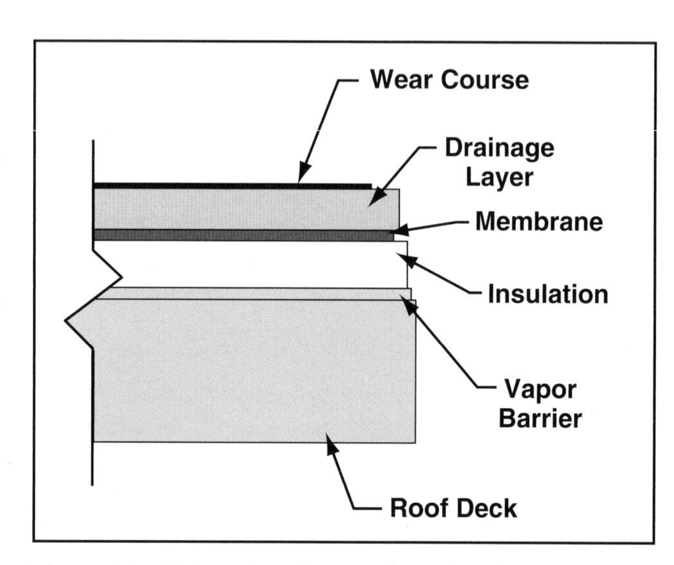

Figure 11.29 The various layers of a typical flat roof.

Vapor Barrier — Watertight material used to prevent the passage of moisture or water vapor into and through walls or roofs or in the case of personal protective equipment that prevents water from penetrating the clothing.

In addition to the roof decking previously described, a flat roof usually consists of several component layers that include the following: **(Figure 11.29)**

- Vapor barrier
- Thermal insulation
- Roofing membrane
- Drainage layer
- Wear course

Vapor Barrier

The vapor barrier is designed to reduce the diffusion of interior moisture into the insulation layer. It is needed when the average outdoor temperature is below 40°F (5°C) and the indoor relative temperature is 45 percent or greater at 68°F (20°C). The vapor barrier consists of a continuous sheet of plastic, aluminum foil, kraft paper laminated with asphalt, asphalt saturated roofing felt, or other material resistant to the passage of water vapor.

Thermal Insulation

The thermal insulation reduces heat loss through the roof. In addition to resisting the flow of heat, the insulation should have resistance to mechanical damage such as gouging, moisture decay, and fire. Insulation can be poured or rigid. Poured insulation materials can be Portland cement or gypsum. Several rigid insulation materials are in use and are listed in **Table 11.1**.

Table 11.1
A Comparison Of Various Rigid-Roof Insulating Materials

Insulating Material	Composition
Cellulose Fiberboard	A rigid, low-density board of wood or sugar cane fibers with a binder
Glass Fiberboard	A rigid, low-density board of glass fibers and a binder
Polystyrene Foam Board	A flammable rigid foam of polystyrene plastic
Polyurethane Foam Board	A flammable rigid foam of polyurethane sometimes faced with felt
Polyisocyanurate Foam Board	A rigid foam of polyisocyanurate sometimes with glass fiber reinforcing and best when combined with materials to increase fire resistance
Cellular Glass Board	A fire-resistant rigid foam of glass
Perlite Board	Fire-resistant granules of expanded volcanic glass and a binder pressed into a rigid board
Lightweight fill with Asphaltic Binder	Lightweight mineral aggregate with asphaltic binder
Composite Insulating Boards	Layers of foam plastic and other materials such as perlite board and glass fiberboard

Membrane Roofing

The membrane of a roof consists of waterproof material that keeps out rain and snow from the interior of the building. The three general categories of membranes used are built-up roof membranes, single-ply membranes, and fluid-applied membranes.

Membrane Roof — Roof covering that consists of a single layer of waterproof synthetic membrane over one or more layers of insulation on a roof deck. Also called single-ply roof.

Built-up membranes. Built-up membranes use several overlapping layers of roofing felt saturated with a bituminous material that may be either tar or asphalt. The layers of roofing felt are cemented together with a hot bituminous roofing cement **(Figure 11.30, p. 322)**. The roofing felt usually is supplied in rolls 3 feet (1 m) wide. The number of layers of roofing felt used varies, but four layers is a common design. The more layers of felt used, the more durable the resulting roof will be. Built-up roofs usually last for 20 years if the manufacturer's specifications are followed.

Single-ply membranes. A single-ply membrane roof consists of a single membrane laid in sheets on the roof deck. The membrane material comes in sheets 10 or 20 feet wide (3.3 m or 6.6 m) and up to 200 feet long (66 m). The membranes are very thin, typically 0.03 to 0.10 inches thick (0.75 mm to 2.5 mm).

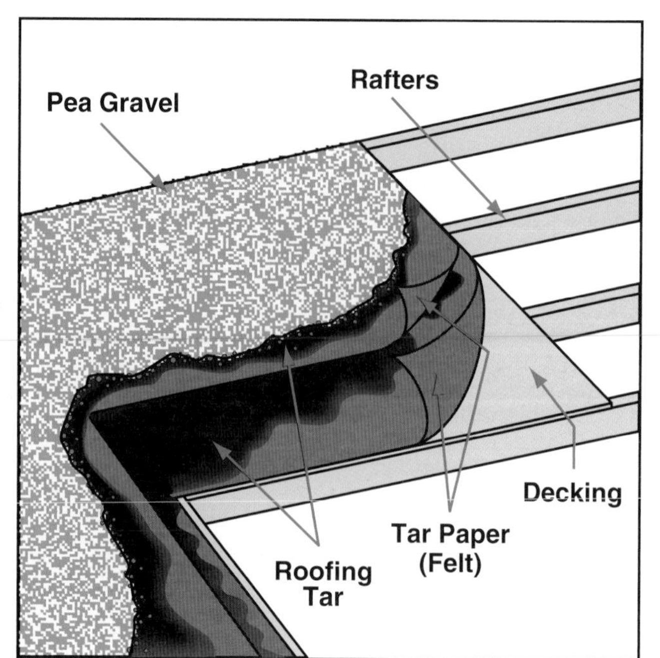

Figure 11.30 A typical built-up, tar-and-gravel roof.

Single-ply membranes are made from several materials. The most common is a synthetic rubber material, ethylene propylene diene monomer (EPDM). Other materials include polyvinyl chloride (PVC) and chlorinated polyethylene (CPE). The single-ply materials can be stretched and consequently will accommodate shifting in a building. They can be applied over decks in new buildings and over existing roofs.

The single-ply membranes are attached to the roof by means of adhesives, gravel ballast, or mechanical fasteners. Using a propane torch to heat the underside of the membrane as it is unrolled will cause it to adhere to the roof. This process has resulted in roof-covering fires that spread rapidly across the entire roof. The burning material liquefies and can drop down through holes in the roof decking onto the contents below.

Fluid-applied membranes. The fluid-applied membranes are useful for buildings with curved roof surfaces such as domes that would be difficult to cover with other materials. The material is applied as a liquid and allowed to cure. Usually several coatings are applied. The materials used include neoprene, silicone, polyurethane, and butyl rubber.

Drainage Layer

The drainage layer permits the free movement of rain water to the roof drains. Depending on the membrane material used, the drainage layer can be the ballast layer in a single-ply roofing system, a drainage fabric, or the aggregate used in a built-up roofing system.

Wear Course — External covering on a roof that protects the roof from mechanical abrasion. The typical tar and gravel roof uses gravel as the wear course.

Wear Course

The wear course protects the roof from mechanical abrasion. It can consist of the aggregate in a built-up roofing system or gravel ballast. Built-up roofs that use gravel as the wear course are commonly known as "tar and gravel roofs." When a gravel surface is used for the wear course, it also increases the resistance of the roof to ignition by flaming brands from adjacent buildings. The wear course may also consist of deck pavers when the roof is used for pedestrian traffic.

Pitched Roof Coverings

From a design standpoint, the roof coverings used on pitched roofs function differently from those used on flat roofs. Water immediately drains from a pitched roof, which minimizes the possibility of it pooling on the roof and working through seams in the roof covering. Because the force of gravity is constantly pulling at the roof, the coverings used on a pitched roof must always be secured to the roof deck or roof support and provision must be made for this on the roof deck **(Figure 11.31)**. Roof coverings used on pitched roofs are generally one of two broad categories - shingle or tile roofs and metal roofs.

Figure 11.31 There are many ways to secure tiles to a pitched roof. These tiles will be nailed. *Courtesy of McKinney (TX) Fire Department.*

NOTE: Technically, thatch, which consists of bundles of reeds, grasses, or leaves, is also a covering for pitched roofs. In North America, its use is limited to a few special situations such as for decorative effect.

Shingles and Tiles

Shingles and tiles are small overlapping units that are relatively easy for workers to handle. Their small size allows for movement between individual units caused by thermal expansion and shifting of the building structural system. Shingles and tiles are available as wood shingles and shakes, asphalt shingles, slate, and clay and concrete tiles **(Figure 11.32)**.

Figure 11.32 Roof tiles and shingles may be asphalt, clay, wood, or slate. *Courtesy of McKinney (TX) Fire Department.*

Wood shingles and shakes. The difference between wood shingles and shakes is in their method of production. Wood shingles are thin, tapered slabs of wood that are sawn from pieces of a tree trunk. Shakes are split from the wood either by hand or by machine and are thicker than shingles. Wood shingles and shakes used in North America are made from red cedar, white cedar, or redwood because of the resistance of these woods to decay.

Asphalt shingles. Asphalt shingles are produced from heavy sheets of asphalt-impregnated felt made from rag, paper, or wool fiber. Asphalt-impregnated fiber glass felt is becoming common. A mineral aggregate is imbedded in the top surface to act as a wearing surface and to provide color. Asphalt shingles are available in several sizes but the most common size is 12 x 36 inches (300 mm x 900 mm).

Slate. Slate is produced from hard rock that has a tendency to split along one plane. This characteristic permits roofing slate to be produced in smooth sheets as thin as 1/16 inches (7 mm) although it may be as thick as 1½ inches (38 mm). Slate is a very durable material and can have a life expectancy of 150 years. Slate is also a heavy material, weighing 8 to 36 pounds per square foot (38 kg/sq m to 175 kg/sq m). Therefore roof framing and decking that is heavier than normal may be required if slate is to be used.

Clay tile. Clay tile is an ancient material known to have been used for thousands of years. It is made by shaping clay in molds and firing it in kilns. It is a dense, hard, and nonabsorbent material and can be used for flat or curved tiles. The curved clay tiles are known as "mission" tiles and are used to create imitation (or genuine) Spanish-style architecture **(Figure 11.33)**.

Concrete tiles. Concrete tiles are made from Portland cement, aggregate, and water. Concrete tiles are frequently made to look like clay tile, slate, or even wood in color and texture. A major advantage of concrete tiles over wood tiles is their greater longevity.

Other tiles. Other materials can be found that, for architectural purposes, are shaped to resemble other roof coverings such as wood shingles. These include porcelainized aluminum, mineral-based materials, and composite materials.

Figure 11.33 Clay tiles are popular in Spanish-style architecture.

Application of Shingles and Tiles

Shingles and tiles are usually attached to the roof with corrosion-resistant nails. Wood shingles are installed in an overlapping manner so that only about one-third of the length of shingle is actually exposed to the weather. Wood shingles can be nailed to a conventional solid deck or to an open deck consisting of wood strips attached to the roof rafters **(Figure 11.34)**. This latter method of construction can produce earlier failure of the roof when a fire occurs in the attic.

Asphalt shingles are usually installed over an underlayment, which is a layer of roofing felt or synthetic covering. The underlayment serves as a cushion and provides protection from wind-driven rain. Slate tiles are provided with prepunched nail holes. The proper installation technique is to have the head of the nail just touching the slate so that the slate is actually hanging from the nail. As with other materials, shingle and tile roof coverings can be applied over a deteriorated existing roof.

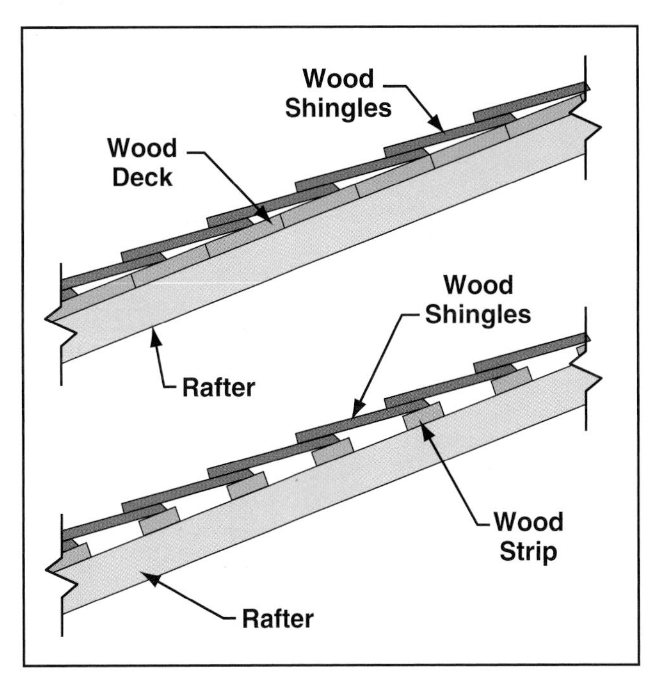

Figure 11.34 Wood shingles can be nailed to a solid deck or an open deck.

Fire Performance

Wood roof shingles and shakes. Wood roof shingles and shakes are popular architecturally because they produce a rustic appearance and may be more resistant to wind damage than asphalt shingles. Their disadvantage is that they pose a serious fire potential. They can be easily ignited by burning brands landing on them from an exposing fire. Once ignited, they can produce embers that spread fire to other roofs. In some parts of the country they have contributed to fires involving entire neighborhoods. For these reasons, some jurisdictions have prohibited their use.

Wood shingles and shakes can be pressure-impregnated with a fire-retardant solution to reduce their combustibility and to meet model code requirements. Experience with the pressure-impregnated wood shingles and shakes indicates that the treatment remains effective after exposure to the elements. Fire-retardant shingles and shakes are shipped to the job site with a paper label identifying them. Once in place, however, identification of fire-retardant shingles or shakes can be difficult. Painting or staining fire retardant shingles or shakes can reduce the effect of the fire retardant, especially if oil-based materials are used.

Asphalt shingles. Asphalt shingles are fundamentally combustible. They tend to drip and run under fire conditions and produce a characteristic heavy black smoke. Asphalt shingles used for roofs are typically produced with a grit surface that reduces their ease of ignition and permits their use under the provisions of building codes **(Figure 11.35)**.

Figure 11.35 A typical roof with asphalt shingles. *Courtesy of McKinney (TX) Fire Department.*

Clay tiles, slate, and cement tiles. Clay, slate, and cement tiles are noncombustible and produce fire-resistant roof coverings that have excellent resistance to flying brands. Flying brands, however, can be blown under tiles such as the Spanish tiles that do not lie flat, and ignite the roof deck. These fires can produce some operational problems of which firefighters should be aware:

- Tiles can become loose and fall from a roof as the deck burns away and nails lose their grip or as firefighters conduct ventilation operations.
- The surface can become slippery, posing a serious fall hazard. Proper equipment and caution should be used at all times.
- The thin pieces of slate are brittle and may have sharp edges, creating a hazard for firefighters below **(Figure 11.36)**.

Some roof coverings, such as wood and asphalt shingles, are used for siding as well as for roofs. When used as siding they can become subject to the heat of an exposing fire **(Figures 11.37 a and b)**.

Metal Roof Coverings

Metal roof coverings make use of several materials including galvanized iron or steel, copper, zinc, aluminum, and lead. Flat or corrugated metal can be used **(Figure 11.38)**. Corrugated sheets of aluminum or steel are widely used on industrial and agricultural buildings. Metal roofs are also found on many residential and commercial buildings.

Figure 11.36 The sharp edges on slate tiles can be a hazard during emergency operations if they start falling on firefighters working below. *Courtesy of McKinney (TX) Fire Department.*

Figures 11.37 a and b When shake shingles are used as siding they present the same fire hazards as shake roofs.

Corrugated roofing sheets are generally strong enough to be installed without decking **(Figure 11.39)**. In these cases the roofing sheets are supported by roof beams or purlins. The spacing between purlins can be from 2 to 6 feet (0.6 m to 2 m).

Flat roofing sheets are nailed to a deck beneath. A layer of roofing felt is placed on the deck beneath the metal sheets. When wood decking is used, the roofing felt increases the fire resistance of the roof because it acts as an insulating layer and protects the wood deck against the heat of external exposing fires.

One problem with metal roofing is the potential for galvanic action when dissimilar metals are in contact – for example, using steel nails with copper sheeting. To prevent this interaction, nails made of the same metal as the sheeting are used. The individual flat sheets of metal roofing are joined at seams that are crimped tight. The seams stand up vertically at the joint between adjacent sheets and make the metal panel roof readily identifiable **(Figure 11.40)**.

Layers of Roofs

When a new roof is installed over an existing one, firefighters can encounter problems performing ventilation. The void spaces created between the new and older roof can also conceal fire spread, creating yet another hazard for firefighters working on top of a building.

Figure 11.38 A corrugated metal roof. *Courtesy of Ed Prendergast.*

Figure 11.39 Sheets of corrugated metal can be installed without decking. *Courtesy of McKinney (TX) Fire Department.*

Figure 11.40 Metal sheets are joined at the seams and crimped tight. *Courtesy of Sturzenbecker Construction Company, Inc.*

Fire Ratings of Roof Coverings

Because of the severe fire danger that combustible roofs can pose, building codes impose restrictions on the combustibilty of roofs of certain buildings, occupancies, or locations within a community. The fire hazards of roof coverings are evaluated by test procedures contained in NFPA® 256, *Standard Method of Fire Tests of Roof Coverings,* also designated ASTM E-108. The test simulates several fire exposure conditions for fires originating *outside* a building. The standard does not evaluate the fire resistance of the structural system supporting a roof or the fire resistance of the roof itself with respect to a fire originating *within* a building.

In the test, samples of roof coverings are attached to a wooden deck measuring 3 feet 4 inches by 4 feet 4 inches (1 m x 1.3 m). The samples are then subjected to the required test procedures. There are six separate test procedures contained in NFPA® 256. They include the following:

- Intermittent flame exposure test
- Burning brand test
- Flying brand test
- Rain test
- Weathering test
- Spread of flame test

The individual test procedures may be repeated from 2 to 15 times on different samples depending on the specific material being tested. For example, 15 samples of wood shakes or shingles may be subjected to the weathering test while another material may be subjected to the burning brand test twice. In addition, if the properties of a specific roof covering material are subject to variation, more than the minimum number of tests contained in NFPA® 256 may be required.

Roof coverings that pass the required test procedures are classified A, B, or C. The three classifications are based on the severity of fire the material can withstand:

- Class A roof coverings are effective against a severe fire exposure.
- Class B roof coverings are effective against a moderate fire exposure.
- Class C roof coverings are effective against a light fire exposure.

Building codes use these three classifications to control the flammability of roofs. Therefore, certain types of construction, such as fire-resistive buildings, may be required to have a Class A or B roof covering. A building code may also use the classifications to restrict the flammability of roofs on buildings in congested areas such as a downtown area. In other parts of a community, roof coverings may be used that do not pass the test procedures and are unclassified.

Laboratories that test roof coverings, such as Underwriters Laboratories Inc., publish a list of roof coverings that have passed NFPA® 256 with their classifications in a manner similar to that of fire-resistance ratings described in Chapter 2.

Rain Roofs

A roof is a building's first line of defense against the elements. As such, roofs deteriorate over the years from exposure to wind, snow, and rain. In some cases the surface of a roof becomes so deteriorated that it cannot be easily repaired. To solve this problem, a second roof, sometimes termed a *rain roof,* can be constructed over the original roof.

Rain Roof — A second roof constructed over an existing roof.

A rain roof presents some special difficulties for the firefighter. A void is created between the rain roof and the original roof beneath, and any concealed space creates problems during fire fighting **(Figure 11.41)**. When a fire enters this space, it can travel undetected in several directions and is exceptionally hard to access and extinguish. The existence of two separate roofs can also impede rapid and effective ventilation.

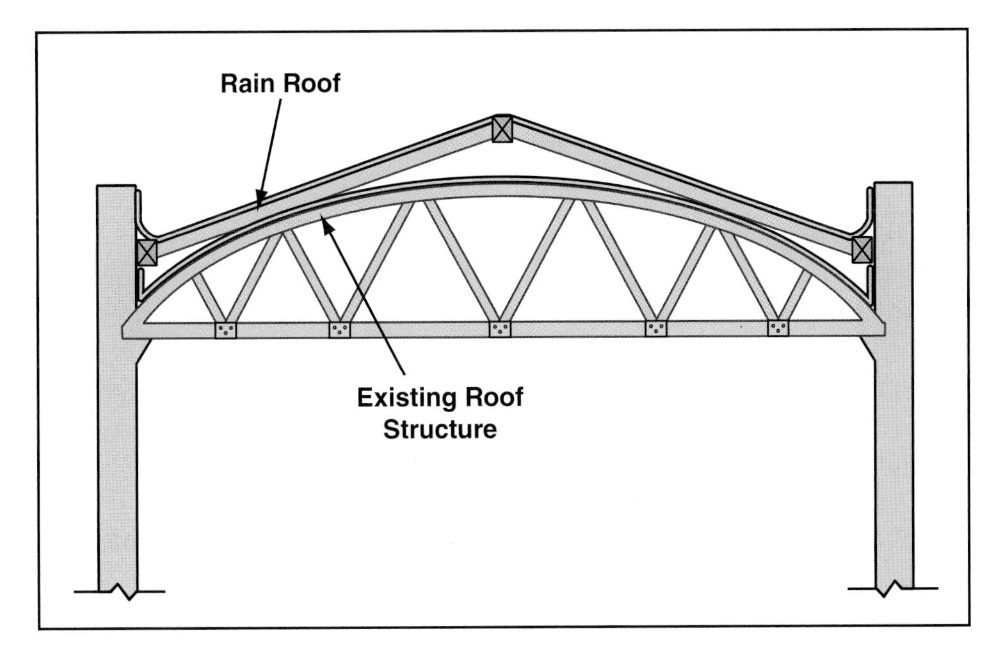

Figure 11.41 Voids are created when a rain roof is installed over an older roof.

Green Roofs

Recent years have seen an increase in designs to protect the environment and conserve energy. One manifestation of the interest in environmental protection has been the development of green roofs. A green roof involves the use of the roof surface of a building for a rooftop garden **(Figure 11.42)**. There are several benefits to this use. One is the increased insulating effects between the building interior and the outside. Probably the greatest benefit, however, is in the increase in air quality due to the oxygen-carbon dioxide exchange of growing plants, particularly in urban areas.

NOTE: Refer to Chapter 1 for a general discussion of green design.

Figure 11.42 What appears to be a park is actually the roof of a six-story parking garage. Note the depth of the vegetation. *Courtesy of Ed Prendergast.*

A green roof can take several forms. A green roof or rooftop garden can vary from the use of potted plants and flower boxes to a layer of earth with growing plants covering a large area of a roof. Green roofs can be developed on existing roofs and in new construction.

A rooftop garden constitutes a dead load on the roof structural system, which must be capable of supporting the load. The layer of earth required for a rooftop garden can vary from a few inches to 1 or 2 feet (0.3 m to 0.6 m). Depending on the depth of the soil, the dead load can vary from 20 pounds per square foot to 150 pounds per square foot (100 kg/sqm to 750 kg/sqm). In new construction the structural engineer can provide for this load in the structural plans just as is done for snow loads. However, when a garden is planned for an existing roof the existing structural system must be analyzed to ensure its adequacy.

Obviously under fire conditions the increased load can hasten structural failure, particularly if the roof is combustible. Green roofs can also interfere with ventilation practices and fire location indicators.

NOTE: Refer to Chapter 3 for more information on loads.

Photovoltaic Roofs

A photovoltaic (solar energy) system produces clean and reliable energy that can be used in a wide range of applications. Photovoltaic cells in panels can be laid on top of a roof or embedded in the roof **(Figures 11.43 a and b)**.

Although solar energy represents a clean source of energy, the electricity generated by the operation of the solar system represents a significant hazard for firefighters. Even if power to the building is shut off, the panels retain a significant amount of electricity. It is not safe to break photovoltaic cells or skylights that are actually solar powered. Furthermore, the panels themselves represent a significant tripping and falling hazard.

In emergency conditions electrical shock, inhalation exposure, falls from roofs, and roof collapse always represent serious safety considerations. For these reasons, it is crucial that fire departments conduct thorough pre-incident planning to identify these structures rather than encountering

Figures 11.43 a and b Solar panels can pose tripping and electrical hazards. *Both photos courtesy of McKinney (TX) Fire Department.*

them during adverse conditions. Solar panels, like other elements on a roof, may not even be visible from the ground on a building with a flat roof. The fire department must preplan for structural emergencies on specific commercial and industrial buildings in their jurisdiction.

Air-Supported Roofs

Air-supported roof structures are often used in sports arenas and at colleges and universities. Air-supported roofs provide protection from the elements and enable year-round use of the area under the roof **(Figure 11.44)**. A typical use with this type of structure would be a sports practice facility. Some industries also use air-supported structures as storage facilities.

Air-supported roofs do not lend themselves to conventional fire fighting tactics. Attempting to ventilate the roof may result in loss of the interior supporting pressure. The roof is curved and will not support the weight of firefighters. Having doors open to provide fire fighting access for lengthy periods could result in the slow deflation of the structure.

NOTE: More information about air-supported structures is contained in Chapter 12.

Roof Openings

The structure of a roof is frequently penetrated for a number of reasons, such as for penthouses, sky lights, vents, and roof hatches. These roof openings are normally provided for purposes other than fire protection, but they can be used in the course of fire fighting for roof access or ventilation.

Penthouse — (1) Structure on the roof of a building that may be used as a living space, to enclose mechanical equipment, or to provide roof access from an interior stairway. (2) Room or building built on the roof, which usually covers stairways or houses elevator machinery, and contains water tanks and/or heating and cooling equipment. Also called a Bulkhead.

Penthouses

Penthouses are small structures erected on the main roof of a building. In some parts of the United States a penthouse is referred to as a "bulkhead." Penthouses are constructed for several purposes such as a stairway enclosure, elevator machinery enclosure, mechanical equipment storage, or additional living space **(Figure 11.45)**. When a stairwell is provided with a rooftop penthouse, firefighters can gain rapid access to the roof to combat fires at the

Figure 11.44 Air-supported structures are often used to enable sports activities year-round.

Figure 11.45 An air-conditioning penthouse on the roof of a building. *Courtesy of McKinney (TX) Fire Department.*

roof level. Unfortunately, a stairwell penthouse can also provide a means of unauthorized access into a building and also may not be desirable from an architectural viewpoint.

Some penthouses may not be directly accessible from the inside of a building and must be accessed from the roof. This is frequently the case with penthouses built for elevator and mechanical equipment. Fires can occur in these spaces, however, so when the roof is beyond the reach of ladders, firefighters must locate a route to the roof. Although some means of access to a roof must always be provided for maintenance purposes, the access may not be readily apparent and may entail climbing up ladders through roof hatches.

CAUTION

Firefighters should not step onto skylights because they may fall through. In addition, skylights may have been covered with a lightweight material and may not be visible.

Skylights — Any of a variety of roof structures or devices intended to increase natural illumination within buildings in rooms or over stairways and other vertical shafts that extend to the roof.

Skylights

Skylights provide natural lighting to the interior of a building. They can be located to serve only the top story of a building or they may be located over the top of an atrium, stairwell, or light shaft **(Figures 11.46 a and b)**. Skylights are sometimes provided with operable glass panes to facilitate normal building ventilation. For safety purposes, building codes require wired glass or tempered glass in skylights. Skylights on modern buildings can also be plastic domes.

Skylights provide a rapid means of ventilating heat and smoke but they usually do not have provision for automatic venting. The glass in a skylight can simply be broken or the skylight housing can be pried up and pushed aside **(Figure 11.47)**.

NOTE: Skylights that feature photovoltaic strips must be approached with caution. Again, pre-incident planning will help remove the guesswork.

Ceilings

The underside of a roof can be left exposed or have a ceiling installed depending on the use and interior design of a building. Warehouse and industrial buildings, for example, are usually built without ceilings. Many other occupancies such as office or residential buildings will have ceilings installed for esthetic effect.

Figures 11.46 a and b Skylights can provide a good way to ventilate a roof but they are also a falling hazard and too weak to stand on. *Courtesy of McKinney (TX) Fire Department.*

Figure 11.47 Some skylights can be tilted open for ventilation.

Ceilings as a distinct building component usually do not play a structural role. However, a ceiling can have a functional role in the design of a building. Ceilings can be designed to control the diffusion of light and the distribution of air in a room. In addition, the space above a ceiling can be used to conceal air conditioning ducts, electrical wiring, and sprinkler piping.

Ceiling materials can be attached directly to the underside of roof joists or trusses or they can be installed at a distance beneath the roof supports creating a considerable concealed space. It is not uncommon for older buildings to have a new ceiling installed beneath an existing ceiling as a means of creating new interior decor.

Figure 11.48 Dropped ceilings can conceal the type of roof structure as well as provide a means for fire spread. *Courtesy of McKinney (TX) Fire Department.*

The concealed space created in fire-resistive or noncombustible construction may conceal the type of roof structure above **(Figure 11.48)**. The extent of fire development in roof spaces is hidden by a ceiling. This uncertainty complicates interior fire fighting.

Summary

Roofs can be constructed in a variety of styles for architectural, functional, and environmental reasons. Roofs can also have several types of coverings. The combustibility of a roof can affect the communication of fire from building to building. The structural system of a roof often creates concealed spaces that are difficult to evaluate, access, and control. A roof typically is not designed to support the same amount of live load as a floor; consequently, roofs should always be viewed very cautiously by firefighters. Some newer types of roofs that feature photovoltaic cells may be too dangerous to support roof operations without excessively endangering firefighters. Because roofs represent falling, tripping, electrical, and collapse hazards, careful preincident planning is necessary to determine as many hazards as possible.

Review Questions

1. On which type of roof is it easiest for firefighters to work?
2. What is a butterfly roof?
3. List several components of a roof deck.
4. What is the difference between wood shingles and shakes?
5. What is a green roof?

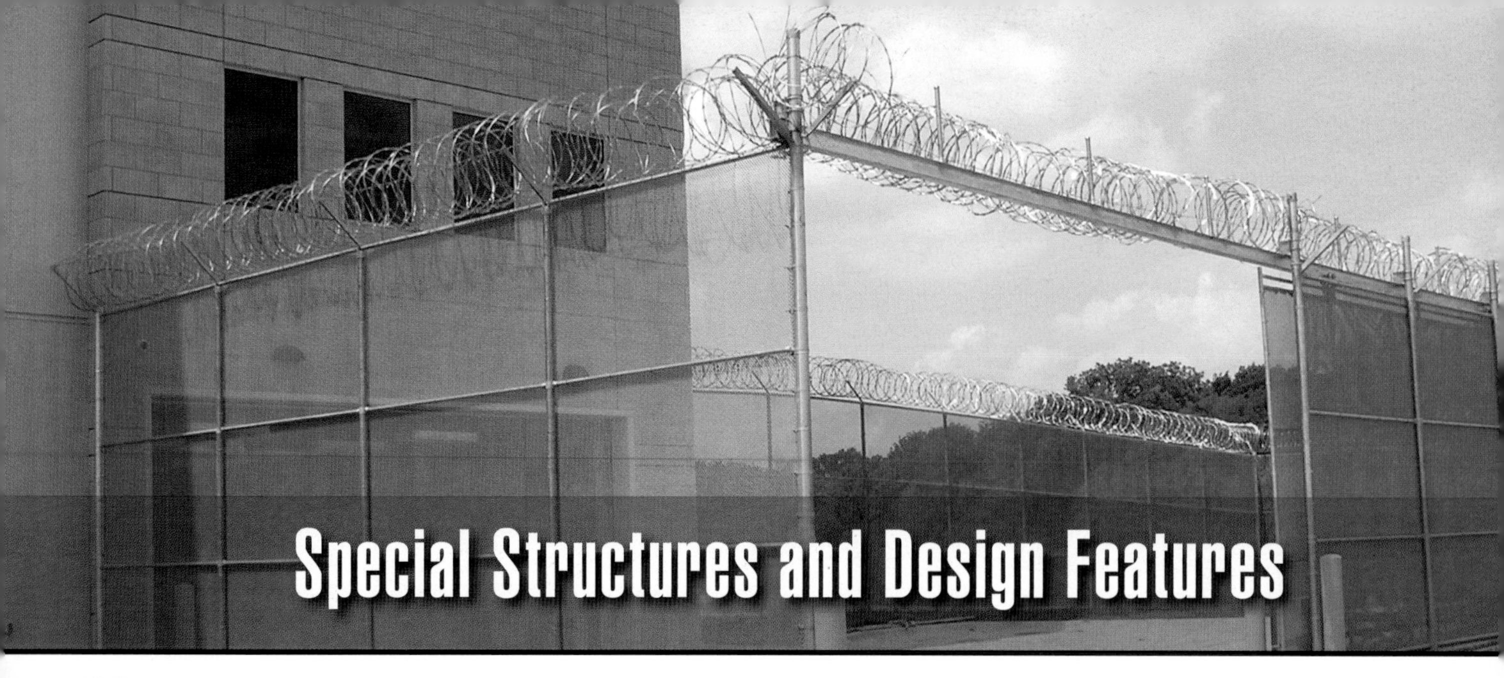

Special Structures and Design Features

Chapter Contents

Divider page photo courtesy of McKinney (TX) Fire Department.

Key Terms

FESHE Objectives

Fire and Emergency Services Higher Education (FESHE) Objectives: *Building Construction for Fire Protection*

3. Analyze the hazards and tactical considerations associated with the various types of building construction.
8. Identify the indicators of potential structural failure as they relate to firefighter safety.

Special Structures and Design Features

Learning Objectives

After reading this chapter, students will be able to:

1. Describe the characteristics of high-rise buildings and their impact on firefighting tactics.
2. Describe the fire protection systems in high-rise buildings and their integration into firefighting tactics.
3. Explain the emergency use of elevators in high-rise buildings during a fire event.
4. Discuss the unique aspects of underground buildings and how they affect firefighting.
5. Identify the usual code requirements for buildings with limited access.
6. Describe the characteristics of membrane structures and their impact on firefighting tactics.
7. Describe the characteristics and construction of covered malls.
8. Identify the primary concerns when managing a fire event in a detention/correctional facility.
9. Discuss the building codes that apply to atriums from a fire safety standpoint.
10. Describe the forces involved in an explosion and the methods to reduce the resultant structural damage.
11. Identify the requirements for areas of refuge for individuals with disabilities.
12. Describe the characteristics of rack storage as it relates to fire spread and firefighting tactics.

Chapter 12
Special Structures and Design Features

Case History

Event Description: The department was operating at a fire on the 8th floor of a residential high-rise. Crews controlled the fire quickly and were using hydraulic ventilation to clear smoke. They were completing a check of the fire floor and floors above the fire for smoke conditions and occupants' welfare by the time the safety officer arrived. A number of occupants had evacuated the building and were congregating in the building lobby or just outside the front entrance. Fire fighting crews were also entering and exiting the building.

As the Safety Officer was walking away from the command post toward the building, crew members on the 8th floor began breaking windows in the fire apartment's bedrooms. This was a corner apartment. Glass fell in large pieces on the A and D sides of the building, nearly striking a number of civilians and firefighters. No warning was given that the windows were about to be broken. Luckily, no one was struck by the falling glass, although one man with two small boys was narrowly missed. An elderly woman walking out from under the entrance canopy was almost struck by shards that bounced off the canopy and showered the area in her travel path.

Lessons Learned: 1. Breaking windows on the upper floors of a building requires careful coordination. 2. Before windows are broken or any debris is released from an upper floor, an all-clear needs to be established. 3. Division commanders and company officers need to be aware of what crews are doing at all times. 4. Constant emphasis on safe practices is a must.

Source: National Fire Fighter Near-Miss Reporting System.

Previous chapters have dealt with basic building construction design, materials, building services, life safety systems, and fire behavior. This chapter addresses special structures and design features that firefighters need to be aware of due to their fire suppression and rescue challenges. These special structures include high-rise buildings, underground structures, windowless buildings, air-supported structures, covered malls, and detection/correctional facilities. Special design features discussed include atriums, rack storage, areas of refuge, and special hazard protection for facilities that house explosive materials.

High-Rise Buildings

In the early part of the 20th century high-rise buildings existed primarily in the larger cities. By the second half of the 20th century high-rises began to be constructed in many medium-sized communities. Today tall buildings may be constructed in almost any community **(Figure 12.1, p. 340)**.

The growing proliferation of high-rise buildings poses a potential problem for almost any fire official; high-rise incidents are among the most challenging the fire department will face. The greater heights of modern buildings are beyond the reach of aerial equipment. Difficulty in gaining access can result in fires burning for extended periods before firefighters can complete staging and begin operations.

In addition, larger buildings contain a greater number of occupants and activities. This in turn results in more frequent calls to high-rise buildings. Emergency operations require a greater level of coordination due to the large numbers of resources and personnel needed.

From a fire protection standpoint, a high-rise building is any building that is beyond the effective reach of fire equipment located at the street level. Model building codes define a high-rise building as a building more than 75 feet (25 m) in height. The height is measured from the lowest level of emergency vehicle access to the floor of the highest occupied story.

Defining a High-Rise

The height at which a building becomes a high-rise by definition varies from jurisdiction to jurisdiction. In some cases a high-rise building may be defined as one that exceeds 80 or 100 feet in height (24.4 to 30.5 m) measured to the roof. In other cases a jurisdiction may, because of limitations in fire fighting resources, define a high-rise building as one exceeding 50 feet (15.2 m) or 5 stories in height.

High-Rise Building — Any building that requires fire fighting on levels above the reach of the department's equipment, often generally given as a building more than 75 feet (25 m) in height.

No matter how a high-rise building is defined, as building height increases, a point is reached where special fire protection problems are created simply by virtue of the building's height. The most obvious aspect of a high-rise building is that exterior means of fire attack and rescue are not possible beyond a certain height **(Figure 12.2)**.

As building height increases, occupant safety and fire fighting become increasingly dependent on the features of the building itself. As the firefighter becomes more dependent on built-in features, pre-incident planning takes on an even more important aspect of fireground operations. These features are described later in this section.

Figure 12.1 A large high-rise building can be like a small town in one area. Responding to structures with many occupants is a significant challenge for any fire department.

Figure 12.2 By definition, a high-rise is taller than the reach of a department's aerial apparatus. *Courtesy of Doug Allen.*

Early High-Rise Buildings

High-rise buildings as they are known today began to be constructed at the end of the 19th century. These buildings were made possible (and practical) by two developments: steel-frame construction and the elevator. It is not practical to build very tall buildings of masonry or wood because of their limitations in strength (See Chapters 7 and 8). Cast iron had been used for structural purposes, but its brittleness made it a somewhat unpredictable material. Without elevators the upper floors of a tall building become virtually inaccessible.

The early high-rise buildings were different in many significant respects from buildings of today. Buildings constructed 75 or 80 years ago did not have the heating, ventilation, and air conditioning (HVAC) systems routinely provided today. Ventilation was provided simply by opening windows. Coincidently, the operable windows facilitated ventilation of smoke when a fire occurred.

Early high-rise buildings also made use of open stairwells and elevator shafts that permitted the vertical communication of products of combustion. Over the years, several disastrous fires demonstrated the need for enclosed stairwells and elevators. The first model building codes that were introduced in the 1920's and 30's required stairs and elevators to be enclosed. Unfortunately, buildings constructed before those dates often did not have to be brought up to standard.

Winecoff Hotel Fire

The 15-story Winecoff Hotel was built in Atlanta, Georgia in 1913. The hotel was designed in a square shape, with stairs and elevator shafts in the center of the building and hotel rooms around the perimeter. The hotel was advertised as fireproof even though it had no sprinkler system, fire alarms, or fire escapes.

On December 7, 1946, the smell of smoke was noticed on the fifth floor of the hotel. The person who sounded the alarm was unaware that the second, third, and fourth floors were already engulfed in flames. To make matters worse, many doors to the stairwells had been propped open, further enabling the fire to travel upward like a chimney. As the stairwells and elevators become impassable, hotel guests had no way to escape. When the fire was finally extinguished, 119 people had died as they leapt to their deaths, were overcome by flames, or suffocated. The Winecoff Hotel fire remains the most deadly fire in US history.

The public outcry after the disaster led to many changes in fire code enforcement. It was determined that local officials could not be relied upon to make responsible decisions about fire safety, and national safety codes were established and strictly enforced.

Fire protection was usually provided only by standpipe systems. Automatic sprinklers and communication systems were not commonly provided until the last quarter of the 20th century. Because the buildings were constructed with fire-resistive materials, automatic sprinklers were not considered necessary.

Although building code requirements for high-rise buildings have changed over the last century, the requirements for modern buildings such as automatic sprinkler systems may not be retroactive; therefore, many older buildings remain without the benefit of modern fire protection features.

Modern High-Rise Buildings

The most significant and obvious feature of modern high-rise buildings is their height. The earliest high-rise buildings were rarely more than 10 or 12 stories. In the 1920's and 1930's several buildings were constructed exceeding 40 stories (New York's Empire State building, which has 102 stories, was constructed in 1930). Today buildings exceeding 60 stories have become commonplace **(Figure 12.3)**. The taller buildings have larger populations, making evacuation under fire conditions more difficult and time consuming.

While early high-rise buildings were usually limited to residential or office use, modern buildings may be used for other purposes, including assembly, institutional, mercantile, and educational purposes. In addition, newer high-rise buildings often have multiple occupancies. Thus, the same building may contain a garage, a hotel, and offices **(Figure 12.4)**.

Contemporary buildings are designed with sophisticated HVAC systems (See Chapter 4) and many building are constructed without operable windows. Ventilation may be accomplished through the fire department's control of the HVAC system. Attempting to ventilate the upper floors of a building by simply

Figure 12.3 High-rises have continued to grow taller, thus complicating emergency response.

Figure 12.4 It is not unusual for high-rises to contain stores, parking garages, and offices. Each type of occupancy presents special challenges during an emergency. *Courtesy of Ed Prendergast.*

breaking out windows poses considerable risk to persons on the street below. The enclosure of stairwells and elevator shafts reduces the upward flow of combustion products. The fire department should arrange to meet someone involved with the engineering staff at the emergency scene for assistance with building operations.

High-Rise Construction

Today, high-rise buildings are of fire-resistive construction. Often a high-rise building will be constructed of a combination of reinforced concrete and a protected steel frame. Tall buildings often have a reinforced concrete core housing the elevator shafts with the remainder of the frame being steel.

The degree of fire resistance required of structural components will be determined by the model building code that has been adopted. Typically, model building codes require 2- or 3-hour fire resistance for the structural frame of a high-rise building depending on the number of stories and occupancy, and 2-hour fire resistance for floor construction. In reinforced concrete construction the floors will be concrete slab. In steel-frame buildings the floors will be lightweight structural concrete placed over corrugated steel.

The fire-resistive construction used in high-rise buildings provides a high degree of structural integrity. Serious fires have occurred in high-rise buildings with only minor damage to the structural system. Significant structural failure in high-rise buildings is extremely rare.

NOTE: The collapse of the World Trade Center Towers in the attacks of September of 2001 occurred as a result of the structural damage caused by the aircraft impact and the large amount of fire fueled initially by the aircraft fuel. For more information on this collapse, the reader is referred to the National Institute of Technology Website at http:/fire.nist.gov/bfrlpubs/.

Fire Protection Systems in High-Rise Buildings

The model building codes require several fire protection features for high-rise buildings. These features always include automatic sprinklers, voice evacuation system, fire department communication system, and often some form of smoke control.

Automatic Sprinklers

Since the mid-1970's, building codes have routinely required that high-rise buildings be equipped with automatic sprinkler systems as well as standpipe systems (see Chapter 4). The sprinkler and standpipe systems are typically supplied from the same vertical riser **(Figure 12.5)**. These systems will be supplied by one or more fire pumps located in basement of the building. In seismic zones, a secondary on-site water supply is required, usually in the form of a storage tank.

Figure 12.5 Sprinkler and standpipe systems are typically supplied from the same riser.

NOTE: See the IFSTA **Fire Detection and Suppression Systems** manual for a complete discussion of automatic sprinkler systems.

The sprinkler system in a high-rise building will be provided with individual floor control valves. Theses valves are important because they permit the rapid shutdown of sprinklers on the floor on which a fire has occurred after the sprinklers have controlled the fire. This permits the sprinkler system to remain in service on the other floors during overhaul operations and reduces the water damage from a fire. These valves are also useful in reducing water damage resulting from a broken sprinkler.

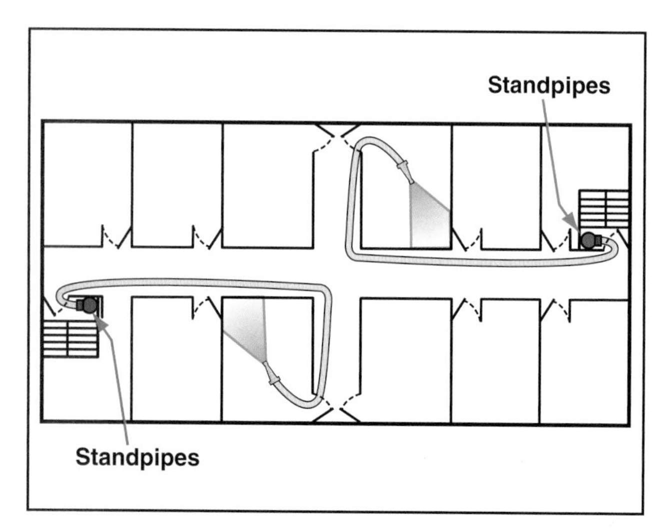

Figure 12.6 To gain complete coverage, standpipes must be properly located to be most effective for fighting fires.

Standpipes

Standpipes are a crucial aspect of fire protection in high-rise buildings. In buildings up to 9 or 10 stories, it might be possible to advance hoselines manually up stairwells. In taller buildings, however, it is very difficult and time-consuming to attempt to advance hoselines in this manner. The taller the building, the more that fire fighting operations are dependent on the availability of standpipes. The standpipe risers and hose valves are located within the stairwells to provide a protected location from which to advance a hoseline **(Figure 12.6)**.

One unavoidable problem with standpipe systems is the variation of pressure with building height (hydrostatic pressure). For example, in a 300-foot (100 m) tall building, the pressure variation due to elevation would be 130 psi (910 kPa). If a minimum 100 psi (700 kPa) pressure is supplied at the top of the riser, the pressure at the base of the riser would be 230 psi (1 610 kPa).

Excessive pressure in the lower portion of a standpipe riser is undesirable because it can make hoselines difficult or dangerous to handle. Therefore, NFPA® 14, *Installation of Standpipe and Hose Systems,* requires pressure-regulating devices to be installed at hose connections when the static pressure at a 1½-inch (38 mm) hose connection exceeds 100 psi (700 kPa) or when the static pressure exceeds 175 psi (1 225 kPa) at larger outlets **(Figure 12.7)**. Pressure-regulating valves can be set according to the jurisdiction's standards. Fire inspectors should verify that these devices are properly set at the time of the original installation. Periodic tests should be conducted to verity their proper maintenance.

Figure 12.7 Pressure-reducing devices are used to manage excess pressure. These devices balance the available pressure within a system against the pressure needed for hoselines.

Meridian Plaza Fire, 1991

On February 23, 1991, a fire occurred in the 38-story Meridian Plaza office building in Philadelphia, heavily damaging the building and resulting in the deaths of three firefighters. This fire is an example of the deadly consequences of fire protection system inadequacy or failure.

Although the building was initially constructed in 1969 without an automatic sprinkler system, one had been partially installed at the time of the fire. At the time of original construction, the primary fire protection features for the upper portions of the building were a dry standpipe riser with fire department connections (FDCs) and a wet-standpipe system with hose supplied by the domestic water system (intended for occupant use).

To facilitate the installation of the automatic sprinklers, the dry standpipe had been converted to wet standpipes supplied by two fire pumps. In addition, pressure-reducing valves were installed at the hose outlets of the standpipes.

The fire began on the 22nd floor of the building. Before it was controlled, it spread down to the 21st floor via convenience stairs and continued up to the 30th floor. Arriving firefighters initially used the building elevators to gain access to the upper floors. Shortly after their arrival, however, a complete electrical failure occurred in the building. This failure not only prevented firefighters from using the elevators and fire pumps, it also forced them to work in a totally darkened building.

When firefighters reached the fire floor, they connected to the former dry standpipe. Their suppression efforts were severely hampered by poor-quality hose streams of limited reach and the crews were forced to use defensive tactics. The availability of water remained low for approximately 4 hours until a sprinkler contractor arrived and adjusted the settings of the pressure-reducing valves on the standpipes. The valves had been improperly adjusted at the time of installation.

Efforts to control the fire became extremely difficult and time-consuming. A 5-inch (125 mm) hoseline was manually advanced up a stairwell in an effort to supply adequate water. The progress of the fire up through the building was essentially unchecked. The three firefighters who died became disoriented in the heavy smoke and darkness.

The fire was finally controlled when it reached the 30th floor where a portion of the automatic sprinkler system had been installed — 10 sprinklers operated.

Lesson Learned: Had the pressure-reducing valves on the standpipe system been properly adjusted at the time of installation and the integrity of the building's electrical system been provided, it is likely that the fire could have been controlled through manual fire suppression efforts.

Fire Alarm Systems

It is not unusual for taller high-rise buildings to have a population of several thousand people; thus, the occupancy of a high-rise can be the equivalent of a small town. For this reason the model building and fire codes require fire alarm systems in high-rise buildings.

Because of the complex nature of a high-rise, the fire alarm systems used in them are more complex and provide more functions than fire alarms used in low-rise buildings. In a high-rise building it is not unusual for a system to include several hundred devices. The fire alarm systems also provide alarm and emergency communication.

Figure 12.8 A two-way fire department connection system is an essential feature for emergency responders at a high-rise.

NOTE: See the IFSTA **Fire Detection and Suppression Systems** manual for a complete discussion of fire alarm systems.

Many modern high-rise buildings have voice alarm systems that automatically sound an alert tone followed by voice instructions on actuation of any detector, waterflow device, or manual pull station. These voice evacuation systems are often zoned by floor. It is not uncommon to evacuate only the floor of origin, the floor above, and the floor below rather than the entire building. These systems will have voice override capability to broadcast further instructions on a selective or all-call basis.

Smoke detection in high-rise buildings typically includes duct detectors arranged to prevent recirculation of smoke to other floors. Smoke detectors are usually provided in elevator machine rooms and elevator lobbies to initiate elevator recall. Some model codes require corridor smoke detectors in residential occupancies. All of these detection devices along with waterflow switches are required to be monitored through the fire alarm system. (Smoke detectors in individual residential units are not monitored through the fire alarm system.)

Another unique fire alarm feature in high-rises is the requirement for a two-way fire department communication system **(Figure 12.8)**. This system operates between the fire command center (central control station) and landings of enclosed exit stairways, areas of refuge, elevators, elevator lobbies, and emergency generator and fire pump rooms. Spare telephone handsets are often provided in the fire command center.

Smoke Control Systems

Prior to the 1970's, few, if any, buildings had any provisions for mechanical smoke control. High-rise buildings constructed before the development of air conditioning systems utilized operable windows for ventilation. During the 1950's, the use of structural steel, and later reinforced concrete, allowed the use of curtain wall construction (see Chapter 9). During the 1970's, the model building codes began requiring breakout panels in exterior walls. However, instead of the breakout panels, the codes typically permitted the use of the building mechanical air handling system to accomplish smoke removal, if the building was completely protected by automatic sprinklers. A common approach using the mechanical equipment was called a "pressure sandwich" concept. This approach exhausted the floor of fire origin and pressurized the floors immediately above and below to contain the smoke to the floor of origin.

Model building codes currently do not require special mechanical smoke removal provisions from the floor of origin. This is based on the premise that the code requirements for shaft construction and the sealing of floor penetrations in conjunction with the automatic sprinklers are adequate to control smoke movement in the building. However, some local or state building codes still require some form of mechanical smoke removal from each floor.

All model building codes require smokeproof exit enclosures in all stairs serving floors 75 feet or higher (see Chapter 4). Entrance to these stairways must be made through an open balcony or a pressurized vestibule **(Figure 12.9)**.

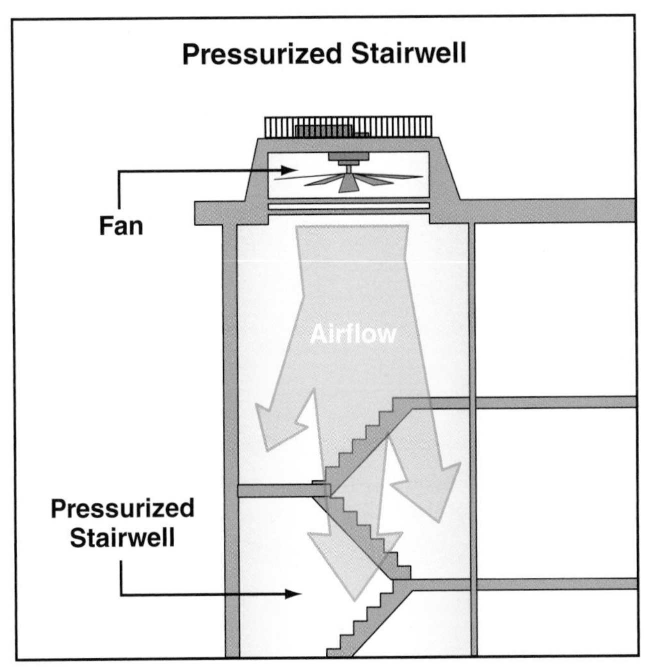

Figure 12.9 A pressurized stairwell incorporates a ventilation system that pushes air into rather than out of the stairwell. This pressurization has the added effect of keeping smoke from a fire in an adjoining floor from entering the stairwell.

Figure 12.10 Firefighters must become familiar with the operation of smoke control panels during pre-incident planning. *Courtesy of McKinney (TX) Fire Department.*

Because of the changes in building codes, especially since the 1970's, it is important that each fire department survey all high-rise buildings in their jurisdiction to determine if these buildings have any smoke-control provisions and how they work **(Figure 12.10)**. This information needs to be incorporated into the pre-incident plans for each high-rise building.

Fire Command Center/Central Control Station

The model building codes require a room or area in a high-rise building to serve as a fire command center. The location of the space must be approved by the fire department. Typically, a fire command center is located on the first floor or level of fire department access **(Figure 12.11, P. 348)**. Often, an enclosed room separated from the remainder of the building by 1-hour fire-rated construction is required. Many jurisdictions require that the room be accessed directly through an exterior door.

Typical features of a fire command center include the following:

- Emergency voice alarm system control panels
- Fire department two-way telephone system panel
- Fire detection and fire alarm system annunciator panel
- Elevator location and status panel **(Figure 12.12, p. 348)**
- Sprinkler valve and waterflow annunciator
- Emergency and standby power status indicators and controls
- Central/status panel for smoke management systems

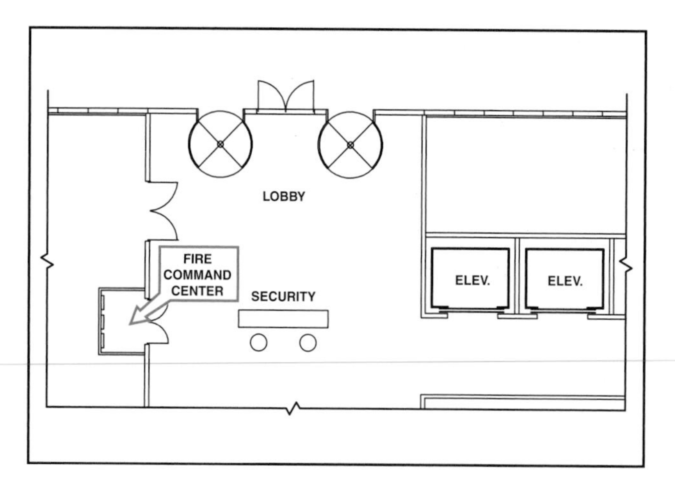

Figure 12.11 Fire command center shown on the floor plan of a building lobby.

Figure 12.12 This room shows the location and status of elevators in a high-rise. *Courtesy of Ed Prendergast.*

- Controls for unlocking stairway doors
- Fire pump status indicators
- Telephone for fire department use with access to the public telephone system

One model building code also requires a work table and a set of building plans.

All building components must be maintained to be sure they are in working order during an emergency. Fire personnel need to do walk-throughs during pre-incident planning and inspection visits to verify that maintenance and repair of all fire protection systems are performed regularly. It should be emphasized that not all fire command centers look and act alike. It is important that panels and other equipment be properly labeled to facilitate use by firefighters.

Fire Extension in High-Rise Buildings

The floor plan of a high-rise building will vary with occupancy and other factors such as site constraints. Many high-rise buildings, especially office buildings and hotels, are designed with a "central core" floor plan. In a central core configuration such building services as elevators, stairwells, and service shafts are grouped in the center of the floor. This arrangement maximizes the amount of space available for development around the periphery of the building.

The fire-resistive construction of high-rise buildings provides a certain degree of inherent compartmentalization and barriers to the vertical extension of fire and smoke. Vertical extension of fire and smoke can occur, however, through floor penetrations such as elevator shafts, stairwells, and utility shafts. For this reason building codes require that vertical shafts be enclosed. Despite the requirement for shaft enclosures, some upward migration of smoke is possible at stairwell doors, utility and elevator shafts, and inadequately protected floor penetrations. For this reason the HVAC systems in high-rise buildings are designed to provide for the management of products of combustion (see Chapter 5).

One means of vertical extension of fire in a high-rise building is by exterior communication from floor to floor. This is particularly likely where a glass curtain wall extends from floor to ceiling. Model buildings codes now have strict requirements for sealing voids where floors intersect with curtain walls.

The horizontal extension of fire in a high-rise building will depend on the extent to which a floor is subdivided by partitions. The partitions between units and the corridor enclosure will act as barriers to fire spread. Some floor plans, such as those found in hotels and apartment buildings, are inherently subdivided; however, in other buildings an entire floor may be open. An example of this would be a floor occupied by a single office tenant.

Emergency Use of Elevators in High-Rise Buildings

Elevators are a potential option for firefighters to reach upper floors in high-rise buildings. The firefighter especially needs to be familiar with Phase I operations (recall) and Phase II operations (override) when confronted with responses to fires in high-rise buildings. A few jurisdictions are currently studying the advantages of providing one or more dedicated elevators for fire department use in high-rise buildings. As noted in Chapter 4, each jurisdiction needs to establish its own policies and procedures for the use of elevators during fires.

Safety During Emergency Elevator Use

The use of elevators by emergency responders during a fire event is always dangerous. The decision to use the elevators must be made carefully and certain guidelines must be followed. Death and injuries can result from the misuse of elevators during a fire.

- Do not use an elevator to travel to the fire floor — stay below the fire floor according to SOPs.
- Maintain the ability to communicate by radio or other means at all times.
- Know the visual signal on the elevator control panel that indicates an impending elevator problem.
- **Be aware that power may fail at any time during a fire.**

NOTE: High-rise buildings for many years have had a minimum of one elevator car on emergency power.

- Never use a fire- or heat-damaged elevator.
- Never use an elevator that has been exposed to water.
- Become familiar with the emergency procedures required to operate elevators by training on actual local systems.

NOTE: The elevator code requires a *shunt trip* where the elevator hoistway or elevator room is protected by automatic sprinklers. This arrangement will shut down the power to the elevator before a sprinkler discharges water on the elevator equipment. As a result the elevator could stop between floors. Therefore, some jurisdictions will not allow these spaces to be protected by automatic sprinklers. The fire department should become familiar with their local requirements.

Phase I Operations

For many years, fire and building officials have recognized the critical importance of elevators in tall buildings while noting their susceptibility to interference by smoke, heat, flame, and water. As a result, codes contain mandatory provisions for the recall of all passenger elevators with vertical travel greater than 25 feet (8.3 m) in the event of fire (called Phase I operations). The automatic recall of elevators to their terminal floor or an alternate floor can be caused by the activation of smoke detectors or sprinkler waterflow alarms. Phase I can also be activated manually by a keyed switch in the terminal floor lobby **(Figure 12.13)**.

Phase I Operation — Emergency operating mode for elevators. Phase I operation recalls the car to a certain floor and opens the doors.

Phase I operation is designed to prevent the deaths of civilians who may find themselves in an elevator that is called to the fire floor, as happened in high-rise fires in the 1970's. Phase I operation automatically stops all the cars that serve the fire floor if they are moving away from their terminal floor (the lobby). It also causes the cars to return nonstop to the lobby or other designated level if the alarm originated from the lobby.

NOTE: The requirement for recall does not apply to freight elevators. Other elevator safety and design features also may not be applicable to freight elevators.

At the lobby, the fire department must account for each car to be certain there are no civilians trapped in a stalled car at or above the fire floor. This can be accomplished by checking the elevator control and information panel that is usually located adjacent to the elevator bank.

Phase I operation opens the car doors and keeps them open when the car reaches the recall floor **(Figure 12.14)**. This makes it easy to see which cars are empty and which may be on another level with people trapped. During Phase I operation, the elevator car's emergency stop and floor selection buttons are rendered inoperative so that car occupants who may be unaware of the fire or emergency cannot stop the car anywhere but at the terminal floor.

Figure 12.13 A firefighter activates Phase I operation.

Figure 12.14 Elevators should be brought to the main elevator lobby.

Phase II Operations

Phase II elevator operation is designed to permit firefighters to use the elevators after they arrive on the scene by overriding the recall feature. The codes specify that all new elevators must be equipped for Phase II operation. Older elevators may not have any provisions for firefighter service. Individual elevator installations must be evaluated during preincident planning to determine their emergency function capabilities.

Phase II Operation — Emergency elevator operating mode that allows emergency use of the elevator with certain safeguards and special functions.

Typically to activate Phase II operation, a firefighter must insert a key in a three-position switch within a car to place that particular car in "fire service" **(Figures 12.15 a and b)**. During this phase an elevator becomes essentially a manually operated elevator. For example, the floor-select buttons within the car remain operable but the floor-call buttons on the individual floors are inoperable. The elevator doors do not open automatically and the operator must push the "Door Open" button in the car.

The electric eye safety, which prevents the doors from closing if there is a person or smoke in the doorway, is disabled during Phase II operations. This is done so the car doors can be closed and the car moved if it inadvertently stops at a smoke-filled floor. The emergency stop button that was inoperable in Phase I should be operable during Phase II. This allows firefighters who feel they may be in trouble to stop the car wherever it is located. Because the car controls are operable only from within a car, it is important that a firefighter remain in the car.

Emergency responders should be aware that some local jurisdictions may have specific, unique requirements for certain occupancies such as correctional facilities where the operational features may be different.

NOTE: Additional discussion of special features at correctional facilities is found later in this chapter.

Figures 12.15 a and b The key is then inserted in the three-position switch inside the elevator car. *Photo b courtesy of McKinney (TX) Fire Department.*

Figure 12.16 The visitor center at Valley Forge National Historical Park is tucked under the edge of George Washington's Revolutionary War encampment field. Photo by Loretta Hall, author of *Underground Buildings: More than Meets the Eye.*

Underground Buildings

The classification as an underground building usually applies to belowgrade buildings or portions of buildings that are deeper than ordinary basements. An underground building is defined by some codes as one in which the lowest level used for human occupancy is 30 feet (10 m) below the main exit that serves that level. Examples of underground buildings that have existed for many years include parking facilities, subway stations, emergency communication/command centers, and storage facilities. With the current trend toward sustainable (green) building design, there are many more underground buildings being built or planned. These occupancies include museums, libraries, academic buildings, laboratories, industrial facilities, and offices.

Underground locations have the advantage of security, relatively constant temperatures, and smaller visual impact **(Figure 12.16)**. Older underground buildings were usually accessed vertically from grade level; however, some modern underground buildings are being built into the side of hills with the primary access being horizontal.

Examples of Underground Buildings in the US

Moscone Convention Center, San Francisco, CA
University of Illinois Library, Urbana, IL
Underground Art Gallery, Brewster, MA
New York Transit Museum, Brooklyn, NY
FedEx, Memphis, TN
Underman Theater, Dallas, TX
Westinghouse Headquarters, Orland, FL
University of Arizona Integrated Learning Center, Tucson, AZ

Emergencies in underground facilities pose very difficult problems for firefighters. Specific difficulties include the following:

- Access to the structure
- Rescue and evacuation of occupants
- Ventilation of heat and smoke
- Water supply and drainage of water from flooding or fire fighting operations **(Figure 12.17)**

The difficulty in venting heat and smoke is probably the greatest single challenge in controlling fires in underground buildings. Normal ventilation tactics usually will not work when the roof is entirely belowgrade. It is critical that these structures be thoroughly pre-planned and that the plans be kept current.

Access to an underground structure may be made in a number of ways. Very large structures may have numerous access points **(Figure 12.18)**. Underground tunnels such as subways may have stairways, ramps, or even railcar access. Knowing how to gain access and how to remove occupants will be crucial for a coordinated emergency response.

Figure 12.17 This underground garage was flooded due to a hurricane. *FEMA/Marty Bahamonde.*

Figure 12.18 The underground tunnel system in downtown Oklahoma City is marked at numerous points, much like a subway system.

Figure 12.19 The Harris Theater for Music and Dance, completed in 2003, was built beside an underground parking garage in Chicago's Millennium Park. The 40-foot-deep, 1,500-seat theater has a street-level entrance pavilion as well as a connection to the adjacent garage. Photo by Loretta Hall, author of *Underground Buildings: More than Meets the Eye.*

Fire fighting can be extremely difficult when a facility is located 100 feet (30 m) or more below grade, and the only access is by means of stairwells or elevators. Evacuation of occupants from underground locations is more difficult than in high-rise buildings because of the greater physical exertion required for occupants going up stairs and the greater potential for a stairwell being filled with smoke from the fire below. Smokeproof enclosures are now required for buildings with levels more than 30 feet (10 m) below the level of exit discharge.

Underground facilities are not built in the same way as an aboveground building. Underground structures must be excavated **(Figure 12.19)**. The structural system of an underground facility is massive compared to the framing systems used for aboveground buildings. However, structural damage may still occur in underground structures when a fire has been of long duration. An example of this is the failure of steel support columns caused by a fire in a subway station. The maximum available fire-resistant protection is typically four hours; therefore, fires exceeding four hours are a serious threat.

The building codes contain special provisions for underground buildings, although they may group underground buildings with windowless buildings. One typical requirement is that the underground portion of the building be of fire-resistive construction. Where buildings have floor levels more than 60 feet (20 m) below the level of exit discharge, separation of each level into two approximately equally sized compartments may be required. Each compartment will be provided with at least one stair and access into the other compartment.

The codes require automatic sprinkler protection for underground buildings, even if a portion of the building extends above the ground and the aboveground portion of the building does not require sprinkler protection. To address the problem of ventilation of smoke and heat, model codes often require a smoke exhaust system. The depth at which a smoke exhaust system becomes necessary as well as specific requirements for any system will depend on the applicable code.

Buildings With Limited Access

Windowless structures, or portions of structures with blank walls or other major obstructions on exterior walls, can present many of the same challenges as underground buildings. Firefighter access for fire suppression and rescue is limited, as well as the ability to quickly and efficiently ventilate the building **(Figure 12.20)**. As a result, building codes generally require that these buildings be fully protected by automatic sprinklers. In addition, emergency access openings are typically required on a minimum of two sides of upper floors of these buildings. These openings must be readily identifiable and operable from both the exterior and interior. The building codes require these openings to be sized to allow for rescue and ventilation operations. However, there is no consistency in the size requirements; therefore, the local code needs to be referenced for specific design details.

NOTE: It should be noted that *readily operable* from the exterior means with use of fire department tools so building security is maintained.

Many public buildings have also been hardened against terrorist attacks or simply wayward drivers. These structures often have concrete pillars and other barriers that make firefighter access more complicated **(Figure 12.21)**.

As with underground buildings, it is important that buildings with limited access be carefully pre-planned and that the pre-incident plans be kept current. Regular inspections of these properties are important to ensure that the emergency access openings are not obstructed on the interior of the building.

Membrane Structures

A membrane structure is a building having its exterior skin consist of a thin "waterproof" fabric. Membrane structures come in various forms including air-inflated, air-supported, membrane-covered cable, and membrane-covered frame. These types of buildings are becoming more common, in part due to economic factors and speed of construction. They are often used to accommodate recreational uses such as tennis courts, athletic fields, and ice skating rinks **(Figure 12.22)**. Several major sports stadiums have been built with air-supported membrane domes over the stadium itself. Membrane structures have been used for such other occupancies as warehouses, casinos, churches, and campus dining facilities. A major advantage of both structure types is that they can be used to provide a large, unobstructed interior space.

Air-inflated and air-supported membrane structures rely on air pressure to form the shape. Membrane-covered cable and frame structures are not pressurized and rely on the cable or frame configuration to form the shape. Model building codes address both these types of structures.

Air-Supported and Air-Inflated Structures

In an air-supported structure, the roof is made of a vinyl-coated polyester fabric material. It is held up by an air machine that keeps a constant supply of air in the structure **(Figure 12.23, p. 356)**. The pressure required is a small

Figure 12.20 This structure will have limited access points due to the fact that it will be a windowless, concrete building. *Courtesy of Dave Coombs.*

Figure 12.21 These concrete planters are actually a form of hardening.

Figure 12.22 This air-supported structure is a sports training facility.

Air-Supported Structure — Membrane structure that is fully or partially held up by interior air pressure.

fraction of atmospheric pressure on the order of 0.35 psi (2.45 kPa). If power is lost the roof may eventually collapse; therefore, a redundant air supply must be provided by a standby generator.

Modern air-supported roofs have computer-controlled air-supply systems that adjust for varying external wind loads. Because the interior pressure is greater than the exterior pressure, cables are frequently used to anchor the roof to the ground or substructure. Normal access or egress is provided through airlocks or revolving doors **(Figure 12.24)**. Because revolving doors may not be permitted to be used as exits by a building code, emergency exits may also be provided.

Air-inflated or air-supported structures have several limitations. They are obviously limited to one story. To comfortably accommodate human occupancy, the interior pressure in an air-supported structure can only be slightly greater than the outside pressure. Air-supported buildings, therefore, make use of a membrane that weighs only a few ounces per square foot for the building skin. This limitation, however, does not apply to air-inflated structures because the occupants are not exposed to any interior pressure.

Another limitation to both types of structures is they cannot be used in situations where fire-rated construction is required. Because of the uplift required to maintain the shape of an air-supported structure, it is necessary to securely anchor the structure to some type of foundation. Various anchoring systems have been developed for this purpose.

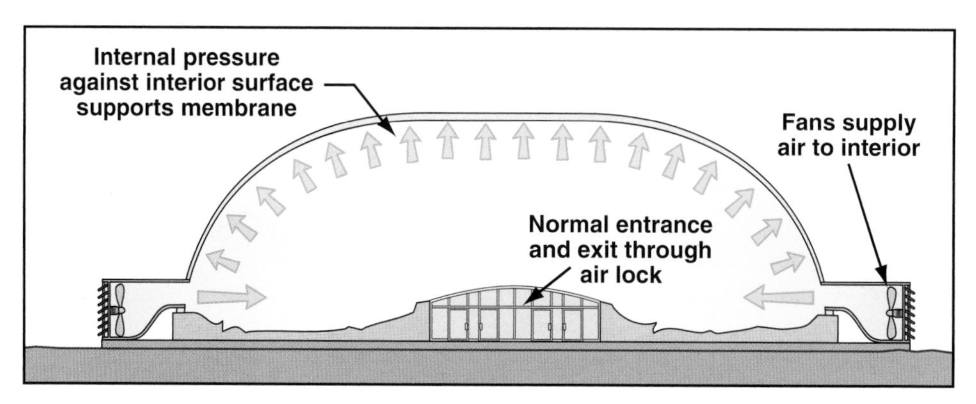

Figure 12.23 The basic concept of an air-supported structure.

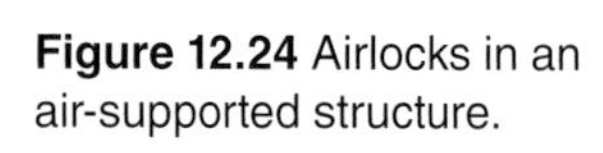
Figure 12.24 Airlocks in an air-supported structure.

Air-inflated and air-supported structures are sometimes vulnerable to high winds and, in some cases, have been blown down in a high wind. Therefore, when exterior winds increase, the interior pressure is increased. Newer structures have computer-controlled fans to automatically adjust to changing atmospheric conditions.

The membrane used in either an air-inflated or air-supported structure may be a limited-combustible or noncombustible material. If the material is limited-combustible, it is required to have a low flame spread and low smoke propagation.

Although most air-supported and air-inflated structures have fire systems to provide initial fire control, fire fighting in these structures can pose some unusual problems. Conventional ventilation of an air-inflated or air-supported building is not possible because firefighters simply cannot open the roof. The fans could probably keep up with the air lost through a small opening in the membrane. However, gaining access to the roof is difficult and dangerous. One offsetting feature, however, is the likelihood that the membrane may self-vent by melting under high temperatures. If there is a serious reduction or total loss of power, the membrane will collapse.

Membrane-Covered Cable and Frame Structures

A membrane-covered cable structure uses a system of masts and cables to provide the support for the membrane covering **(Figure 12.25).** Similarly, membrane-covered frame structures utilize a rigid frame system, usually with columns, to support the membrane **(Figure 12.26)**.

These types of membrane construction are more commonly used than air-inflated or air-supported structures. They are well suited for many uses; however, they cannot be used to provide a large unobstructed interior space because of the necessary support system.

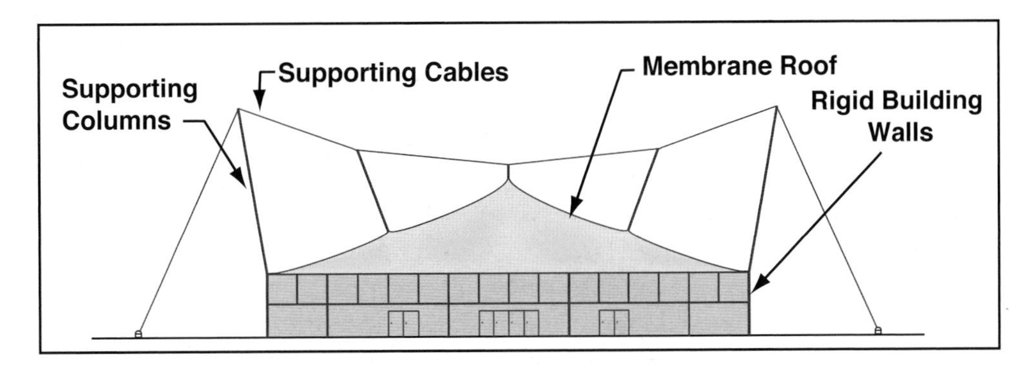

Figure 12.25 Membrane structure using a cable support system.

Figure 12.26 A membrane structure with a rigid frame.

As with air-inflated or air-supported membrane structures, membrane-covered cable and frame structures cannot be used where fire-rated construction is required because membranes do not have a fire-resistance rating. Some membranes are considered noncombustible and the remainder are required by code to have a low frame spread.

There are several obvious fire fighting considerations associated with membrane-covered cable or membrane-covered frame structures. These include ventilation and early building collapse if the cable or frame system does not have a fire-resistance rating. As with air-inflated and air-supported structures, one offsetting feature, however, is the likelihood that the membrane may self-vent by melting under high temperatures.

Covered Mall Buildings

The covered mall building or simply "the mall" consists of a building with numerous individual tenants that face a common covered pedestrian way **(Figure 12.27)**. Technically, the covered pedestrian way is defined as the *mall*. The advantage of the covered mall design is that customers may do their shopping in a comfortable and protected environment.

Shopping malls are often developed in a suburban community on the edge of a larger city. Sometimes, however, malls are incorporated into redevelopment projects in downtown metropolitan areas as part of mixed-use projects. A suburban fire department accustomed to dealing with residential buildings and small- or medium-sized individual stores may now be confronted with a shopping mall of several hundred thousand square feet and 100 or more stores under one roof. In fact, the shopping mall can be described as comparable to the business district of a medium-sized city under one roof.

A shopping mall often contains restaurants, movie theater complexes, and professional offices as well as retail stores. To the fire officer, it is as though several commercial blocks were roofed over and the streets on which they

Figure 12.27 A typical covered mall with several stories.

fronted were turned into pedestrian ways not accessible to emergency vehicles. In some large cities streets have given way to multi-block projects where the original streets have been built over.

Configuration and Construction

Shopping malls are constructed in various sizes and configurations and typically consist of one to three levels. When they contain more than one level, covered malls will have multiple openings between the levels. Building codes require that the mall be a minimum of 20 feet (6.1m) in width. This space permits the use of open storefronts while still allowing for pedestrian exiting.

Malls are usually designed with one or more large, well-known perimeter stores that are known as *anchor stores*. In some cases, the mall may include a hotel as an anchor. The anchor buildings serve to heighten the overall commercial appeal of the mall. The anchor buildings are usually operated by nationally known chains and owned and managed as separate entities. The anchor stores may be separated from the smaller stores by a fire wall; however, codes permit unprotected openings between the anchor stores and the mall.

While some shopping malls may be built of fire-resistive construction, many are of noncombustible or wood-joisted construction. In recent years, major developers of covered malls have preferred unprotected noncombustible protection. (Current codes do not permit covered malls to be constructed of wood-joisted construction.)

Sections of covered malls may have been constructed at different times; mixed construction is definitely a possibility. Current codes require malls to be fully sprinklered; however, it is possible to find older covered malls that are only partially sprinklered or even nonsprinklered. It is also possible to find in older covered malls that the anchor stores are sprinklered and the smaller stores are nonsprinklered.

Codes require individual stores within a shopping mall to have 1-hour fire-resistive separations from each other. However, one characteristic of covered malls is that the individual storefronts are not separated from the mall itself by any type of fire-resistive construction. The storefronts usually are separated from the mall only by show windows and either a security gate or swinging glass doors **(Figure 12.28)**. During business hours, the gates or glass doors are open. Products of combustion from a fire occurring within an individual store can readily communicate into the mall.

Figure 12.28 Security gates can permit products of combustion to travel out into the main areas of the mall. They can also make fire department access more difficult.

For many years, most codes required a smoke control system in all covered mall buildings, regardless of the number of stories. Current codes have eliminated this requirement in one-story shopping malls; however, a smoke control system may be required in two- and three-story covered mall buildings. (See applicable code.)

Figure 12.29 Hose outlets are required at major entrances to malls.

Access Considerations

Because the smaller stores face into the mall, access to an individual store by responding fire companies may be slowed because apparatus cannot drive directly up to the front of the store. Furthermore, a store may be some distance from a mall entrance, necessitating a long hose lay. To compensate for this condition, codes now require fire department hose outlets in stairs, at major entrances to the mall, and at entrances from the mall to corridors and passageways **(Figure 12.29)**.

Another problem characteristic of covered malls is the periodic turnover of smaller tenants. This often results in vacant stores and stores undergoing renovation. When stores are remodeled, the work is typically done while other stores in the mall are open for business. The stores that are being renovated will have all the hazards associated with construction (see Chapter 13). In addition, the sprinkler zone protecting that store may be shut off at times during the renovation. This usually results in the sprinklers being shut off in nearby stores.

As a result of economic changes in neighborhoods, there have been instances where covered malls that originally included primarily retail tenants have been converted to office malls. This change should not have a negative impact on the fire protection challenges of a covered mall because office occupancies are considered to be light-hazard and would be typically covered by the covered mall code provisions.

Detention and Correctional Facilities

Buildings housing detention and correctional facilities differ significantly from most occupancy uses. First of all, the occupants (inmates) are confined, typically in cells. Although they may be mobile, they are unable to evacuate in case of a fire until a lock or several locks are opened. In addition, when the occupants are evacuated, they must be relocated to a secure area because they cannot be allowed to escape.

A detention facility is generally a smaller temporary holding facility before release or transfer of inmates to a larger correctional facility. However, some inmates may be kept in a detention facility for several months or longer.

Defend in Place — Procedures taken to shelter persons from harm during an emergency without evacuating them from a structure. Used especially in hospitals and prisons.

For larger correctional facilities located in remote areas, evacuation to a secure area can be achieved by directing inmates into outdoor exercise areas. However, there are some correctional facilities located in metropolitan areas where large secured outdoor areas are not available. This situation is also the case in many local detention facilities. Where evacuation is a last resort, it is necessary to apply the concept of *defend-in-place* similar to the approach used for hospitals.

Model building code requirements for fire protection and life safety features for detention and correctional facilities vary depending on the level of restraint required at the facility. These features may include remote release of door locks, smoke compartments, areas of refuge on either side of smoke barriers, automatic sprinklers, smoke control systems, and both manual and automatic fire alarm systems **(Figures 12.30 a and b)**. Newer correctional facilities will likely have a central control center where locking devices are monitored and controlled. However, current codes still allow the use of keys under certain conditions. Fire codes have strict flammability requirements for furnishings, including mattresses, to reduce the fire hazard exposure to inmates (**Figure 12.31**).

Until recent years, there was a reluctance to install automatic sprinklers in detection and correctional facilities because of the possibility of inmates committing suicide by hanging themselves from sprinklers. About 25 years ago, the automatic sprinkler industry responded to this concern by developing a "breakaway" sprinkler. Now many of the newer facilities have automatic sprinklers installed. This fire protection feature specifically addresses the concept of defend in place (**Figure 12.32**).

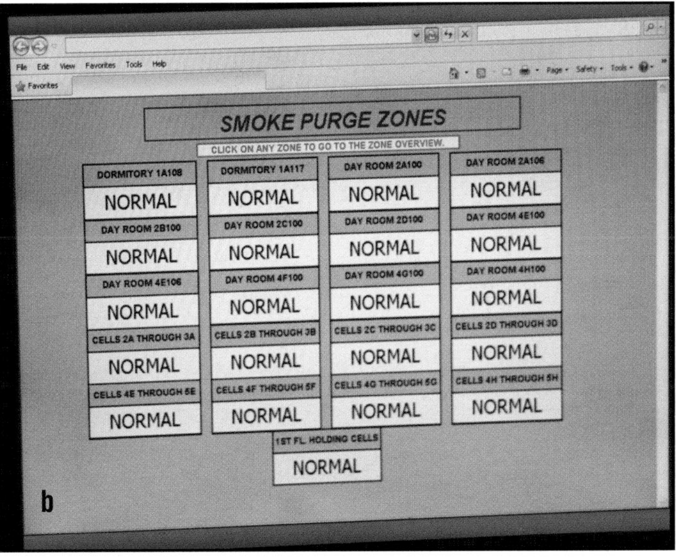

Figures 12.30 a and b (a) This sallyport door can only be opened electronically. Note the smoke alarm to the side. (b) Smoke purge zones can be quickly monitored in this jail's fire control room.

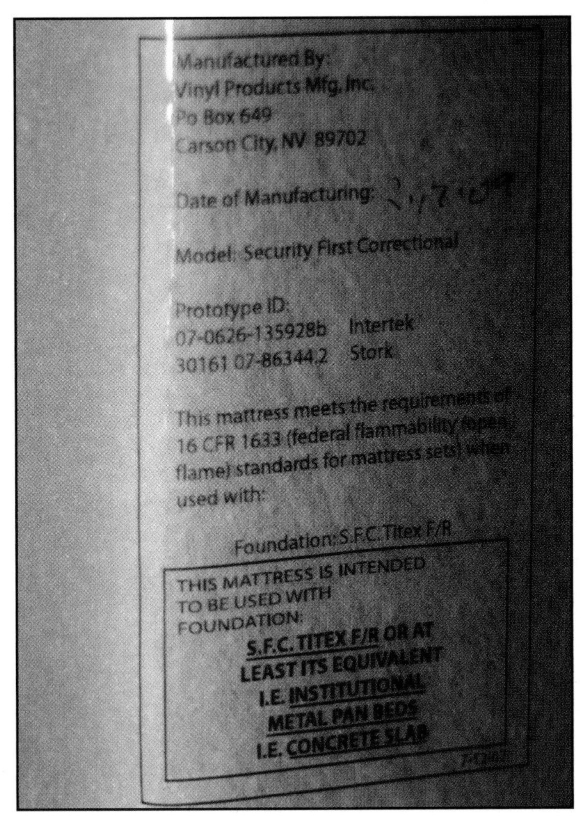

Figure 12.31 Flammability codes extend to mattresses in order to reduce hazards.

Figure 12.32 A breakaway sprinkler installed in a jail.

Figure 12.33 In this jail, inmates can be moved from one secure area to another during an emergency.

Another key element of the defend-in-place concept is the creation of one or more compartments on each floor, separated by a fire-rated smoke barrier **(Figure 12.33)**. This allows the inmates to be moved horizontally from the compartment of fire origin to an adjacent compartment on the same floor. Therefore, in most cases, it will not be necessary to utilize stairs and evacuate inmates.

Older detention and correctional facilities may not have automatic sprinklers installed and may be dependent on keys to open locks. It is important for the responding fire department to work closely with the agencies operating these facilities to develop a comprehensive plan of action to guide firefighters during emergency response.

Atriums

An *atrium* is a large vertical opening extending through two or more floors of a building that is not used for building services such as enclosed stairs, elevators, or building utilities (**Figure 12.34**). In an architectural sense, an atrium can have a roof or a ceiling or simply be open to the atmosphere. However, the code definition of an "atrium" refers to a covered vertical opening.

Atrium — Open area in the center of a building, extending through two or more stories, similar to a courtyard but usually covered by a skylight, to allow natural light and ventilation to interior rooms.

Atriums are a very popular architectural design feature. An atrium can provide light and ventilate the interior of a building. In contemporary design practice, their appeal lies mainly in the openness they provide within a building. For low-height buildings, an atrium often extends from the first floor up to the roof. In taller buildings, it is common for an atrium to extend only partway up through a building **(Figure 12.35)**.

From a fire safety standpoint, an atrium poses the same potential as other vertical openings for communication of heat and smoke up through a building **(Figures 12.36 a and b)**. There is often a desire to have some floors of a building open to the atrium without any physical separation. In addition, the floor level of an atrium is frequently occupied by combustible furnishings or other contents. Therefore, a fire in floors open to the atrium or on the atrium floor itself has the potential to impact occupants of the other open floors - even upper floors that are otherwise enclosed.

Typical model code requirements for atriums include automatic sprinkler protection. Automatic sprinklers may only be required for those floors that are connected by the atrium. The building codes have a basic requirement that an atrium be enclosed with 1-hour fire-rated construction or a combination of glass and automatic sprinklers. However, codes usually make provision for elimination of the 1-hour enclosure for up to three stories or more when certain conditions are met.

A smoke control system is required to vent the products of combustion to the outside whether floors are enclosed or not. Until recently, the exhaust capacity was determined by the volume of the atrium and the unenclosed floors connected to the atrium. Current codes base the exhaust capacity to maintain the smoke layer at a specified height above the highest walking level serving the exit system. The required exhaust capacity is based on the magnitude of the expected fire and the height to the bottom of the desired smoke layer. The design of the system, therefore, requires a thorough engineering analysis. Several different methods of venting atriums can be found, depending on when a particular building was built. Providing adequate fire protection in a building with an atrium is complicated by the height of many atriums, which can extend

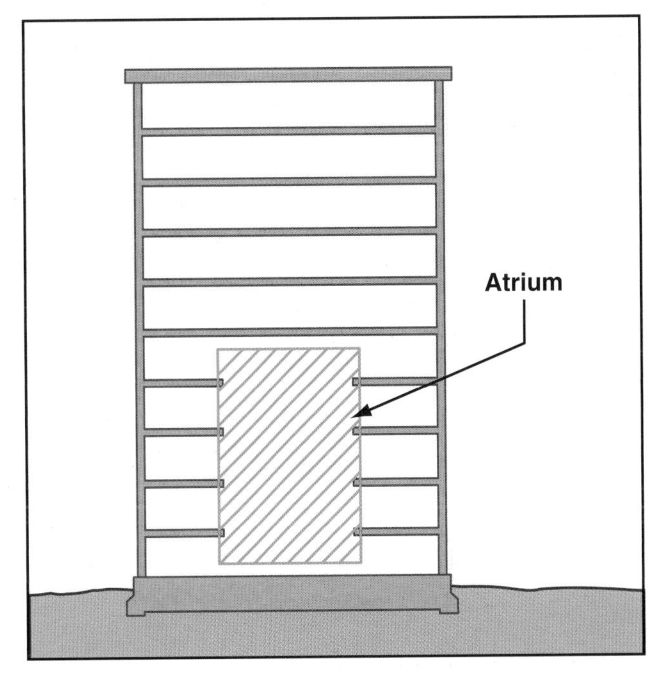

Figure 12.35 An atrium may extend only part of the way through a building.

Figure 12.34 Large, open areas like this university atrium require an active smoke-control system because of the extended length of time it may take to evacuate the structure and gain access to a fire.

Figures 12.36 a and b (a) An atrium can provide a path for the vertical communication of heat and smoke. (b) Note the travel of smoke as the fire protection systems are being tested in this new structure. *Photo courtesy of McKinney (TX) Fire Department.*

up to 50 stories or more. A vertical space of that height creates several specific problems. For example, when sprinklers are installed at an atrium ceiling 50 or more feet (16.6 m) above the floor, they will be less effective because of the longer time it takes for them to operate — if they operate at all. Some codes now waive the requirements for sprinklers where the ceilings of atriums are more than 55 feet (18.3 m) above the floor.

Explosion Venting of Buildings

Chapter 3 discussed the various types of forces that can be exerted on a building. One type of force that was not discussed was the internal force within a building that results from an explosion. An explosion inside a building creates an outward pressure on the building and its structural components **(Figure 12.37)**. In cases of severe explosions, the building can be literally torn apart. Exterior structural components such as glass and bricks can be hurled a considerable distance.

For most occupancies, internal explosions are not taken into account in the design of a building. In specialized industrial occupancies, such as where flammable liquids are processed or where combustible dusts are produced, structural design provisions can and should be made to reduce the structural damage due to an explosion.

Types of Explosions

Explosion — A physical or chemical process that results in the rapid release of high pressure gas into the environment.

An *explosion* can be defined as an event that produces a rapid release of energy. This sudden release of energy produces outward pressures, often referred to as blast waves. What distinguishes an explosion from such other occurrences as ordinary combustion is the speed with which the process occurs **(Figure 12.38)**.

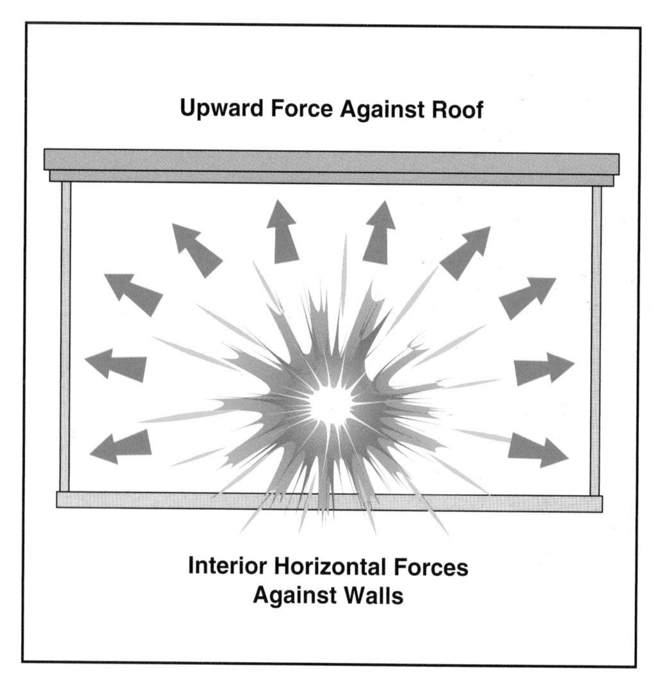

Figure 12.37 An explosion within a building creates internal forces that the building is not designed to withstand.

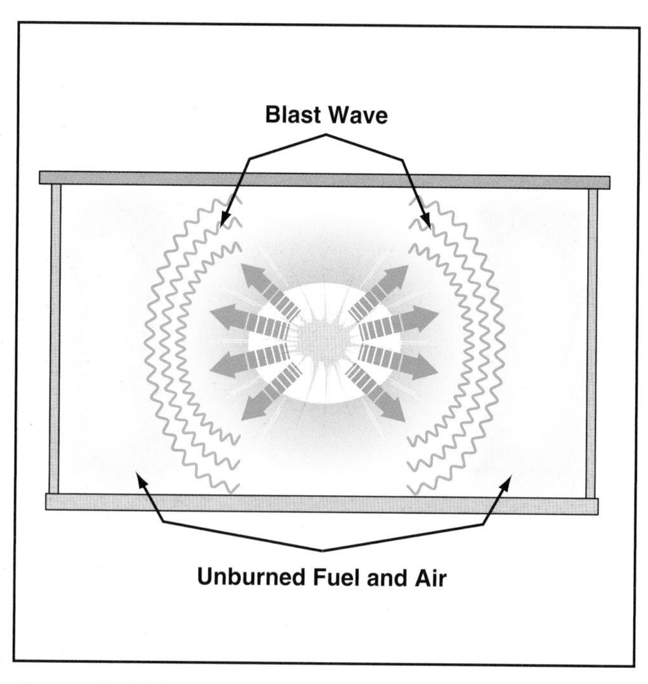

Figure 12.38 In a detonation, the blast wave moves faster than the speed of sound.

Explosions in buildings occur in a number of ways. Common examples are as follows:

- Explosions involving a chemical reaction such as the rapid combustion resulting from the ignition of a mixture of air and a flammable vapor, or ignition of a mixture of air and dust **(Figure 12.39)**. Dust explosions often occur in grain elevators and in milling operations involving sawdust.
- Explosions resulting from uncontrolled chemical reactions in processing plants or decomposition of unstable compounds.
- Boiler explosions — a purely physical event in which no chemical reaction occurs.

The damage an explosion may inflict on a structure depends on the maximum pressure developed, the rate of pressure rise, the duration of the peak pressure, and the resistance of the confining structure.

Containment and Venting

Two general methods can be employed to reduce the structural damage from an explosion: containment or venting.

Containment. In containment, the building enclosure is constructed with adequate reinforcement to contain the pressure resulting from an explosion without failure. Containment is usually expensive because it normally requires reinforcement beyond what is necessary for ordinary structural design purposes. The maximum pressure reached during an explosion may be as high as ten times the atmospheric pressure, or more.

Venting. Explosion venting is designed to quickly relieve the pressure produced by a explosion before it causes excessive damage. Ideally, an explosion vent would be open at all times. However, there are very few industrial operations that can be carried out without some kind of an enclosure.

Figure 12.39 Metro-Dade (FLTF-1) Task Force began performing search and rescue operations inside the Humberto Vidal Building following a gas mainline explosion. A view of the destroyed building. *FEMA/ Roman Bas.*

Explosion vents include louvered openings, hangar-type doors, wall panels, windows, or roof vents **(Figure 12.40)**. Some types of explosion vents can be purchased ready to install. Others are custom designed for a specific location. Vent closures must be designed to operate at as low an internal pressure as practical; however, they must still be designed to remain in place when subjected to the forces of external winds.

Because vent panels must operate quickly, they must be relatively light. If the explosion panels are too heavy, their inertia will slow the speed at which they operate resulting in a faster rate of internal pressure rise. Ideally, explosion vent panels should not weigh more than 3 pounds per square foot (14.6 kg/m^2). Several different materials, such as lightweight corrugated steel or aluminum sheets, are often used in industrial buildings for explosion vents.

One method to help ensure rapid operation of vent panels is to attach the panels with reduced-diameter bolts, which are designed to break under the force of an explosion **(Figure 12.41)**. The panels also may be hinged at one side and fastened at the other so that they swing open in an explosion.

Explosion vents must be of an adequate size to vent the pressure of an explosion. For a given type of vent, a larger vent area results in a lower pressure within the building. The vent area must be large enough to keep the pressure of an explosion below that which would cause structural damage. No simple rule exists with respect to the required size of explosion vents. The required explosion vent area is a function of the size and strength of the structure, the expected forces of the potential explosion, and the type of vent. Determination of the actual vent area, therefore, requires an engineering analysis.

It is important to remember that the venting of an explosion is a means of limiting structural damage. Explosion venting is not a substitute for the prevention of explosions and it does not provide for the protection of personnel within a building. The pressure wave developed within the space

Figure 12.40 Buildings that contain unstable liquids or hazardous processes are required to be equipped with explosion vents.

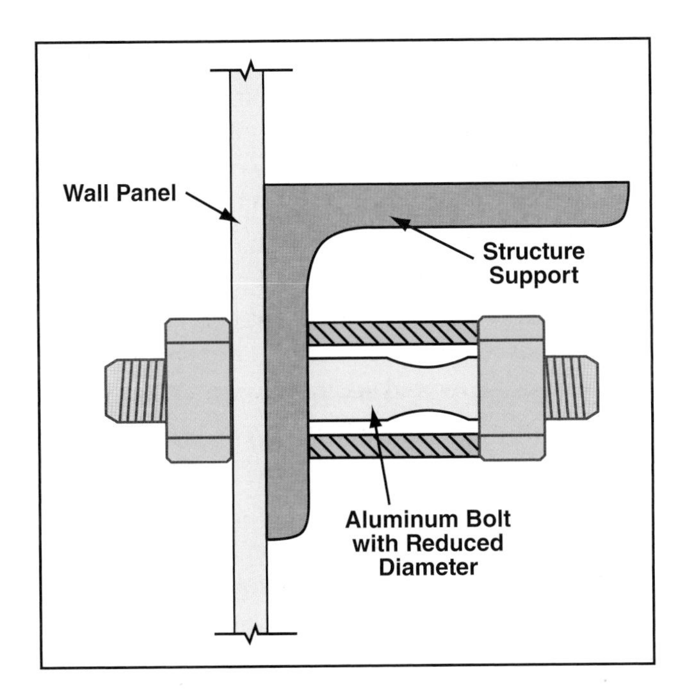

Figure 12.41 A bolt with a reduced cross section is designed to fail in tension under the force of an explosion.

may be great enough to cause death or injury — even when it is vented — because the pressure wave may come into contact with personnel before the vents open.

When explosion vents operate, the force is carried beyond the opening. Therefore, in the design of vents, consideration must be given to what lies in the path of the pressure wave in the area beyond the vent opening.

Areas of Refuge

As a result of the adoption of the Americans With Disabilities Act (ADA) of 1990, there has been a significant effort to provide buildings that are readily accessible and usable for individuals with disabilities. A key element of this effort has been the recognition of the need for special provisions to alert and evacuate disabled persons if the need arises. For example, if a fire alarm evacuation system is required by code in an occupancy or building, it must provide adequate audible and visual means to alert occupants who are visually or hearing impaired.

Area of Refuge — (1) Space in the normal means of egress protected from fire by an approved sprinkler system, by means of separation from other spaces within the same building by smokeproof walls, or by virtue of location in an adjacent building. (2) Area where persons who are unable to use stairs can temporarily wait for instructions or assistance during an emergency building evacuation.

Once an alarm is initiated, there must be provisions to provide accessible means of egress, especially for occupants who are unable to readily use stairs. Some codes reduce or eliminate the requirements in buildings that are fully protected by automatic sprinklers. When required, accessible means of egress typically consist of areas of refuge on a floor in conjunction with stairs or elevators, or a combination of both. In recent codes, there is a requirement for two accessible means of egress from a building, with both routes being continuous to ground level. For those buildings with unusual site or configuration restraints, there may be a need for an area of refuge at the level of stair or elevator discharge.

An area of refuge for a stairway serving as an accessible means of egress can be located with the stairway on a landing or from an adjacent vestibule. Generally, the area of refuge is designed to accommodate one or two wheelchairs, depending on the occupant load served. The wheelchair space(s) must not obstruct other occupants using the stairway or vestibule for egress. The design of the space must also consider the presence of automatic sprinkler or standpipe risers. Elevator lobbies are often used for areas of refuge to utilize an elevator as an accessible means of egress.

A smoke- and heat-free environment is essential to protect occupants while they are awaiting rescue. Areas of refuge located in an enclosed exit stair are inherently protected by the very nature of the construction requirements for a stair enclosure. Likewise, required elevator lobbies are similarly protected. However, vestibules outside of stairways that provide areas of refuge need to be separated from the rest of the floor by a smoke barrier. The typical smoke barrier consists of a minimum 1-hour, fire rated enclosure.

NOTE: Refer to Chapter 4 for additional information on stairs and elevators.

Occupants of an area of refuge could potentially be left unattended or go unnoticed. Therefore, areas of refuge must be provided with some type of two-way communication system connected to a constantly attended location. This system must be designed to accommodate persons with any type of disability. Also, areas of refuge need to be well identified from the exterior of the space and provided on the interior with adequate instructions on the use of the space and the communication system.

Rack Storage

Industry has made use of various types of racks for storing all types of commodities for several decades. The use of multiple-tier racks greatly increases the efficiency of warehouse operations because it permits greater utilization of a building's interior volume. Rack storage varies from the simple use of forklifts positioning pallets in the racks to specialized automated warehouses in which unmanned pickers or stackers handle large commodities **(Figure 12.42)**. Such industrial technology permits very precise control of inventory. Storage racks can vary from two or three tiers with a total height of just 12 feet (4 m) to in excess of 100 feet (33 m).

Normally, storage racks are structurally independent of the building in which they are located, and are often bolted to the floor **(Figure 12.43)**. In some instances, however, the rack system provides part of the structural support for the building. Because racks consist of unprotected steel members, they may collapse under fire conditions. This situation not only adds to the difficulty of fire suppression, but can also affect the stability of the entire structure when the racks are used as part of the building structural support.

The use of racks for storage has created some very difficult fire protection problems. Although rack storage is highly efficient, it results in a very high density of storage. When the stored materials are combustible, such as paper or containers of flammable liquid, a high fire load is produced **(Figure 12.44)**. Furthermore, the racks are frequently arranged with narrow aisles, which makes access by firefighters very difficult.

Storage racks are arranged with several horizontal tiers. While racks may be in single rows, the racks are usually arranged back to back (double row racks). Warehouses may have multiple row of racks wider than 12 feet (4 m) or with aisles narrower than 3.5 ft (1.1 m). These configurations create two problems. First, the penetration of water from overhead sprinklers is obstructed by the intervening tiers of storage. Second, flue spaces may be created that permit vertical communication of fire through the racks.

The design of automatic sprinklers for rack storage is very specialized **(Figure 12.45, p. 370)**. The required level of sprinkler protection is determined by the height and style of racks, the commodity being stored, and the type of containers or palletizing used for the storage. Sprinklers are often installed at the ceiling and within high-rack configurations. However, the effectiveness of in-rack sprinklers can be reduced if the sprinkler discharge is obstructed by the material being stored.

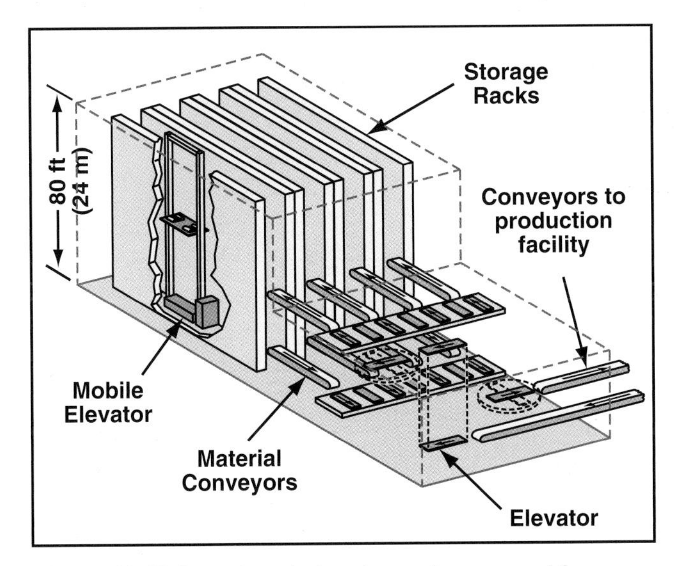

Figure 12.42 An automated rack warehouse used for the storage of automobile bodies. *Courtesy of Gala and Associates.*

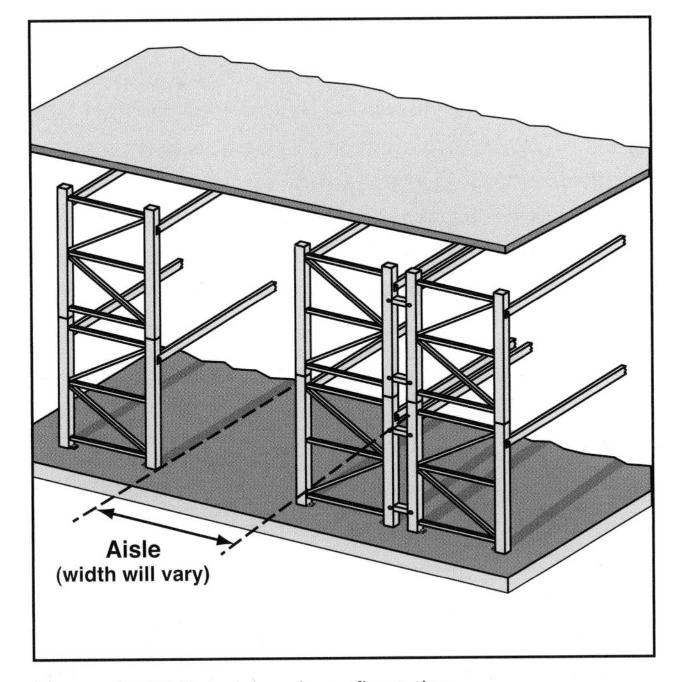

Figure 12.43 Storage rack configuration.

Figure 12.44 This warehouse is filled with books and therefore represents a high fire load.

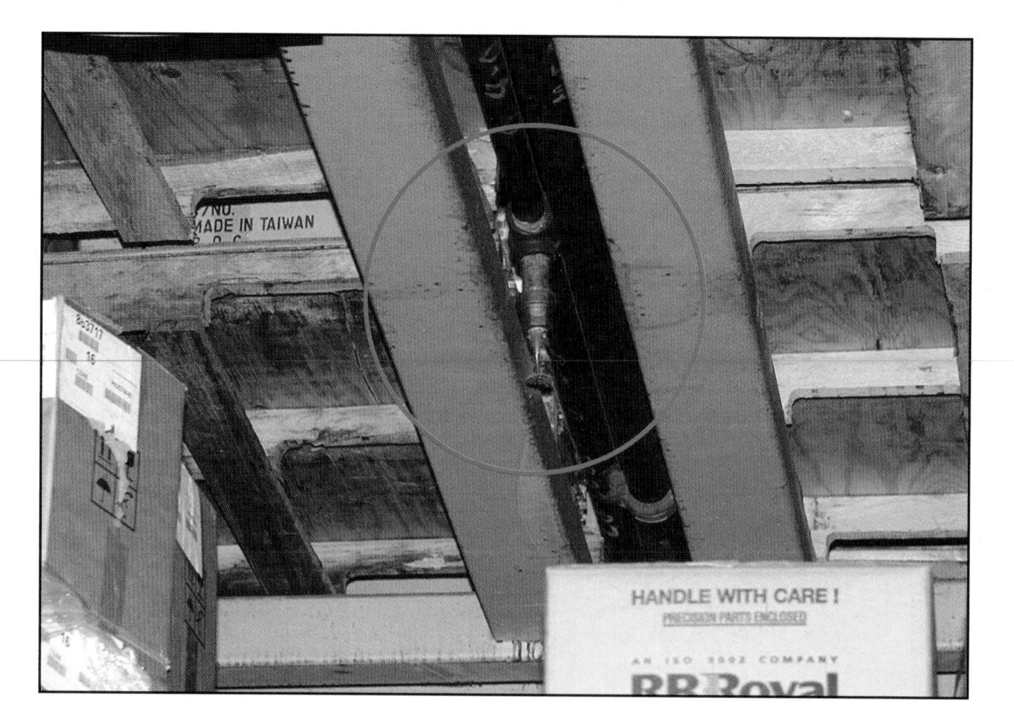

Figure 12.45 Stored materials must not be placed where they can block sprinklers.

Newer sprinkler technology has led to the development of control-mode and suppression-mode sprinklers. These sprinklers are used to control or suppress fires in rack storage, usually without in-rack sprinklers. However, the use of these sprinklers not only needs to take into account the materials stored and rack arrangement, but also any obstructions created by the building elements such as beams, trusses, ventilation and heating ducts, and slope of the roof.

As with all other information regarding building construction, fire loads, and suppression systems, pre-incident planning is of the utmost importance for fire suppression and overhaul operations.

Summary

While many buildings have similar fire suppression and rescue challenges, others have special characteristics or design features that present unique challenges. This chapter addresses several building types or design features that fall into that category. Many smaller communities may not have any of these special structures within their jurisdiction now; however, they could be constructed at any time. If currently faced with these buildings or fire-suppression challenges, it is important that firefighters recognize and prepare for the effects they will have on fire suppression efforts. It is highly recommended that pre-incident plans be prepared for all these locations.

Review Questions

1. Under which type of building construction do today's high-rise buildings fall?
2. What is a Phase I operation in the emergency use of elevators?
3. Why is it sometimes necessary to defend in place?
4. How does an atrium differ from an elevator shaft?
5. What are the limitations of air-supported structures?

References

1. *International Building Code 2006*, International Code Conference, Washington, D.C.
2. *Fire Protection Handbook*, 19th ed., National Fire Protection Association®, Quincy, MA.
3. NFPA® 68, *Guide for Venting Deflagrations*, 2002 ed., National Fire Protection Association®, Quincy, MA.
4. NFPA® 5000, *Building Construction and Safety Code*, 2003 ed., National Fire Protection Association®, Quincy, MA.
5. The American Society of Mechanical Engineers, ASME A17.1 – 20007/CSA B44-07, *Safety Code for Elevators and Escalators.*
6. NFPA® 13, *Standard for the Installation of Sprinkler Systems.*
7. NFPA® 14, *Standard for the Installation of Standpipes and Hose Systems*
8. NFPA® 25, *Standard for the Inspection, Testing, and Maintenance of Water-Based Fire Protection Systems.*
9. NFPA® 72, *National Fire Alarm and Signaling Code.*

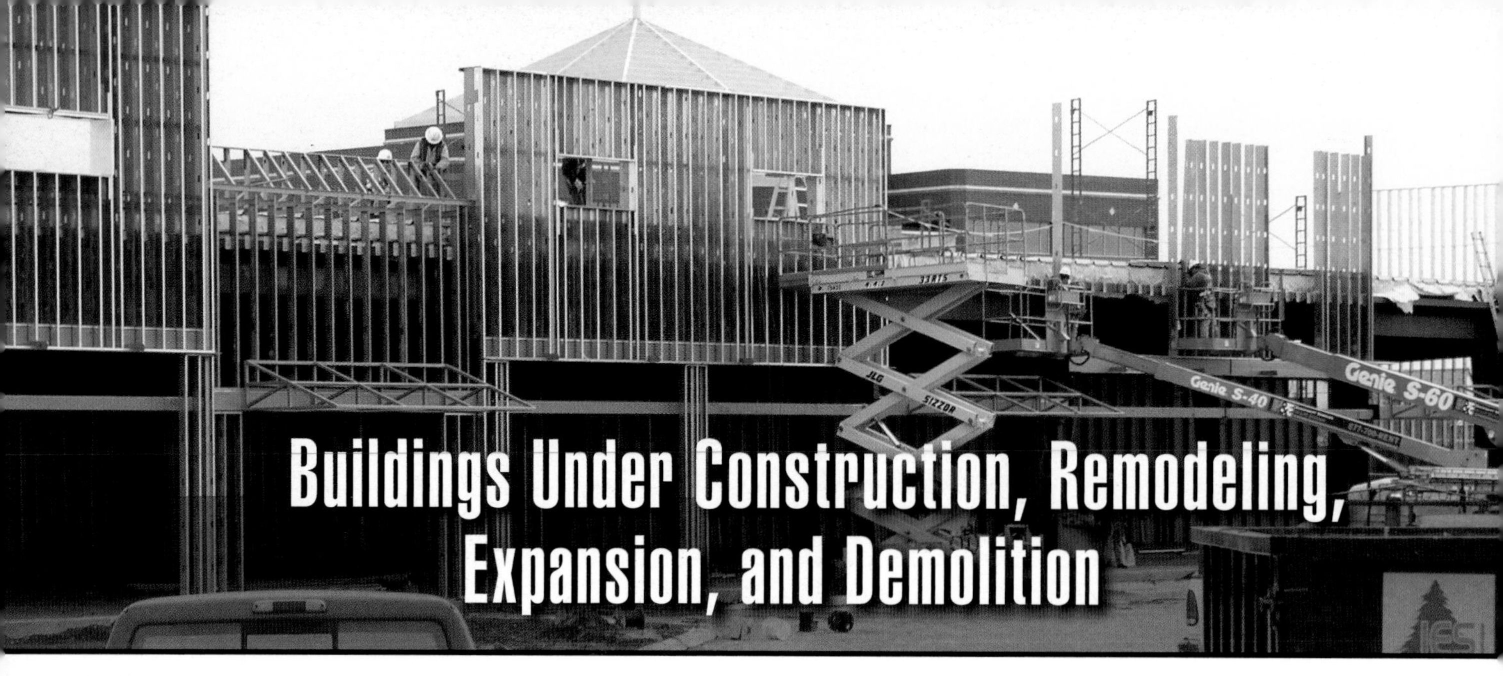

Buildings Under Construction, Remodeling, Expansion, and Demolition

Chapter Contents

Divider page photo courtesy of McKinney (TX) Fire Department.

Key Terms

FESHE Objectives

Fire and Emergency Services Higher Education (FESHE) Objectives: *Building Construction for Fire Protection*

1. Describe building construction as it relates to firefighter safety, building codes, fire prevention, code inspection, firefighting strategy, and tactics.
3. Analyze the hazards and tactical considerations associated with the various types of building construction.
8. Identify the indicators of potential structural failure as they relate to firefighter safety.

Buildings Under Construction, Remodeling, Expansion, and Demolition

Learning Objectives

After reading this chapter, students will be able to:

1. Describe the impact of conditions found at construction sites on fire fighting tactics.
2. Discuss the methods of providing fire protection at construction sites.
3. Identify and discuss the hazards associated with building remodeling and renovation as they impact fire fighting.
4. Describe the impact building expansion projects have on life safety systems in the existing building.
5. Describe the hazards presented by buildings being demolished as they relate to fire fighting tactics.

Chapter 13
Buildings Under Construction, Remodeling, Expansion, and Demolition

Case History

Event Description: Firefighters responding to a multiple-alarm fire at a construction site were positioned in the bucket of an aerial platform. Suddenly, salamander-type heating units and propane tanks stored at the site exploded under the apparatus. This action propelled tanks and flaming salamander units into the air traveling directly toward the bucket. The ruptured tanks traveled over the bucket with a margin of only 2 to 3 feet from the personnel in the bucket. Fast repositioning and ducking down in the bucket were the only ways firefighters avoided injury.

Lesson Learned: Consider construction sites as high-caution areas. Apparatus placement is critical; getting stuck in one position is dangerous. This can be prevented by evaluating contents of structures that are under construction.

Source: National Fire Fighter Near-Miss Reporting System.

The study of the fire behavior in buildings typically focuses on the building as a completed structure. Often, little attention is paid to fire protection needs during the period of construction. However, the construction process creates unique fire situations and poses special problems for firefighters. Similar situations also apply to buildings being remodeled, renovated, expanded, or demolished. This chapter discusses a number of these problems as well as what building and fire codes are doing to address these issues. This knowledge will benefit the firefighter when responding to emergencies at these locations.

Tactical Problems of Construction Sites

The complexity of a construction project obviously varies with the complexity of the building being constructed. A one-story mercantile building of several thousand square feet can be completed in a matter of a few weeks but a high-high-rise building can take as long as three years to complete.

A building can change in shape and size on a regular basis as it is remodeled. In addition, the interior building configuration changes as interior floors, walls, elevator shafts, and stairwells are completed. The changing configuration of a large construction project complicates the task of responding firefighters. Frequent site visits by first-due fire companies are necessary to keep firefighters familiar with a project.

Construction Site Access

One problem firefighters are likely to face is gaining access to the construction site. Construction sites are frequently surrounded by fences or barricades for security and public safety **(Figures 13.1 a and b)**. Construction sites also involve excavations that make access difficult and driving hazardous **(Figures 13.2 a and b)**. Gates in fences and barricades may be locked at night and on weekends to prevent unauthorized access, and there may not be a watchman present to open the gates.

Emergency vehicle access up to the actual building may be difficult. Completion of paved driveways and parking areas may be one of the last phases of a project **(Figure 13.3)**. Fire codes have requirements for providing adequate access roads, either temporary or permanent, before construction begins.

Figures 13.1 a and b Fencing around larger construction sites can hamper access during an emergency. *Photo b courtesy of McKinney (TX) Fire Department.*

Figures 13.2 a and b Excavation sites, stored equipment, debris, and unfinished portions of construction can make driving, access, and staging hazardous. *Both photos courtesy of McKinney (TX) Fire Department.*

Often a large construction project is located a considerable distance from a paved street or highway. Reaching the construction site can involve long hose lays from existing hydrants located on adjacent streets. This situation is especially true for industrial complexes, sports complexes, and covered mall buildings. The fire codes require that fire hydrants be located close to buildings with a water supply capable of meeting the required fire flow for the premises. The codes also allow the fire department to require the installation of necessary mains and hydrants before construction begins.

Large building projects are typically complex and often built in stages. It is important that the fire department meet with the owner/developer and general contractor before construction starts to coordinate the installation of fire department access and water supply **(Figures 13.4 a and b)**. If it is not possible to initially provide permanent roadways and water supply, a strategy needs to be developed among all parties to ensure that adequate temporary provisions are in place during the course of construction.

Fires frequently occur in the upper floors of high-rise construction projects. It is just as necessary for firefighters to be able to reach the upper floors of a building under construction as it is for them to do so in a completed building. However, it can be far more difficult and dangerous to go up into an uncom-

Figure 13.3 Hydrants may not be operational on a construction site. In this community, hydrants are painted silver before they are in service. *Courtesy of McKinney (TX) Fire Department.*

Figures 13.4 a and b Both of these sites will present significant access problems at this stage – particularly at night. *Both photos courtesy of McKinney (TX) Fire Department.*

Figure 13.5 Construction elevators can be used to gain access during an incident. *Courtesy of Ed Prendergast.*

pleted structure. The normal elevators used in a building are usually not in service until near the completion of a project. It may be necessary to use a construction elevator to gain access to the upper floors in a high-rise project. A construction elevator is a temporary elevator usually erected on the outside of a building for use by construction workers **(Figure 13.5)**. It is removed when the project is finished.

Construction elevators are manually operated. During normal work hours, an operator is stationed on the car, and the construction elevator can be used by firefighters to reach the upper floors. During nonworking hours an operator is not provided. It is common to find that the construction site watchman is unfamiliar with the operation of the construction elevator and will be unable to assist firefighters in its operation.

It is also common to disconnect the power to a construction elevator during nonconstruction hours. During site visits, firefighters should determine what provision exists for operating the construction elevator during nonworking hours.

It is important to note that construction projects may make use of a material hoist as well as a construction elevator. Material hoists are intended only to transport material. They do not have the same safety features as a construction elevator and should never be used to transport personnel. Likewise, the scaffolding systems used during construction are also not designed to be used by firefighters to gain access to upper floors. These systems are often extensive enough that they alone present a tactical challenge for firefighters **(Figures 13.6 a and b)**.

Figures 13.6 a and b Scaffolding systems can be extensive and present a tactical challenge of their own. They are not safe for use by firefighters. *Both photos courtesy of McKinney (TX) Fire Department.*

Newer building codes require that a minimum of one lighted stairway be provided when building construction reaches a height above four stories or 50 feet (16.6 m). This can be a temporary stair if one or more of the permanent stairways are not to be erected as the construction proceeds.

If a fire occurs at the very top of a construction project, it will be necessary to use stairs beyond the elevator to reach the fire. Because the permanent stairs may not be complete, firefighters may have to use temporary construction stairs or construction ladders. This can be a very difficult and dangerous undertaking, especially when a fire occurs several hundred feet up.

Figure 13.7 Unfinished or uncovered floors are an obvious hazard, particularly at night in smoky conditions. *Courtesy of McKinney (TX) Fire Department.*

A building under construction will have openings in floors for shafts and stairwells **(Figure 13.7)**. Contemporary safety standards require that these openings as well as the outside edges of floors have barricades to keep workers from falling. However, a danger to firefighters still exists, particularly under the conditions of limited lighting on a construction project at night. This danger is compounded under fire conditions where smoke may also compromise visibility.

Fire Hazards at Construction Sites

The construction process tends to be inherently chaotic. Fire hazards, such as temporary wiring and combustible debris, exist at construction projects that would never be tolerated in a finished building.

Electrical Wiring

Temporary electrical wiring is often installed on construction projects for lights and power equipment **(Figures 13.8 a and b)**. The temporary wiring can become a source of ignition because it is subject to

Figures 13.8 a and b Temporary wiring is a shock and access hazard on large and small sites. The wiring in the larger structure could easily present a hazard to arriving vehicles as well as firefighters. *Photo b courtesy of McKinney (TX) Fire Department.*

being moved and rearranged in the daily course of the work. Even when it is carefully installed, temporary wiring is subject to mechanical damage from work being performed in the immediate vicinity.

Heating

When a project takes more than a year for completion, it will involve working through the winter. Temporary heating is typically provided during cold periods to provide a more comfortable environment for workers. Temporary heat may also be needed for the curing of concrete (see Chapter 10) and protection of the fire sprinkler and standpipe systems if they are being installed in the building.

Temporary heat may be provided by several means including natural gas, kerosene, and propane. The heating systems used at construction sites become a cause of fires not only because of the temporary nature of their installations, but also because they are frequently left unattended **(Figure 13.9)**. Unattended propane heaters have ignited the combustible formwork used to support concrete. Fires have also occurred on construction sites when temporary heaters have been blown over by the wind.

BLEVE —Acronym for Boiling Liquid Expanding Vapor Explosion.

When propane is used as the fuel on a construction site, the tanks are susceptible to mechanical damage. One hundred-pound (45.4kg) size cylinders are commonly used at construction sites, and a number of extra cylinders are always stored on the site. Propane tanks can become exposed to the heat of a fire, and a BLEVE can occur. When kerosene is used as fuel, the fuel storage tanks are susceptible to sparks from welding and exposure to fires that may occur in the construction debris.

Welding

Welding is a hazard commonly encountered on construction sites **(Figure 13.10)**. Welding is used in the erection of steel-frame buildings and is also used on reinforced concrete buildings to weld reinforcing bars (see Chapter 10). Sparks from welding operations can fall through an incomplete structure to ignite combustible materials. The sparks can also ignite residual fuel vapors around liquid fuel storage tanks.

Fire Watch — Usually refers to someone who has the responsibility to tour a building or facility on at least an hourly basis, look for actual or potential fire emergency conditions, and send an appropriate warning if such conditions are found.

Fire prevention codes usually require that a fire watch be provided during and after welding or cutting with torches. In addition to welding, plumbers use torches in other operations such as the soldering of pipe joints. The person assigned fire watch responsibilities should be provided with an extinguisher and should have no other duties. However, it is not uncommon for a fire watch to be omitted as an economy measure.

Flammable Liquids

Much of the work on a construction site is done with heavy equipment such as excavators, cranes, lifters, and forklifts. In order to keep equipment supplied with fuel, diesel fuel and gasoline likely to be stored onsite **(Figure 13.11)**. Like all flammable liquids, diesel fuel and gasoline will contribute to a fire if they are spilled.

Combustible Debris

As construction progresses, a large amount of combustible debris is generated at construction sites **(Figure 13.12)**. At large sites, several truckloads of debris may be removed each day. On a high-rise project, a temporary chute is typically provided on the building exterior so that debris from the upper floors can be dumped into a container on the ground. If a fire occurs in the container, it can communicate up the chute into the building.

Structural Integrity

A common problem encountered at construction sites is that the structural fireproofing may not be complete **(Figures 13.13 a and b, p. 382)**. Although a building may be designed as a fire-resistive building, it does not necessarily have the structural integrity of a fire-resistive building while under construction. In steel-framed buildings, the materials used to provide the fire-resistive insulation need to be installed like any other aspect of the structural system. In concrete buildings, the ultimate fire resistance of the concrete cannot be ensured until the concrete has cured **(Figure 13.14, p. 382)**.

Figure 13.9 Temporary heating at construction sites can be a hazard, particularly if systems are left unattended. *Courtesy of Ed Prendergast.*

Figure 13.10 Sparks produced during welding operations present a fire hazard that should be controlled through a work environment that follows applicable codes.

Figure 13.11 This diesel tank is well stored, but would be a hazard if it were knocked over or punctured.

Figure 13.12 If this combustible debris should ignite, it could easily communicate fire to the adjoining structure.

Figures 13.13 a and b Until structural fireproofing is installed, these buildings have little protection against fire. *Both photos courtesy of McKinney (TX) Fire Department.*

Figure 13.14 Even concrete structures will not have full fire resistance until they are cured properly. *Courtesy of McKinney (TX) Fire Department.*

The wood formwork used in the placing of concrete poses a special danger because it contributes fuel to a fire. In addition, if the formwork is destroyed, freshly placed concrete will collapse to the floors below. Therefore, it is imperative that firefighters not be positioned under burning concrete formwork.

Fire Protection

Like other building systems, such fire protection systems as automatic sprinklers, standpipes, pumps, and hydrants, must be installed before they can provide protection. During construction, firefighters are often confronted with an unprotected structure filled with the hazards and combustible materials previously described, but with unfinished fire protection systems **(Figures 13.15 a and b)** . In some cases, the sprinkler system may be one of the last building components to be placed in service.

Temporary Fire Protection

On large, long-duration construction projects, some level of temporary fire protection must be provided. The installation of interim fire protection is especially critical on high-rise projects, but it is also appropriate for expansive low-rise projects such as regional shopping centers (see Chapter 12).

The most common temporary fire protection measure is the installation of standpipes with outlets. It also can include automatic sprinklers when sufficient progress has been made after the installation of a water supply. Automatic sprinklers sometimes are placed in service to protect completed portions of the building that are used to store construction materials, or to protect the construction offices that may be located within the building.

The most efficient method of providing fire protection on a construction project is to make use of the permanent fire protection systems as they are installed **(Figure 13.16)**. A separate temporary system can be installed, but this has the disadvantage of increasing construction costs.

It is especially important on high-rise buildings, and low-rise buildings more than three stories, that the standpipe risers be extended up as construction progresses. Building and fire codes typically require that standpipes be extended before the construction reaches 40 feet (13.3 m) above the lowest level of fire department access. The top hose outlets should be within one story of the uppermost level having a secure floor. This requires coordination on the part of the contractor installing the system. On some projects, this may require that two standpipe risers be available so that one can be maintained in service while the contractor extends the other.

Figures 13.15 a and b Planned fire protection system may not yet be operational. Departments must perform pre-incident planning to develop strategies for responding to these structures. *Both photos courtesy of McKinney (TX) Fire Department.*

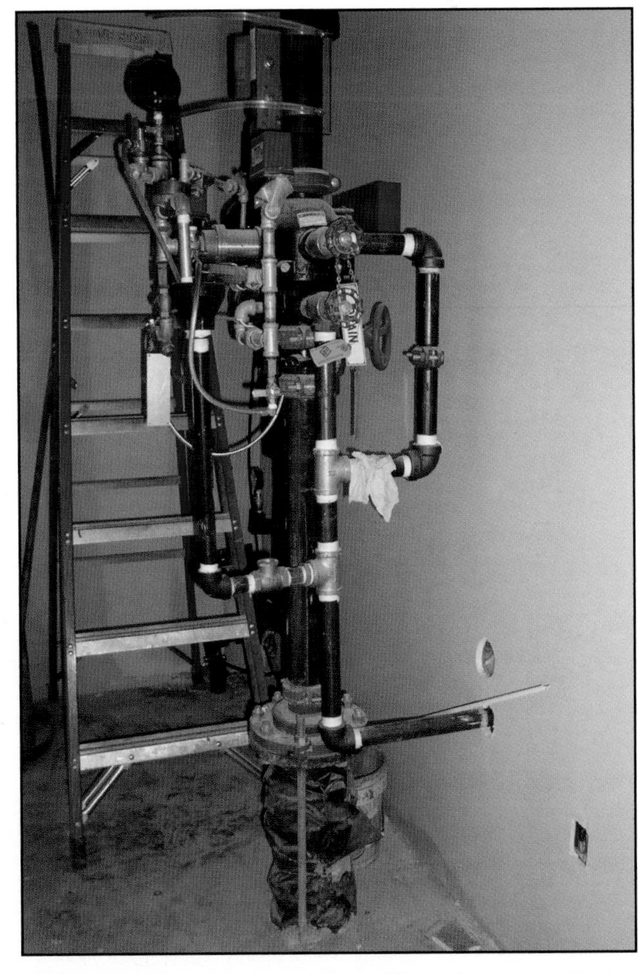

Figure 13.16 The fire department will need to ensure that all newly installed fire protection systems are operational before relying on them during an emergency. *Courtesy of McKinney (TX) Fire Department.*

When standpipes are installed in building projects, they cannot be maintained wet during freezing weather. Dry standpipes supplied through fire department connections must then be used.

There are practical difficulties with maintaining dry standpipes on construction sites. Workers, for example, have a tendency to try to use standpipes as a source of water for construction purposes. Workers may open a hose valve to get water and then leave the valve open when they get no water. If the fire department connection is charged for a fire, water will flow out any open valve. This will reduce the available water for hose lines and may limit or delay fire fighting operations.

The fire department connection can be obstructed by construction materials, trucks, or barricades. The hose connections can be damaged or even stolen. If a dry standpipe is used during cold weather, the riser must be drained after use to prevent freezing. All of this necessitates frequent inspection by fire companies or fire prevention personnel to ensure that the system will be available when needed.

Fire Extinguishers

Fire extinguishers can be very useful at construction sites when workers are trained to use them. However, fire extinguishers often are stolen from a construction site. When theft of extinguishers is a problem, barrels of water with buckets can sometimes be substituted.

Hazards of Remodeling and Renovation

It is normal for a building to be periodically remodeled or renovated. The renovations can be either for purposes of modernization or to accommodate the desires or needs of new occupants. The extent of remodeling can vary from relatively simple to major construction projects. Renovation can consist of replacing bathroom fixtures, or it can consist of extensive structural and architectural alterations **(Figures 13.17 a and b)**.

The remodeling and renovation of buildings can introduce unusual hazards. In some respects, the remodeling of a building can be more hazardous than new construction operations. This is because the remodeling often takes place

Figures 13.17 a and b Remodeling projects can have all the hazards of construction sites. Photo a *Courtesy of McKinney (TX) Fire Department; photo b courtesy of Dave Coombs.*

in one portion of a building while the remainder of the building continues to be occupied. For example, a hotel may elect to renovate one floor of a building while the other floors are still occupied to minimize the effect on revenues. Renovation of an occupied building often involves the erection of construction barricades that can obstruct exits or increase the exit travel distance from the occupied portion of the building.

Remodeling and renovation introduces many hazards such as trash, temporary wiring, and open flames. Propane torches, for example, may be used to remove the adhesive used for floor coverings. When occupancy continues on other floors, it is especially important to have a fire guard patrol the construction area after work stops for the day.

The dangers of remodeling can be aggravated by the need to shut down automatic sprinklers in the area being remodeled. If sprinklers must be shut off in the construction area, first aid hose stations and extinguishers must be provided. A renovation project should be planned in such a way that the interruption to sprinkler protection is minimized. In addition, the number of sprinklers shut off should be limited to those in the area affected.

If a portion of a sprinkler system has been shut off for remodeling, it is imperative that the restoration of the system be verified when the project is finished. Leaving a portion of a sprinkler system shut off occurs frequently in a project of long duration. Sectional control valves are sometimes located in obscure corners of old warehouse and factory buildings and can be forgotten at the end of a project. Fortunately, in newer buildings with sprinkler systems, building and fire codes require valves controlling the water supply for sprinkler systems with more than 20 sprinklers to be monitored at a constantly attended location.

Expansion of Buildings

Occasionally a building will undergo major expansion. These projects are typically located in suburban areas, or on a campus-type complex, where adjacent open space is available. Expansions of existing buildings often are major construction projects in themselves **(Figure 13.18)**. Therefore, these projects present both the concerns faced in new construction along with some of the concerns faced in major remodeling projects. These include construction site access, site fire hazards, and structural integrity.

Figure 13.18 The expansion of this university building was larger than the original building and occupied nearly a city block.

Some examples of building expansions include regional shopping centers, industrial buildings, warehouses, convention centers, airports, and research facilities. A common theme in most of these projects is the need to keep the existing portion of the building in operation. Therefore, it is critical to maintain an adequate level of life safety to occupants in the existing portion while providing the necessary protection to the construction site.

Often the new construction will impact one or more of the exits for the existing building. Temporary measures must be taken to provide for the continued protection of the occupants in the existing building. These measures may include providing new exits or by providing temporary fire-rated "tunnels" through

Figure 13.19 Redirected traffic often allows less road space to accommodate emergency vehicles.

the new construction zone until occupants can reach a safe location. Arrangements for these provisions need to be developed prior to the start of construction. The fire inspector assigned to the project must meet with the owner and general contractor to assure that an acceptable approach is in place. First-due fire companies should be made aware of these arrangements and visit the site to become familiar with the project.

Another concern is the need to maintain all life safety systems provided in the existing building. Most of these expansions will involve buildings that are fully protected by automatic sprinklers. Those buildings with large occupant loads will have a fire alarm evacuation system and possibly a smoke control system installed. The new construction will often impact the existing water supply system on the premises including mains, hydrants, and fire department connections. As with the exiting system, temporary provisions will likely be needed to keep existing fire protection systems operational. Again, the local fire inspector must meet with the owner and contractor to ensure that measures are in place before construction begins. Also, first-due fire companies should be made aware of any temporary arrangements and visit the site to become familiar with the project **(Figure 13.19)**.

Measures that may be required to maintain the water supply include temporary mains, hydrants, and fire department connections. It may be necessary to provide temporary risers and temporary bulk sprinkler piping until the new construction is complete.

Demolition of Buildings

The demolition of a building is a process that is even more chaotic than its construction, and fires on demolition sites are very common. Major fires can occur that involve buildings of wood-joisted or wood-frame construction. When a building is being wrecked, there is little inclination to spend money to protect it. It may be argued that it is irrelevant to protect such a building from fire. However, a fire involving a building being demolished can expose neighboring structures and creates an environmental threat; therefore, the community must be protected **(Figure 13.20)**.

There are many things that firefighters should remember when fires occur in buildings being demolished. In the demolition process, it is normal for materials that are salvageable or that have scrap value to be stripped from buildings. This can include portions of the fire protection system such as standpipe hose valves and fire department connections. Removals can also include architectural artifacts such as ornate elevator shaft gates, bath tubs, stair railings, and even marble slabs used for stairs. The wrecking contractor may carry out the stripping of buildings, or the salvageable materials simply may be stolen by vandals.

Removal of scrap steel often involves cutting with torches. This procedure is a very common cause of fires. Unauthorized removal of materials, of course, frequently takes place at night and obviously with no regard for safety.

As a building is demolished, its structural integrity is gradually lost. With almost each swing of a wrecking ball or the removal of each beam, the integrity of the building is diminished. If a fire occurs, structural collapse is accelerated.

While major fires in buildings being demolished usually involve combustible structures or structures with combustible framing, fire-resistive or noncombustible buildings can have some unpleasant surprises. For example, a building may contain a fuel oil tank with residual oil. Or, there may be an accumulation of combustible debris at the bottom of an elevator shaft.

Figure 13.20 Buildings in the process of demolition generate large quantities of combustible materials. Nearby structures can be at risk if the operation doesn't go as planned.

To a firefighter, a building being demolished can present a virtual maze consisting of floor openings, an unstable structural system, inoperative fire protection systems, and unknown hazardous materials. Obviously, such a structural mess is not worth a firefighter's life. Exterior fire fighting tactics from a safe distance are the best course of action.

Situations exist where firefighters must enter buildings being wrecked. These sometimes arise when it is necessary to rescue demolition workers who become injured or even trapped in the course of demolition. Ironically, scavengers pilfering materials frequently become victims of their own activities and need to be rescued from a building.

Summary

All completed and occupied buildings create a need for the firefighter to know what to expect when arriving at the building during an emergency response. Many target hazards have pre-incident plans and others have been the subject of company inspections. The same respect must be given to buildings that are under construction, being remodeled or expanded, or are being demolished. These buildings are often full of surprises. This chapter has discussed many of the concerns that the firefighter must be aware of when responding to these locations.

Review Questions

1. Why do firefighters encounter problems regarding construction site access?
2. What fire hazards exist at construction sites?
3. What is the most efficient method of providing protection on a construction project?
4. What factors can increase the dangers of remodeling?
5. Why is it important to protect a building under construction from fire?

References

1. *International Building Code® 2006*, International Code Council®, Washington, D.C.
2. *International Fire Code® 2006*, International Code Council®, Washington, D.C.
3. NFPA® 5000, *Building Construction and Safety Code*, 2003 ed., National Fire Protection Association, Quincy, MA.

Chapter Contents

Divider page photo courtesy of FEMA News Photo.

Key Terms

FESHE Objectives

Fire and Emergency Services Higher Education (FESHE) Objectives: *Building Construction for Fire Protection*

1. Demonstrate an understanding of building construction as it relates to firefighter safety, buildings codes, fire prevention, code inspection and firefighting strategy and tactics.
8. Identify the indicators of potential structural failure as they relate to firefighter safety

Non-Fire Building Collapse

Learning Objectives

After reading this chapter, students will be able to:

1. Describe the forces of nature that can result in partial or total building collapse.
2. Discuss the building code requirements to minimize the effect of the forces of nature on building stability.
3. Describe the scenarios that result in human-caused building collapse.

Chapter 14
Non-Fire Building Collapse

Case History

Event Description: Our community received major flood damage as a result of Hurricane Dennis in 2005. Our specialty rescue squad answered several water-related calls in the surrounding area in agreement with our auto-aid dispatch. On one incident, a dead-end road development was flooded under the cover of morning darkness. About six homes were eventually consumed from a nearby river's overflow. Nine residents had to be rescued from the homes. Water 5 feet deep (1.6 m) rapidly flowed down the street between these houses. Our Swift Water Technicians assisted the local jurisdiction's rescue teams that were already in operation.

The initial responding Engine Company made a decision to attempt water entry to evacuate nearby houses. This crew's level of training is unknown, but it was apparent they were not sufficiently trained in swiftwater rescue. The crew had entered the water wearing a Type V Rescue PFDs over full turnout gear. One firefighter became trapped inside a house by rapidly rising water. Self-rescue would have been very difficult for this individual due to weight and bulk of his turnout gear. This firefighter was eventually rescued by boat after the residents were taken to higher ground.

Lesson Learned: Every swiftwater training class emphasizes the dangers of wearing turnout gear near the water's edge. If equipment is placed on the apparatus with untrained personnel, safety may become secondary to the feeling of helplessness. To prevent this, proper training is recommended. If it is unavailable, awareness of the hazards must be a training priority.

Source: National Fire Fighter Near-Miss Reporting System.

Responding firefighters, from recruits to senior officers, must be alert to the potential for building collapse during a fire. Firefighters are also called upon to respond to partial or full collapse of buildings that are *not* involved in a fire. The causes of nonfire building collapse can be put into two broad categories: nature-caused and human-caused. For many of these events, such as earthquakes and explosions, there is no forewarning. For other events, the potential for building collapse can be anticipated, but the specific location will not be known until the event reaches exposed structures. Examples include tornadoes and hurricanes.

Whenever possible, specific precautions will be identified for each type of event. When confronted by a partially collapsed building, firefighters must assume that the remaining building could be seriously weakened. Use caution when performing search and rescue operations when facing this condition.

This chapter discusses many causes of non-fire building collapse, including earthquakes, mudslides, floods, snow, poor construction methods, and explosion. Applicable building code requirements that minimize the impact of the forces of nature will be presented as appropriate.

NOTE: Refer to IFSTA's **Rescue** and **Technical Rescue for Structural Collapse** manuals for more information on structural collapse.

Nature-Caused Collapse

There are several forces of nature that can result in partial or total building collapse. Most fire departments will be faced with at least one of these perils. The following are some concerns common to most of these events:

- Loss of local or area-wide water supplies.*
- Rupture of local or area-wide gas distribution mains.*
- Loss of local or area-wide electrical distribution system.*
- Loss of local or area-wide landline communications.
- Impaired access roads **(Figure 14.1)**.

  NOTE: Utilities and roads may be impaired for days or even weeks.
- Multiple building collapses.
- Multiple incidents involving rescue of trapped occupants.
- Fires following event due to accumulation/ignition of vapors from ruptured gas mains/lines.
- Impairment of building life safety systems.
- Fires occurring days/weeks later when utilities are restored, especially electrical services.
- Arson fires in subsequent days when owners realize that property damage was not covered by the insurance policies in effect.

> * It is imperative that the fire department work closely with the appropriate agencies when faced with the loss of these utilities. For safety reasons, there is a need to ensure that the electric power is off and that gas mains are shut down. It may be necessary to isolate some areas of the water systems so local main breaks do not deplete water supplies in other, less-impacted areas.

Earthquakes

Many areas are susceptible to earthquakes. In simple terms, an earthquake results from the energy released by a sudden shift in the earth's crust that produces seismic waves. (See Chapter 3 for additional discussion of earthquakes.) In the United States and Canada, the West Coast, including Alaska and Hawaii are particularly vulnerable; however, earthquakes do occur in other parts of both countries. For example, significant earthquakes occurred elsewhere in the U.S. in the 1800's: one in the Mississippi River Valley centered near New Madrid, Missouri and one in Charleston, South Carolina. There have been two high-magnitude earthquakes in Eastern Canada; one in Charlevoix, Quebec (1663) and another in the Grand Banks, South of Newfoundland (1939). Two of the more recent significant U.S. earthquakes were the Loma Prieta in the San Francisco Bay Area, CA in 1989 and the Northridge in the San Fernando Valley of Los Angeles, CA in 1994 **(Figure 14.2)**.

Ground motion associated with major earthquakes can cause extensive structural damage to buildings that are near earthquake fault lines, often resulting in partial or total building collapse. Unreinforced masonry construction is highly susceptible to seismic events. Buildings constructed prior to the 1930's are generally more at risk than newer buildings. This is because of the gradual introduction of mandatory seismic design requirements in model building codes during the 1930's to the 1950's. These requirements continue to be strengthened as geologists and engineers study the performance of buildings during major seismic events around the world.

Figure 14.1 Natural-caused events often result in inaccessible roads. Here DOT crews in Kapaau, HI (2006) work to clear debris from the Lolou Lookout Road after a series of earthquakes resulted in a landslide. *FEMA/Adam DuBrowa.*

Figure 14.2 Many apartments in this hillside complex were destroyed or heavily damaged by the 6.7 magnitude Northridge earthquake (1994). *FEMA/Robert A. Eplett.*

Numerous factors contribute to the extent of building damage and potential collapse during earthquakes. One important factor is the location of the building relative to the responsible fault and the epicenter of the earthquake. The soil conditions beneath the building, especially the potential for liquefaction of loose, sandy soil, are another critical factor.

Model building codes now have stringent requirements for seismic design. These requirements vary depending on the expected ground motion in the area where the building is to be located. (Maps or tables showing these values are usually in the applicable code itself.) Seismic design considerations include lateral bracing and other features to provide sufficient resistance to seismic motion in order to minimize damage and avoid collapse. However, the requirements for a critical building are usually higher so that the building will be able to continue to function. These buildings include hospitals, command centers, and other important public uses.

Currently, there is no proven method to predict an earthquake. Fire departments in areas subject to earthquakes should have a general plan in place for responding to incidents after the event. The fire department should coordinate with the local building department to identify in advance buildings that may be more likely to suffer major damage and possible collapse during a major seismic event.

Firefighters should always be aware that major earthquakes will likely be followed by aftershocks, some of which may be nearly as strong as the original event. Therefore, partially collapsed buildings are vulnerable to further failure. Even with severely collapsed buildings, aftershocks may cause rubble to shift and threaten firefighters engaged in search and rescue operations. Even if a building has not collapsed, a minor aftershock may be strong enough to topple a weakened cornice or other damaged building element **(Figure 14.3)**.

Figure 14.3 There is substantial earthquake damage in Pioneer Square, part of Seattle's historic district, after an earthquake in 2001. *FEMA/Kevin Galvin.*

Other challenges firefighters may encounter after earthquakes include fires, loss of water supply, rupture of gas mains, loss of electric power, and impaired roads. Building collapse, either total or partial, will likely involve multiple buildings over a widespread area.

Landslides, Subsidence, and Sinkholes

The potential for landslides occurs nearly everywhere. Landslides are defined as the movement of rock, earth, or debris down a slope **(Figure 14.4)**. Urban expansion has increased the potential for this geological event because there are more cuts into hillsides to accommodate building development. Although landslides are more likely to occur during heavy rains or in conjunction with such events as earthquakes, they may occur anytime. A recent example is the landslide in the La Jolla section of San Diego, CA in October, 2007. Prior to this event, there had been little or no rain for months. The landslide resulted in red-tagging (indicating unsafe to enter) nine homes with lesser damage to numerous other homes. Landslides also occurred in La Conchita, California in 2005 after winter storms **(Figure 14.5)**.

In some instances there may be forewarning of an eventual major landslide. Examples of early warning signs include damage to underground utilities, cracking in roads, and minor slides in the area. On hillsides that are prone to landslides, the addition of groundwater through landscape watering or a broken water main can be the trigger to a major landslide.

Figure 14.4 Lake Delton, WI, June 23, 2008 - Houses fall into the emptied Lake Delton as the side banks collapse. Recent rains have created flooding and a dam break which caused the lake to drain. *FEMA/Robert Kaufmann.*

Figure 14.5 Damaged buildings are surrounded by mud in La Conchita, California, where winter storms caused fatal landslides that damaged private property and roads. *FEMA/John Shea.*

There are no specific model building code requirements to address landslides. Environmental impact reports should identify this issue and zoning regulations typically address it. Developers are often required to provide adequate drainage and other measures to ensure ground stability. However, soil reports may not identify all existing conditions. The fire service needs to develop a close working relationship with public works agencies to ensure notification when problems are noted in the field or where high-risk areas have been identified by historical evidence. It is interesting to note that there was a previous landslide in 1961 in the vicinity of the landslide in La Jolla, CA described earlier.

As with earthquakes, landslides may cause failure of gas and waterlines, along with electrical outages. Firefighters must also be aware of the potential for continued landslide activity after arriving at the scene. For example, caution must be exercised in parking apparatus to avoid unsafe areas.

Subsidence — Sinking or settling of land due to various natural and human-caused factors such as removal of underground water or oil.

Land subsidence is similar to landslides in that the ground gives way, but usually in the form of gradual sinking. If this occurs under or near buildings, damage or eventual collapse of the building may occur. The primary cause of this phenomenon in the U.S. is the removal of large amounts of underground water. This has occurred in almost every state. Sometimes the cause is the pumping of oil over many years beneath built-up areas. This has occurred in California, Louisiana, and Texas. Subsidence is usually gradual and more-or-less uniform when the cause is the removal of an underground liquid. There are instances in the San Joaquin Valley of California where the ground over time has subsided 30 feet (10 m) or more.

Sinkhole — A natural depression in a land surface formed by the collapse of a cavern roof. Generally occurs in limestone regions.

Another form of ground collapse is the sinkhole. Like landslides, sinkholes are usually sudden events **(Figure 14.6)**. This type of failure represents an extreme form of subsidence. While sinkholes more often occur in the Midwest and Eastern/Southeastern states, especially Florida, they can occur in all states and throughout most of Canada. They are caused by the collapse

Figure 14.6 San Diego, CA (1998) — Extensive reengineering to replace a storm drain in a giant sinkhole after El Nino rains collapsed the original pipe to an estimated 850-foot long hole. *FEMA/Dave Gatley.*

of the ground surface into a belowground cavity. These cavities are usually created by the interaction of ground water with rock formations that are water soluble. However, they can occur over areas that have been subjected to underground mining, or more commonly in urban areas as the result of water main breaks.

Determining the potential for subsidence and sinkholes is often done by examining historical records. The fire service should be aware of these areas and establish SOPs for responding to incidents of this type. The same precautions that apply to landslides apply to these events.

Windstorms, Tornadoes, and Hurricanes

High winds, especially these associated with tornados and hurricanes, often result in catastrophic regional events. Damage to buildings is usually extensive; in the case of tornadoes many structures may be completely destroyed **(Figure 14.7)**. With hurricanes, there is the added factor of water surge along immediate adjacent coastal areas that causes additional building damage and destruction **(Figure 14.8, p. 400)**.

Unlike earthquakes and sinkholes, which usually occur without warning, modern weather forecasting usually enables early warnings about the potential for high winds, tornadoes, and hurricanes. However, the paths of these storms are often unpredictable. One resource available to fire departments to assist them in monitoring weather conditions is by contacting the National Oceanographic and Atmospheric Administration (NOAA) National Weather Services website, www.nws.noaa.gov.

Windstorms can occur throughout all 50 states and Canada. Although tornadoes are more likely in the Midwest, South-Central, and Southeastern states and in the prairie/plains areas of Southern Canada, they have been known to occur in most states and provinces. Hurricanes generally occur only along the Gulf States, Eastern Seaboard, and Hawaii. However, the remnants of hurricanes have moved into other states as tropical storms and heavy rainfall.

Figure 14.7 Oklahoma, May 4, 1999 — Urban Search and Rescue teams worked to recover missing persons after the devastating tornado killed 38 people and destroyed more than 1,500 houses. *FEMA/Andrea Booher.*

Model building codes include wind loads as part of the structural design requirements. The codes specify minimum design wind speeds that are to be considered in the design analysis (see Chapter 3). In wind-borne debris regions, such as coastal areas subject to hurricanes where the basic wind speed is 110 mph (48 m/s) or greater, window glazing is required to be impact-resistant. Residents may also employ special brackets, also known as wind clips, to provide additional reinforcement as storms approach **(Figure 14.9)**. For example, southern Florida locally adopted more stringent requirements after Hurricane Andrew in 1992.

NOTE: A downburst featuring 70 mph (112 km/h) winds partially collapsed the roof of a Dallas Cowboys training facility in May 2009.

Many existing buildings were constructed before more stringent requirements appeared in model building codes. Therefore, heavy structural damage can be expected as a result of major windstorms, tornadoes, and hurricanes **(Figure 14.10)**. General concerns facing firefighters include collapse of mul-

Figure 14.8 Hancock County, Miss., September 23, 2005 — The storm surge of Hurricane Rita severely affected the gulf coast of Mississippi. *FEMA/ Mark Wolfe.*

Figure 14.9 These brackets, or wind clips, help to protect this coastal Gulf Breeze (FL) house from the wind and storm surge of Hurricane Ivan in 2004. *FEMA Photo/Mark Wolfe.*

Figure 14.10 The Windemere Condominiums in Orange Beach, Ala., show the vast fury of Hurricane Ivan's 130 mph winds and 30-foot swells in 2004. *FEMA/Butch Kinerney.*

tiple buildings, fires, loss of water supplies, and loss of electrical power. These types of catastrophic events require a highly coordinated multiagency regional plan for coordinated response.

Snow and Water

Building collapse can occur due to the force associated with accumulated snow, water, or a combination of both. This type of collapse is usually associated with major weather events but may occur after a series of storms **(Figure 14.11)**. These types of collapses often happen without warning and are usually isolated events. A classic example was the 1979 collapse of the roof of the

Figure 14.11 Approximately 3,000 square feet of this building collapsed due to snow/ice load. *Courtesy of West Allis (WI) Fire Department.*

Kemper Arena in Kansas City, MO, which caused some of the walls to fail. This was caused by a storm with 70 mph (112 km/h) winds and heavy rains, which overwhelmed the roof drainage system. Similarly, in January 1978, the roof of the Hartford Civic Center in Hartford, CT collapsed due to the added weight resulting from a heavy snowstorm.

As with the potential for earthquakes and windstorms, model building codes require that snow and rain loads be addressed as part of the structural design of the building. The codes specify the snow loads expected throughout the U.S. and Canada (see Chapter 3). These values range from zero in certain portions of the Sun Belt states to 300 psf in Whittier, AK.

Another cause of building collapse is from floods. Floods can be simply described as water flowing where it is not normally expected. This threat exists in all areas and takes on many forms. Examples include overflow of a river bed or body of water resulting from too much rainfall, either over an extended period or from a brief intense downpour **(Figure 14.12)**. Other examples include breaches of dams or levees, or water surge along coastal areas in conjunction with high winds, especially those associated with hurricanes **(Figure 14.13)**. Coastal areas along the West Coast of the U.S. and Canada, Alaska, and Hawaii are subjected to the possibilities of tsunamis caused by earthquakes.

Damage resulting from floods occurs in several ways. Water can undermine foundations causing the building to partially or totally collapse. Or, in the case of storm surges in coastal areas, the house can be swept off its foundation and deposited in whole or part hundreds of feet from where it originally rested.

Flood loads are another required aspect of structural design considerations for a new building. U.S. model building codes base their requirements for flood loads on maps of flood hazard areas prepared by the Federal Emergency Management Agency (FEMA). The requirements vary depending on whether

Figure 14.12 Downtown East Grand Forks, MN after the Red River flooded the town in 1997. *Photo courtesy of FEMA/David Saville.*

Figure 14.13 The remains of a building in New Orleans after Hurricane Katrina (2005). *Photo courtesy of District Chief Chris E. Mickal, NOFD Photo Unit.*

or not the building is also subject to high-velocity wave action. High-velocity wave action would apply to coastal areas prone to hurricanes. Provisions for buildings in these locations require simultaneous consideration of wind and flood loads on all components. Foundations and structures are required to resist flotation, collapse, and lateral movement.

Human-Caused Collapse

There are numerous scenarios for human-caused building collapse. Several of these are described in the following sections.

Inadequate Structural Design

Building codes have established stringent structural design requirements. The structural design, including calculations, is typically reviewed during the plan review process before construction begins. The codes also require inspections during construction to ensure that the proper materials and methods of construction are being used. Despite all these safeguards, occasionally an inadequate design will slip through, or there will be a field change that negatively impacts the integrity of the design. When this happens, a partial or total collapse may occur.

NOTE: For more information on plans review, consult the IFSTA **Plans Examiner for Fire and Emergency Services** manual.

Change in Building Use

As discussed in Chapter 3, the structural design of a building includes consideration of the maximum live loads associated with the expected occupancy of the building. If the use changes over the years, it is possible that the new use could result in higher live loads than the original design can accommodate, eventually resulting in structural failure. An example is when the upper levels

Figure 14.14 Changes in occupancy occur frequently, some of which can significantly affect structural or fire loads. *Courtesy of Ed Prendergast.*

of retail or office buildings are used for concentrated storage **(Figure 14.14)**. Although the concern of increased loads may be identified during review of an application for a change in occupancy, all too often these changes occur without the knowledge of the building or fire department.

Poor or Careless Construction Methods

Until a building's structural frame is completed, the building will not have the level of structural stability that it will have when the frame is finished. Therefore, there is always a potential for collapse due to poor or hurried construction techniques. There are several causes of structural collapse during the course of construction.

Temporary Loads. Collapse can result when temporary loads on the structural members exceed final design loads due to careless stockpiling of heavy building materials on upper floors. An example of this occurred in Los Angeles in December 1985 when a 21-story building under construction collapsed, resulting in the deaths of three workers. Structural steel was stockpiled on one bay on an upper floor. The load of the steel was twice the design load, causing three beams to fail.

Sequencing. Another cause of collapse is improper sequencing of the construction process. By some accounts, this was a contributing factor in 1987 in Bridgeport, CT when a housing project collapsed during construction, killing 28 workers. At the time of the failure, construction of the shear walls was several floors below the lift-slab erection operation.

Weakness of building frame. A third cause of failure during construction results from temporary weakness of the building frame. This danger especially applies to poured-in-place concrete structures. (See Chapter 10.) This comes about due to the length of time required for concrete to cure and develop its ultimate design strength. If the construction progresses too fast, the supporting elements may not have sufficient strength to support the new load.

> An example of building frame weakness occurred in March 1973 during the construction of a 26-story condominium building in Bailey's Crossroads, VA. The building collapsed, resulting in the death of 14 workers and the injury of 35 others. The collapse was attributed to the removal of shoring before the concrete had cured sufficiently to support the loads above.

Instability of building frame. A fourth cause is the instability of the building frame during construction. Until the final members of the structural frame are in place, the frame is vulnerable to vertical (gravity) and horizontal loads. The use of temporary bracing is common to all types of building construction **(Figure 14.15)**. The structural engineer or contractor is responsible for identifying and providing temporary bracing. If the bracing is not adequate, failure

Figure 14.15 During construction, it is common for major structural elements to be braced. If bracing is not done carefully or is removed too soon, collapse can result. *Photo courtesy of McKinney (TX) Fire Department.*

can occur. One example of inadequate temporary bracing was the collapse in 1987 of a grandstand addition to the Husky Stadium at the University of Washington, Seattle, WA. Several of the guy lines being used to provide support during the construction were removed before all of the structural frame members were in place. Fortunately, there was early warning of the failure and no injuries or deaths occurred.

Event Description: Firefighters were to called to a structure fire in a single-story residential building. The air was heavy with fog and moisture and smoke was banked down around the outside of the house. Two firefighters pulled a 1¾-inch line and headed to the B/C side of the structure where fire and smoke were showing. They decided to make a rapid exterior attack while waiting for backup crews. The lieutenant had completed a 360-degree sizeup of the house and noted utilities, etc. The lower electrical drop to the masthead on the C/D corner of the roof was intact.

What the firefighters on the hoseline didn't know was that there was an illegal second drop from a power pole to the structure on the B/C side that had burnt through and was on the ground. The backyard was overgrown with grass and weeds and the two firefighters either stepped on or dragged the hoseline over the downed power line. They were engulfed by a bright white light. They dropped the hose and managed to walk out on the C/D side. They never saw the downed line because it was hidden and they assumed there was only one electrical drop.

Lessons Learned: Even when firefighters try to make safe decisions, the unexpected can happen. The weather, smoke and environmental conditions played huge roles in this incident, but so did firefighter complacency with regard to potential hazards. They had poor lighting and walked into a bad situation; however, they did keep their heads and walk away in a different direction from where they had come.

Source: National Fire Fighter Near-Miss Reporting System.

Poor or Careless Demolition Methods

The common means of performing demolition include both piecemeal and controlled collapse. Piecemeal demolition is performed by using hand tools or machines. Controlled collapse is performed by using a crane with a demolition ball, hydraulic pusher arms, wire rope pulling, or explosives. Both piecemeal and controlled collapse require a well-thought-out plan. Such planning is especially critical when the building being demolished is close to other structures. A primary goal of the demolition process must be to protect nearby structures.

When primary structural members are being removed in a piecemeal demolition, there is always the potential for unexpected collapse. Often this approach will require temporary bracing or props to ensure that the remaining structure will be able to support the load imposed by workers, equipment, and removed debris (building materials) temporarily stored in the building.

With piecemeal demolition, an unexpected collapse will likely involve rescue because it is likely to occur when the building is occupied by workers involved in the actual demolition. On the other hand, a well-planned and executed controlled collapse should result in a pile of debris at the ground to be subsequently removed. Rescue normally should not be an issue. However, under either piecemeal or controlled collapse, things can go wrong, and if they do there is always the potential for impact on nearby buildings.

Figure 14.16 A view of the destroyed Humberto Vidal building in San Juan, Puerto Rico following a gas explosion (1996). Here the Metro-Dade (FLTF-1) Task Force begins to perform search and rescue operations inside the building. Photo courtesy of *FEMA/Roman Bas.*

A rising concern with vacant buildings, whether scheduled for demolition or not, are the problems of vandalism and scavenging. The latter activity is being accelerated by the increased price of wood and metals. Amateur scavenging may result in weakening of the structure or even an unwanted building collapse. Scavenging activities during building demolition may also lead to an unexpected collapse.

Explosions

Explosions – whether accidental or deliberate -- frequently result in building collapse. Often, accidental explosions are the result of natural gas leaks within a building. The ignition of these leaks can produce a powerful force on the structural frame. Except in the rare instance that explosion venting has been provided (see Chapter 13), these forces can cause major structural failure **(Figure 14.16)**.

Other common sources of explosions in buildings include boiler furnaces, gasoline vapors, finely powdered dust, storage of fireworks or blasting agents, and in today's society, the methamphetamine laboratories ("meth labs") found in single or multifamily residential properties. Deliberate explosions are a rising threat due to the increase of terrorist activities throughout the world. This is in addition to the deliberate use of explosives associated with gang activities and other organized crime acts. Explosions may or may not be followed by fire.

Other Causes

Other unexpected events occasionally cause a building to collapse. One example is the collision of a motor vehicle with a structure, which typically results in a partial collapse. Less common, although more dramatic, is an aircraft crash into a structure. When this occurs, it is almost always followed by a fire due to the fuel carried in the airplane fuel tanks.

Some older buildings have collapsed due to age and deterioration **(Figure 14.17)**. An example of deterioration was the partial collapse of Pier 34 and the nightclub it supported in Philadelphia, PA on May 18, 2000. The collapse resulted in the death of three patrons of the nightclub and injury to 36 patrons and employees. The pier, built in 1909, reportedly failed due to the deterioration of the old piles supporting the deck of the pier.

Figure 14.17 More than 20 tarps being used to remove water from roof leaks were discovered in this warehouse. The owner and insurance company had been attempting to repair leaks for several weeks without notifying the fire department or utilities. Energized power lines were in the building. *Courtesy of West Allis (WI) Fire Department.*

Summary

There are numerous causes of building collapse other than as a result of fire. Although communities are not likely scenes for some nature-caused emergencies, firefighters and communities need to be prepared for events that may occur. Modern building codes have reduced the vulnerability of buildings to nature-caused collapse, but most jurisdictions have structures that were built before the requirements came about, or before they were strengthened.

Even where modern codes are in effect, mistakes can be made in the design plan review process or on the job site that result in a partial or total collapse. Always remember when responding to partial building collapse that the remaining structure is likely to be weakened. Therefore, the firefighter must exercise extreme care when performing search and rescue operations in these situations.

Review Questions

1. Why are pre-1930's buildings more susceptible to collapse due to an earthquake than those built since the 1930's?
2. What is a landslide?
3. How does snow cause building collapse?
4. How can a change in a building's use cause it to collapse?
5. What should be a primary goal of the demolition process?

References

1. Carper, Ken, *Beware of Vulnerabilities during Construction*, Daily Journal of Commerce, March 25, 2004.
2. International Building Code 2006, International Code Council, Washington, D.C., Chapter 16.
3. National Fire Protection Association® 5000, Building Construction and Safety Code, 2003 ed. National Fire Protection Association®, Quincy, MA, Chapter 35.
4. *National Building Code of Canada* 2005, Canadian Commission on Building and Fire Codes/National Research Council of Canada, Ottawa, ON.

Appendix

Contents

Appendix A

Divider page photo courtesy of McKinney (TX) Fire Department.

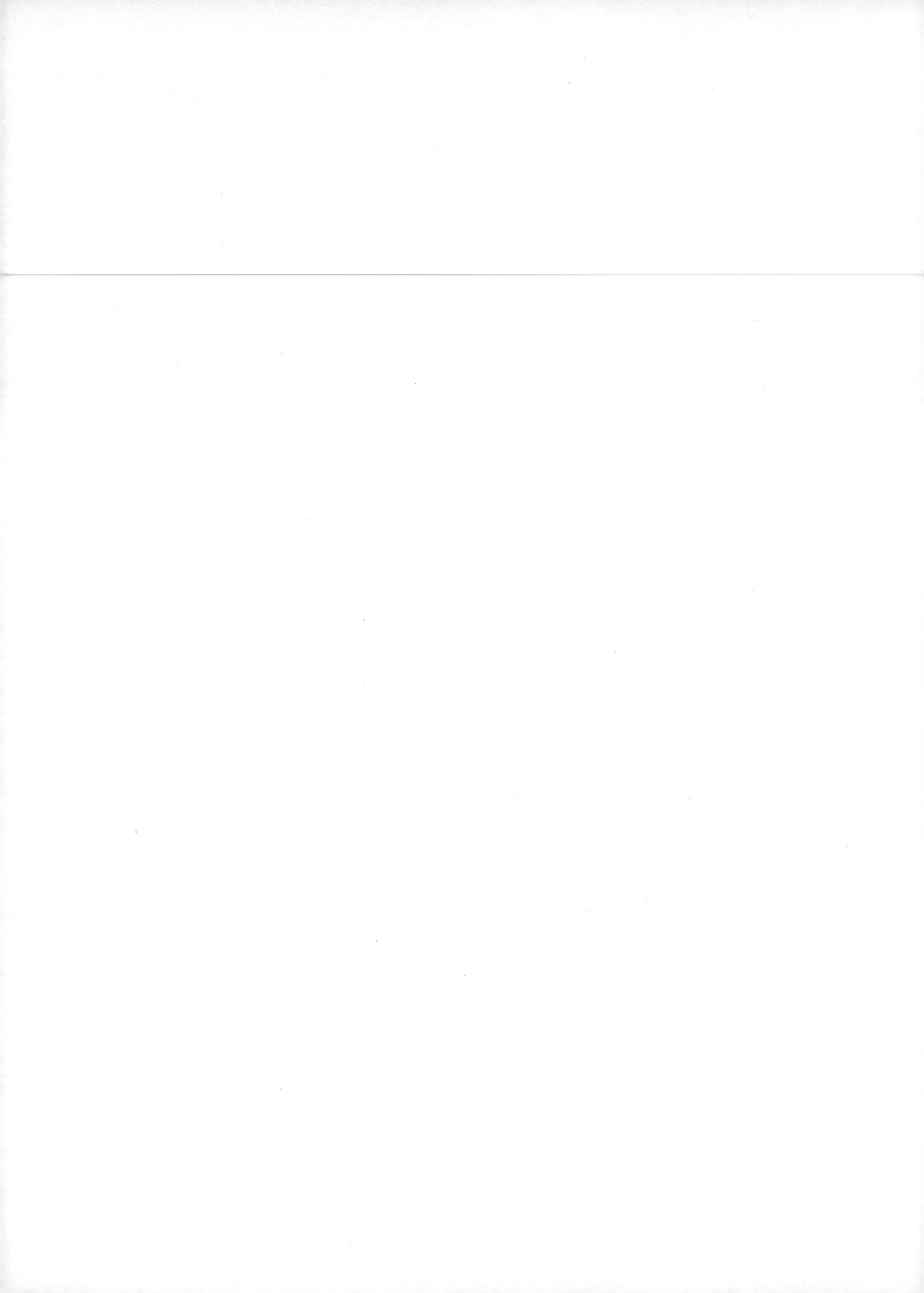

Appendix A

Correlation FESHE *Building Construction for Fire Protection* Course Outline

FESHE Objective	Chapter Reference
1. Describe Building construction as it relates to firefighter safety, building codes, fire prevention, code inspection, firefighting strategy, and tactics.	1, 7, 8, 9, 10, 11, 12, 13, 14
2. Classify major types of building construction in accordance with a local/model building code.	2, 5
3. Analyze the hazards and tactical considerations associated with the various types of building construction.	2, 4, 5, 12, 13
4. Explain the different loads and stresses that are placed on a building and their interrelationships.	3, 5
5. Identify the function of each principle structural component of typical building design.	3, 4, 6
6. Differentiate between fire resistance, flame spread, and describe the testing procedures used to establish ratings for each.	5, 11
7. Classify occupancy designations of the building code.	2
8. Identify the indicators of potential structural failure as they relate to firefighter safety.	7, 8, 9, 10, 11, 12, 13, 14
9. Identify the role of GIS as it relates to building construction.	1

Glossary

Glossary

A

Aesthetics — Branch of philosophy dealing with the nature of beauty, art, and taste.

Admixture — Ingredients or chemicals added to concrete mix to produce concrete with specific characteristics.

Aggregate — Gravel, stone, sand, or other inert materials used in concrete. These materials may be fine or coarse.

Air-Supported Structure — Membrane structure that is fully or partially held up by interior air pressure.

Ambient Temperature — Temperature of the surrounding environment.

Americans with Disabilities Act (ADA) of 1990 - Public Law 101-336 — A federal statute intended to remove barriers — physical and otherwise — that limit access by individuals with disabilities.

Area of Refuge — (1) Area where persons who are unable to use stairs can temporarily wait for instructions or assistance during an emergency building evacuation. (2) Space in the normal means of egress protected from fire by an approved sprinkler system, by means of separation from other spaces within the same building by smokeproof walls, or by virtue of location in an adjacent building.

Assembly — All component or manufactured parts necessary for and fitted together to form a complete machine, structure, unit, or system.

Atrium — Open area in the center of a building, extending through two or more stories, similar to a courtyard but usually covered by a skylight, to allow natural light and ventilation to interior rooms.

Axial Load — Load applied to the center of the cross-section of a member and perpendicular to that cross section. It can be either tensile or compressive and creates uniform stresses across the cross-section of the material.

B

Balloon-Frame Construction — Type of structural framing used in some single-story and multistory wood frame buildings wherein the studs are continuous from the foundation to the roof. There may be no fire stops between the studs.

Bar Joist — Open web truss constructed entirely of steel, with steel bars used as the web members.

Beam — Structural member subjected to loads, usually vertical, perpendicular to its length.

Bearing Wall — Wall that supports itself and the weight of the roof and/or other internal structural framing components such as the floor beams above it.

BLEVE — Acronym for Boiling Liquid Expanding Vapor Explosion.

Blind Hoistway — Used for express elevators that serve only upper floors of tall buildings. There are no entrances to the shaft on floors between the main entrance and the lowest floor served.

Board of Appeals — Group of people, usually five to seven, with experience in fire prevention, building construction, and/or code enforcement legally constituted to arbitrate differences of opinion between fire inspectors and building officials, property owners, occupants or builders.

Bowstring Truss — Lightweight truss design noted by the bow shape, or curve, of the top chord.

Building Code — Body of local law, adopted by states, counties, cities, or other governmental bodies to regulate the construction, renovation, and maintenance of buildings.

Butterfly Roof — V-shaped roof style resembling two opposing shed roofs joined along their lower edges.

C

Cantilever — Projecting beam or slab supported at one end.

Cast-in-Place Concrete — A common type of concrete construction. Refers to concrete that is poured into forms as a liquid and assumes the shape of the form in the position and location it will be used.

Cementitious — Containing or composed of cement. Has cementlike characteristics.

Column — Vertical supporting member.

Column Footing — Square pad of concrete that supports a column.

Compartmentation Systems — Series of barriers designed to keep flames, smoke, and heat from spreading from one room or floor to another; barriers may be doors, extra walls or partitions, fire-stopping materials inside walls or other concealed spaces, or floors.

Composite Panels — Produced with parallel external face veneers bonded to a core of reconstituted fibers.

Concentrated Load — Load that is applied at one point or over a small area.

Concrete Block — Also known as concrete masonry units (CMU). The most commonly used concrete block is the hollow concrete block.

Conflagration — Large, uncontrollable fire covering a considerable area and crossing natural fire barriers such as streets; usually involves buildings in more than one block and causes a large fire loss. Forest fires can also be considered conflagrations.

Convection — Transfer of heat by the movement of heated fluids or gases, usually in an upward direction.

Convenience Stair — Stair that usually connects two floors in a multistory building.

Course — Horizontal layer of individual masonry units.

Curing — Maintaining conditions to achieve proper strength during the hardening of concrete.

Curtain Boards — Vertical boards, fire-resistive half-walls, that extend down from the underside of the roof of some commercial buildings and are intended to limit the spread of fire, heat, smoke and fire gases.

Curtain Wall — Nonbearing exterior wall attached to the outside of a building with a rigid steel frame. Usually the front exterior wall of a building intended to provide a certain appearance.

D

Damping Mechanism — Structural element designed to control vibration.

Dead Load — Weight of the structure, structural members, building components, and any other feature permanently attached to the building that is constant and immobile. Load on a structure due to its own weight and other fixed weights.

Dielectric — Nonconductor of direct electric current. Term usually applied to tools that are used to handle energized electrical wires or equipment.

Draft Stops — Dividers hung from the ceiling in large open areas that are designed to minimize the mushrooming effect of heat and smoke. Also called Curtain Boards and Draft Curtains.

Duct — (1) Channel or enclosure, usually of sheet metal, used to move heating and cooling air through a building. (2) Hollow pathways used to move air from one area to another in ventilation systems.

Ductile — Capable of being shaped, bent, or drawn out.

Dynamic Load — Loads that involve motion. They include the forces arising from wind, moving vehicles, earthquakes, vibration, falling objects, as well as the addition of a moving load force to an aerial device or structure. Also called Shock Loading.

E

Eccentric Load — Load perpendicular to the cross-section of the structural member but does not pass through the center of the cross section. An eccentric load creates stresses that vary across the cross-section and may be both tensile and compressive.

Equilibrium — Condition in which the support provided by a structural system is equal to the applied loads.

Explosion — A physical or chemical process that results in the rapid release of high pressure gas into the environment.

Exposure — Structure or separate part of the fireground to which the fire could spread.

F

Failure Point — Point at which material ceases to perform satisfactorily. Depending on the application this can be breaking, permanent deformation, excessive deflection, or vibration.

Fascia — (1) Flat horizontal or vertical board located at the outer face of a cornice. (2) Broad flat surface over a storefront or below a cornice.

Fire Cut — Angled cut made at the end of a wood joist or wood beam that rests in a masonry wall to allow the beam to fall away freely from the wall in case of failure of the beam. This helps prevent the beam acting as a lever to push against the masonry.

Fire Door — A specially constructed, tested, and approved fire-rated door assembly designed and installed to prevent fire spread by automatically closing and covering a doorway in a fire wall during a fire to block the spread of fire through the door opening.

Fire Escape — (1) Means of escaping from a building in case of fire; usually an interior or exterior stairway or slide independently supported and made of fire-resistive material. (2) Traditional term for an exterior stair, frequently incorporating a movable section, usually of noncombustible construction that is intended as an emergency exit. It is usually supported by hangers installed in the exterior wall of the building.

Fire Load — The amount of fuel within a compartment expressed in pounds per square foot obtained by dividing the amount of fuel present by the floor area. Fire load is used as a measure of the potential heat release of a fire within a compartment. Also known as Fuel Load.

Fire Partition — Fire barrier that extends from one floor to the bottom of the floor above or to the underside of a fire-rated ceiling assembly. A fire partition provides a lower level of protection than a fire wall. An example is a one-hour rated corridor wall.

Fire Resistance Directory — Directory that lists building assemblies that have been tested and given fire-resistance ratings. Published by Underwriters Laboratories, Inc.

Fire Resistance Rating — Rating assigned to a material or an assembly after standardized testing by an independent testing organization that identifies the amount of time a material or assembly of materials will resist a typical fire as measured on a standard time-temperature curve.

Fire Retardant — Any substance, except plain water, that is applied to another material or substance to reduce the flammability of fuels or slow their rate of combustion by chemical or physical action.

Fire Stop — Solid materials, such as wood blocks, used to prevent or limit the vertical and horizontal spread of fire and the products of combustion in hollow walls or floors, above false ceilings, in penetrations for plumbing or electrical installations, in penetrations of a fire-rated assembly, or in cocklofts and crawl spaces.

Fire Wall — Fire-rated wall with a specified degree of fire resistance, built of fire-resistive materials and usually extending from the foundation up to and through the roof of a building, that is designed to limit the spread of a fire within a structure or between adjacent structures.

Fire Watch — Usually refers to someone who has the responsibility to tour a building or facility on at least an hourly basis, look for actual or potential fire emergency conditions, and send an appropriate warning if such conditions are found.

Flame Spread Rating — Numerical rating assigned to a material based on the speed and extent to which flame travels over its surface.

Floating Foundation — Foundation for which the volume of earth excavated will approximately equal the weight of the building supported. Thus, the total weight supported by the soil beneath the foundation remains about the same, and settlement is minimized because of the weight of the building.

Footing — (1) That part of the building that rests on the bearing soil and is wider than the foundation wall. (2) Base for a column.

Fuel Loading — Amount of fuel present expressed quantitatively in terms of weight of fuel per unit area. This may be available fuel (consumable fuel) or total fuel and is usually dry weight.

G

Gabled Roof — Style of pitched roof with square ends in which the end walls of the building form triangular areas beneath the roof.

Gambrel Roof — Style of gabled roof on which each side slopes at two different angles; often used on barns and similar structures.

Generator — Auxiliary electrical power generating device. Portable generators are powered by small gasoline or diesel engines and generally have 110- and/or 220-volt capacities.

Gentrification — Process of restoring rundown or deteriorated properties by more affluent people, often displacing poorer residents.

Girder — Large, horizontal structural member used to support joists and beams at isolated points along their length.

Glazing — Glass or thermoplastic panel in a window that allows light to pass.

Glulam — Short for glue-laminated structural lumber.

Gravity — Force acting to draw an object toward the earth's center; force is equal to the object's weight.

Green Design — Term used to describe the incorporation of such environmental principles as energy efficiency and environmentally friendly building materials into design and construction.

Grillage Footing — Footing consisting of layers of beams placed at right angles to each other and usually encased in concrete.

Gusset Plates — Metal or wooden plates used to connect and strengthen the intersections of metal or wooden truss components roof or floor components into a load-bearing unit.

Gypsum Board — Widely used interior finish material. Consists of a core of calcined gypsum, starch, water, and other additives that are sandwiched between two paper faces. Also known as Gypsum Wallboard, Plasterboard, and Drywall.

Interstitial Space — In building construction, refers to generally inaccessible spaces between layers of building materials. May be large enough to provide a potential space for fire to spread unseen to other parts of the building.

Intumescent Coating — Coating or paintlike product that expands when exposed to the heat of a fire to create an insulating barrier that protects the material underneath.

H

Header Course — Course of bricks with the ends of the bricks facing outward.

Heat of Combustion — Total amount of thermal energy (heat) that could be generated by the combustion (oxidation) reaction if a fuel were completely burned. The heat of combustion is measured in British Thermal Units (Btu) per pound or calories per gram.

Heat of Hydration — During the hardening of concrete, heat is given off by the chemical process of hydration.

Heat Release Rate (HRR) — Total amount of heat produced or released to the atmosphere from the convective-lift fire phase of a fire per unit mass of fuel consumed per unit time.

Heating, Ventilating, and Air Conditioning (HVAC) System — Heating, ventilating, and air-conditioning system within a building and the equipment necessary to make it function; usually a single, integrated unit with a complex system of ducts throughout the building. Also called Air-Handling System.

High-Rise Building — Any building that requires fire fighting on levels above the reach of the department's equipment, often generally given as a building more than 75 feet (25 m) in height.

Hip Roof — Pitched roof that has no gables. All facets of the roof slope down from the peak to an outside wall.

I

Inertia — The tendency of a body to remain in motion or at rest until it is acted upon by force.

Interior Finish — Exposed interior surfaces of buildings including, but not limited to, fixed or movable walls and partitions, columns, and ceilings. Commonly refers to finish on walls and ceilings and not floor coverings.

K

Kinetic Energy — The energy possessed by a moving object.

L

Lamella Arch — A special type of arch constructed of short pieces of wood called lamellas.

Light-Frame Construction — Method for construction of wood-frame buildings. Replaced the use of heavy timber wood framing.

Lintel — Support for masonry over an opening; usually made of steel angles or other rolled shapes singularly or in combination.

Live Load — Force placed upon a structure by the addition of people, objects, or weather.

Load — Any effect that a structure must be designed to resist. Forces of loads, such as gravity, wind, earthquakes, and soil pressure, are exerted on a building.

Lumber — Lengths of wood cut and prepared for use in construction.

M

Mansard Roof — Roof style with characteristics similar to both gambrel and hip roofs. Mansard roofs have slopes of two different angles, and all sides slope down to an outside wall.

Masonry — Bricks, blocks, stones, and unreinforced and reinforced concrete products.

Mat Foundation — Thick slab beneath the entire area of a building. A mat foundation differs from a simple floor slab in its thickness and amount of reinforcement.

Means of Egress — Safe, continuous path of travel from any point in a structure to a public way; the means of egress is composed of three parts: the exit access, the exit, and the exit discharge.

Membrane Ceiling — Usually refers to a suspended, insulating ceiling tile system.

Membrane Roof — Roof covering that consists of a single layer of waterproof synthetic membrane over one or more layers of insulation on a roof deck. Also called Single-Ply Roof.

Membrane Structure — Structure with an enclosing surface of a thin stretched flexible material. Examples of these are a simple tent or an air-supported structure.

Monitor Roof — Roof style similar to an exaggerated lantern roof having a raised section along the ridge line, providing additional natural light and ventilation.

Mortise — Notch, hole, or space cut into a piece of timber to receive the projecting part (tenon) of another piece of timber.

Mushrooming — Tendency of heat, smoke, and other products of combustion to rise until they encounter a horizontal obstruction. At this point they will spread laterally until they encounter vertical obstructions and begin to bank downward.

N

Noncombustible — Incapable of supporting combustion under normal circumstances.

O

Oriented Strand Board (OSB) — Construction material made of many small wooden pieces (strands) bonded together to form sheets, similar to plywood.

P

Parapet — Portion of the exterior walls of a building that extends above the roof. A low wall at the edge of a roof.

Penthouse — (1) Structure on the roof of a building that may be used as a living space, to enclose mechanical equipment, or to provide roof access from an interior stairway. (2) Room or building built on the roof, which usually covers stairways or houses elevator machinery, and contains water tanks and/or heating and cooling equipment. Also called a Bulkhead.

Phase I Operation — Emergency operating mode for elevators. Phase I operation recalls the car to a certain floor and opens the doors.

Phase II Operation — Emergency elevator operating mode that allows emergency use of the elevator with certain safeguards and special functions.

Pier — Load-supporting member constructed by drilling or digging a shaft, then filling the shaft with concrete.

Piles — Used to support loads, piles are driven into the ground and develop their load-carrying ability either through friction with the surrounding soil or by being driven into contact with rock or a load-bearing soil layer.

Pipe Chase — Concealed vertical channel in which pipes and other utility conduits are housed. Pipe chases that are not properly protected can be major contributors to the vertical spread of smoke and fire in a building. Also called Chase.

Platform Frame Construction — Type of framing in which each floor is built as a separate platform and the studs are not continuous beyond each floor. Also called Western Frame Construction.

Plywood — Wood sheet product made from several thin veneer layers that are sliced from logs and glued together.

Portland Cement — Most commonly used cement consisting chiefly of calcium and aluminum silicates. Portland cement is mixed with water to form a paste that hardens and is, therefore, known as a hydraulic cement.

Post-Tensioned Reinforcing — Technique used in post-tensioned concrete. Reinforcing steel in the concrete is tensioned after the concrete has hardened.

Precast Concrete — Method of building construction where the concrete building member is poured and set according to specification in a controlled environment and is then shipped to the construction site for use.

Preincident Planning — Act of preparing to handle an incident at a particular location or a particular type of incident before an incident occurs. Also called Prefire Planning, Preplanning, Prefire Inspection, or Preincident Inspection.

Pretensioned Reinforcing — Used in pretensioned concrete. Steel strands are stretched between anchors producing a tensile force in the steel. Concrete is then placed around the steel strands and allowed to harden.

Protected Stair Enclosure — Stair with code required fire-rated enclosure construction. Intended to protect occupants as they make their way through the stair enclosure.

Pyrolysis — Thermal or chemical decomposition of fuel (matter) because of heat that generally results in the lowered ignition temperature of the material. The pre-ignition combustion phase of burning during which heat energy is absorbed by the fuel, which in turn gives off flammable tars, pitches, and gases. Pyrolysis of wood releases combustible gases and leaves a charred surface.

R

Rafter — Inclined beam that supports a roof, runs parallel to the slope of the roof, and to which the roof decking is attached.

Rain Roof — A second roof constructed over an existing roof.

Rated Assembly — Assemblies of building components such as doors, walls, roofs, and other structural features that may be, because of the occupancy, required by code to have a minimum fire-resistance rating from an independent testing agency. Also called Labeled Assembly.

Refuse Chute — Vertical shaft with a self-closing access door on every floor; usually extending from the basement or ground floor to the top floor of multistory buildings.

Rise — Vertical distance between the treads of a stairway or the height of the entire stairway.

Roof Covering — Final outside cover that is placed on top of a roof deck assembly. Common roof coverings include composition or wood shake shingles, tile, slate, tin, or asphaltic tar paper.

Run — The horizontal measurement of a stair tread or the distance of the entire stair length.

S

Sawtooth Roof — Roof style characterized by a series of alternating vertical walls and sloping roofs that resembles the teeth of a saw. This type of roof is most often found on older industrial buildings to provide light and ventilation.

Scissor Stairs — Two sets of crisscrossing stairs in a common shaft; each set serves every floor but on alternately opposite sides of the stair shaft. For example, one set would serve the west wing on even-numbered floors and the east wing odd-numbered floors, while the other set would serve floors opposite to the first set.

Seismic Forces — Forces developed by earthquakes. Seismic forces are the some of the most complex forces exerted on a building.

Setback — Distance from the street line to the front of a building.

Sheathing — (1) Covering applied to the framing of a building to which siding is applied. (2) First layer of roof covering laid directly over the rafters or other roof supports. Sheathing may be plywood, chipboard sheets, or planks that are butted together or spaced about 1 inch (25 mm) apart. Also called Decking or Roof Decking.

Shed Roof — Pitched roof with a single sloping aspect, resembling half of a gabled roof.

Shell Structure — Rigid, three-dimensional structure having an outer "skin" thickness that is small compared to other dimensions.

Shelter in Place — Having occupants remaining in a structure or vehicle in order to provide protection from a rapidly approaching hazard (fire, hazardous gas cloud, etc...). Opposite of evacuation.

Shoring — General term used for lengths of timber, screw jacks, hydraulic and pneumatic jacks, and other devices that can be used as temporary support for formwork or structural components or used to hold sheeting against trench walls. Individual supports are called shores, cross braces, and struts.

Sinkhole — A natural depression in a land surface formed by the collapse of a cavern roof. Generally occurs in limestone regions.

Skylights — Any of a variety of roof structures or devices intended to increase natural illumination within buildings in rooms or over stairways and other vertical shafts that extend to the roof.

Smoke-Control System — Engineered system designed to control smoke by the using mechanical fans to produce airflows and pressure differences across smoke barriers to limit and direct smoke movement.

Smokeproof Enclosures — Stairways that are designed to limit the penetration of smoke, heat, and toxic gases from a fire on a floor of a building into the stairway and that serve as part of a means of egress.

Spalling — Expansion of excess moisture within concrete due to exposure to the heat of a fire resulting in tensile forces within the concrete, causing it to break apart. The expansion causes sections of the concrete surface to violently disintegrate, resulting in explosive pitting or chipping destruction of the material's surface.

Spec Building — (short for speculation). Building built without a tenant or occupant.

Static Load(s) — Loads that are steady, motionless, constant, or applied gradually.

Steiner Tunnel Test — Unofficial name for the test used to determine the flame spread ratings of various materials.

Subsidence — Sinking or settling of land due to various natural and human-caused factors such as removal of underground water or oil.

Surface System — System of construction in which the building consists primarily of an enclosing surface and in which the stresses resulting from the applied loads occur within the surface bearing wall structures.

Surface-To-Mass Ratio — The ratio of the surface area of the fuel to the mass of the fuel.

T

Tenon — Projecting member in a piece of wood or other material for insertion into a mortise to make a joint.

Tension — Those vertical or horizontal forces that tend to pull things apart; for example, the force exerted on the bottom chord of a truss.

Thermal Radiation — The transmission or transfer of heat energy from one body to another body at a lower temperature through intervening space by electromagnetic waves similar to radio waves or X-rays.

Tilt-Up Construction — Type of construction in which concrete wall sections (slabs) are cast on the concrete floor of the building and are then tilted up into the vertical position. Also known as Tilt-Slab Construction.

Torsional Load — Load offset from the center of the cross section of the member and at an angle to or in the same plane as the cross-section. A torsional load produces a twisting effect that creates shear stresses in a material.

Toxicity — Ability of a substance to do harm within the body.

Truss — Structural member used to form a roof or floor framework. Trusses form triangles or combinations of triangles to provide maximum load-bearing capacity with a minimum amount of material; often rendered dangerous by exposure to intense heat, which weakens gusset plate attachment.

U

Underpinning — Process of strengthening an existing foundation.

Underwriters Laboratories, Inc. (UL) — Independent fire research and testing laboratory with headquarters in Northbrook, Illinois that certifies equipment and materials. Equipment and materials are approved only for the specific use for which it is tested.

V

Vapor Barrier — Watertight material used to prevent the passage of moisture or water vapor into and through walls or roofs or in the case of personal protective equipment that prevents water from penetrating the clothing.

Veneer — Surface layer of attractive material laid over a base of common material; for example, a veneered wall (faced with brick) or a veneered door (faced with a thin layer of hardwood).

Volatility — Ability of a substance to vaporize easily at a relatively low temperature.

Voltage — The electrical force that causes a charge (electrons) to move through a conductor. Sometimes called the electromotive force (EMF). Measured in volts (V).

W

Wear Course — External covering on a roof that protects the roof from mechanical abrasion. The typical tar and gravel roof uses gravel as the wear course.

Wildland/Urban Interface — Line, area, or zone where structures and other human development meet or intermingle with undeveloped wildland or vegetative fuels.

Wythe — Single vertical row of multiple rows of masonry units in a wall, usually brick.

Index

Index

T

U